Advances in
MICROBIAL ECOLOGY
Volume 4

ADVANCES IN MICROBIAL ECOLOGY

Sponsored by International Commission on Microbial Ecology,
a unit of International Association of Microbiological Societies
and the Division of Environmental Biology of the
International Union of Biological Societies

Editorial Board

A Continuation Order Plan is available for this series. A continuation order will bring delivery of each new volume immediately upon publication. Volumes are billed only upon actual shipment. For further information please contact the publisher.

Advances in
MICROBIAL ECOLOGY

Volume 4

Edited by

M. Alexander

Cornell University
Ithaca, New York

PLENUM PRESS · NEW YORK AND LONDON

The Library of Congress cataloged the first volume of this title as follows:

Advances in microbial ecology. v. 1—
  New York, Plenum Press c1977—
    v. ill. 24 cm.
  Key title: Advances in microbial ecology, ISSN 0147-4863

  1. Microbial ecology—Collected works.
QR100.A36   576'.15   77-649698

Library of Congress Catalog Card Number 77-649698
ISBN 0-306-40493-1

© 1980 Plenum Press, New York
A Division of Plenum Publishing Corporation
227 West 17th Street, New York, N.Y. 10011

Printed in the United States of America

Contributors

B. Ben Bohlool, Department of Microbiology, University of Hawaii, Honolulu, Hawaii 96822

C. Furusaka, Institute of Agricultural Research, Tohoku University, Sendai, Japan

Henry S. Lowendorf, Department of Agronomy, Cornell University, Ithaca, New York 14853

Birgitta Norkrans, Department of Marine Microbiology, Botanical Institute, University of Göteborg, Carl Skottsbergs Gata 22, S-413 19 Göteborg, Sweden

Edwin L. Schmidt, Departments of Microbiology and Soil Science, University of Minnesota, St. Paul/Minneapolis, Minnesota 55108

John D. Stout, Soil Bureau, New Zealand Department of Scientific and Industrial Research, Lower Hutt, New Zealand

Robert L. Tate III, University of Florida Agricultural Research and Education Center, Belle Glade, Florida 33430

I. Watanabe, The International Rice Research Institute, Los Baños, Laguna, Philippines

Preface

The literature in microbial ecology is growing rapidly. Journals in many countries dealing with microbiology, ecology, environmental sciences, and environmental technology are publishing an ever-increasing number of papers, and these reports are providing microbial ecologists with a wealth of information. This body of data is now so large and the research is published in so many journals and monographs that maintaining an overview of the development of the field grows more difficult. The role of *Advances in Microbial Ecology* thus becomes more obvious with time.

The articles in the present volume encompass an array of topics appropriate to the development of the discipline of microbial ecology. Both terrestrial and aquatic ecosystems are subjects of attention, and a variety of microbiological groups come under review. Furthermore, methodological problems and approaches are not overlooked.

The ecology of protozoa, constraints on their populations, and their role in nutrient cycling and energy flow are considered by J. D. Stout. A unique microenvironment is discussed by B. Norkrans, the surface microlayer of aquatic ecosystems, and Dr. Norkrans presents information on a field that has blossomed in the last few years. The subject of the review by H. S. Lowendorf is the genus *Rhizobium*, a group of bacteria whose importance has grown as the cost of fuel for production of nitrogen fertilizers and ultimately for protein production has increased. Two special terrestrial ecosystems are also considered in the present volume, flooded soils that are common in much of Asia for rice production and Histosols, a related ecosystem in that it is flooded but one that is dominated by organic rather than inorganic materials. I. Watanabe and C. Furusaka are the authors of the former review, and R. L. Tate III is the author of the latter. B. B. Bohlool and E. L. Schmidt present a thorough evaluation of immunofluorescence as a technique for study of the ecology of microorganisms.

The Editorial Board of *Advances in Microbial Ecology* and the sponsors of the series hope to maintain the standards and the direction of the earlier volumes. *Advances* is designed to serve an international audience and to provide reviews written by scientists from various countries. The various reviews that are published are selected because they include the basic and applied aspects of microbial ecology and properties of diverse ecosystems and various microorganisms. We believe that the series has been useful in accomplishing this goal, and we express our thanks to our colleagues for their contributions, comments, and suggestions.

M. Alexander, Editor
K. C. Marshall
T. Rosswall
H. Veldkamp

Contents

Chapter 3

Factors Affecting Survival of Rhizobium in Soil

Henry S. Lowendorf

Chapter 4

Microbial Ecology of Flooded Rice Soils

I. Watanabe and C. Furusaka

Chapter 5
Microbial Oxidation of Organic Matter of Histosols
Robert L. Tate III

Chapter 6

The Immunofluorescence Approach in Microbial Ecology

B. Ben Bohlool and Edwin L. Schmidt

The Role of Protozoa in Nutrient Cycling and Energy Flow

JOHN D. STOUT

1. Introduction

Nutrient cycling and energy flow are centered on photosynthesis and plant growth, for plant tissue forms the greater part of the earth's biomass. But all organisms participate, and their role is determined not simply by their biomass but by their catalytic reaction in different ecosystems. The major nutrient cycles are the carbon cycle, in which the organic energy cycle is implicit, the nitrogen cycle, the sulfur cycle, and the phosphorus cycle. The major energy cycles are the solar cycle and the hydrological cycle that is not only the source of the major part of living matter but also provides the medium for all organic cycles. The geometry of nutrient cycling is determined by the nature and distribution of sources and sinks, the most important being the atmosphere, the ocean and soil, and the availability of the major, minor, and trace elements, and other growth factors.

Cycling is a function of reaction rate, and turnover is determined by pool size and residence time. The greatest difficulty lies in the identification and measurement of cycling (Payne and Wiebe, 1978). The determination of isotopic ratios has provided a valuable guide to nutrient cycling, and fortunately stable and/or radioactive isotope species of all the major nutrient elements exist.

The role of protozoa in nutrient cycling and energy flow is determined by their bionomics. The distinctive features of protozoa are their small size, their high rate of reproduction, often through a complex life history, the high conversion efficiency of nutrients to new cell tissue, and their potentially high metabolic

JOHN D. STOUT • Soil Bureau, New Zealand Department of Scientific and Industrial Research, Lower Hutt, New Zealand.

rates. They occupy a wide range of ecological niches, but their role is generally in association with a range of other microorganisms that function together as a microcosm. In such a system, the components tend to be soluble and insoluble nutrients, saprobionts assimilating the available nutrients, and micropredators accelerating turnover rates by constant grazing on the saprobiont populations.

In this chapter, nutrient dynamics in free-living microbial ecosystems, at both the cell and population levels, and the determination of turnover rates in communities of different population structure are discussed. The ecological constraints of the microhabitat limiting turnover rates and the structure and dynamics of the population will be examined in relation to nutrient cycling and energy flow.

2. Nutrient Dynamics

2.1. Nutrient Cycles and Energy Flow

The major factor affecting nutrient dynamics is whether a transformation is energy yielding or energy demanding. The two most important transformations are oxidation–reduction and the interchange between the organic and inorganic pools, the cell and the substrate. The energetics of the cell are centered in the hydrolysis and synthesis of ATP, and the major expenditure of energy and nutrients is in cell maintenance, including motility and feeding behavior, and cell synthesis and division (Fig. 1). On this conceptually simple scheme of microbial bioenergetics is based the mechanism of the major nutrient cycles. In the simplest case, that of a single cell, bioenergetics may be centered on the cell membrane, separating the organic from the inorganic cycles. In this case, nutrient cycling and energy flow can be related to simple parameters of cell size, cell mass, and the rate of metabolic reactions. The limitations of monaxenic culture may similarly be reduced to simple parameters. Where controversy exists is over the effects of food chains, particularly predation, on the rate of nutrient cycling. Opinions have fluctuated between those who consider that direct cycling by primary saprophytes provides the most rapid transformations and those who consider that a more complex food chain, involving prey–predator relations, accelerates mineralization and hence cycling. Because they constitute the most notable microbial predator in most natural microcosms, this controversy has centered particularly on the role of protozoa as micropredators.

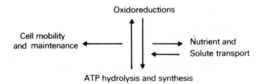

Figure 1. Bioenergetic relationships (after Garland, 1977).

Nutrition and microbial interaction involving protozoa have been recently reviewed (Curds, 1977), and in this section only the literature directly pertaining to nutrient and energy cycling is covered.

2.1.1. Single Populations

2.1.1a. Carbon. Carbon is central to nutrient cycling both because of its relation to energy and because, apart from water, it comprises the greatest bulk of protoplasm. Two aspects are important: the respiration or fermentation of carbon, and the relation of assimilation to carbon loss; and secondly, the intrinsic rate of natural increase. Unicellular organisms, and protozoa in particular, differ from metazoa, both heterotherms and homotherms, in their metabolic rate and their intrinsic growth. Whereas for a heterotherm of comparable size, the metabolic rate is 8.3 times greater than that of the protozoan, its intrinsic growth rate is only about twice as great. Thus, the more complex organism requires a greater proportion of assimilated energy for maintenance and consequently is less efficient as a secondary producer (Fenchel, 1974). Put more simply, the same nutrient and energy resources will sustain a much larger protozoan population than metazoan population, and similarly nutrients and energy cycled through a protozoan population will be conserved more efficiently than if cycled through a metazoan population. The peculiar status of protozoa in nutrient cycling turns on this point. The second point of ecological importance is the dependence of the intrinsic growth rate on cell size, smaller protozoa having a shorter generation time than larger protozoa (Fenchel, 1968b; Finlay, 1977). The actual metabolism of carbon substrates by protozoa in most decomposition processes does not differ biochemically from that of other organisms, the same mechanisms and pathways operate as in many bacterial and animal cells (Stout, 1974), and as with other organisms, their metabolism is regulated by availability of substrate and environmental conditions. There are, however, intrinsic differences in metabolic rates between the different protozoan groups and between different species of the same group. Lee and Muller (1973) discussed metabolic activity of some salt-marsh foraminifera which have a relatively high rate of metabolism but show considerable variation between species. The testacea, on the other hand, appear generally to have low metabolic activity, though some species may be comparable to the foraminifera. Although the small amebae have quite a high rate of respiration, ciliates appear to have a much higher rate, although both are dependent upon an adequate food supply (Stout and Heal, 1967), and the cysts have a much lower respiratory rate than the trophic cells. Protozoa, of course, may also assimilate CO_2 in the fermentation of glucose to succinic acid (Peak and Peak, 1977; van Niel *et al.*, 1942), thus constituting a sink as well as a source of carbon flow.

2.1.1b. Nitrogen. Normally most protozoa assimilate amino acids and excrete ammonia. Some, such as *Chlamydomonas*, may be able to assimilate nitrate (Nichols and Syrett, 1978). When protein or an amino acid is used as an energy source, the ammonia released is excreted, but starved cells may assimilate ammonia as endogenous carbon sources are metabolized (Harding, 1937a,b; Doyle and Harding, 1937).

   2.1.1c. Phosphorus. Because of its importance in both the energy and nutrient cycles, the metabolism of phosphorus by protozoa is of key importance (Matheja and Degens, 1971). The first search for labile phosphorus in protozoa was by Needham *et al.* (1932), with negative results, but Kandatsu and Horiguchi (1962) showed the presence of ciliatine, 2-amino ethylphosphonic acid (with its C—P bond) in the ciliate *Tetrahymena pyriformis.* Now the role of the nucleophosphatide pools in both the ciliate *Tetrahymena* and the small ameba, *Acanthamoeba*, is well worked out. Study of changes in the ADP, ATP, and AMP levels of *Acanthamoeba* during different stages of development have shown differences in concentration between the logarithmic growth phase and the stationary phase, when encystation takes place (Jantzen, 1974; Gessat and Jantzen, 1974). Encystment is accompanied by a depletion of total adenosine phosphate to about 85% because of the depletion of ATP. During development, the energy charge becomes stabilized between 0.58 and 0.81, characteristic of the different modes of encystation. During this cell cycle, changes also occur in the cytoplasmic membranes and alkaline phosphatase activity (Pauls and Thompson, 1978), and in the carbon cycle, in the activity of cytochrome oxidase, dehydrogenase, and catalase, associated with changes in the nucleophosphatides (Edwards and Lloyd, 1977a). Such changes are also correlated with cyanide sensitivity of the respiration (Edwards and Lloyd, 1977b). Edwards and Lloyd (1978) also showed oscillations in pool levels of ATP, ADP, and AMP during the 7- to 8-hr cell cycle (Fig. 2). Roti and Stevens (1975) showed that encystment was associated with the breakdown and synthesis of DNA and that inhibition of encystment was correlated with inhibition of phosphate incorporation. Respiration rates, adenine nucleotide levels, and heat production were measured during exponential growth of *Tetrahymena pyriformis* by Lloyd *et al.* (1978), who found that the ATP pool oscillated in phase with respiratory activity. The ADP and AMP pools oscillated in phase with each other but out of phase with ATP. Values calculated for the energy charge were low (0.2-0.5). Echetebu and Plesner (1977) also followed the nucleotides during cell division (Fig. 3). *Tetrahymena* was grown in synchronized culture, and concentrations peaked 1 hr after the heat shock during the first division cycle and reached a minimum in the second cycle 2 hr after the heat shock. There was thus a complete turnover of adenine phosphates during this period. Similar results were reported by Lloyd *et al.* (1978), who followed O_2 uptake, heat production, and adenine nucleotide levels in exponential and synchronized cultures of *T. pyriformis.* They also found that the O_2 uptake and ATP pool oscillated in phase but not the ADP or AMP pools. Values calculated for adenylate charge were low, increasing from less than 0.2 to more than 0.4 in the mid-exponential phase of growth. There is thus a contrast in behavior of the adenylates between the ameba and the ciliate, and in both the

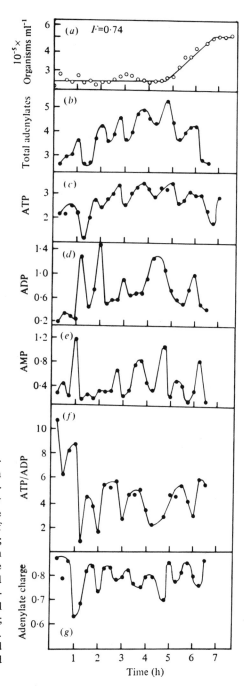

Figure 2. Energy metabolism in synchronous *Acanthamoeba*. Changes in adenine nucleotide pool levels and adenylate charge values in a synchronously dividing culture of *A. castellanii*. The synchronous culture contained 10% of the exponentially growing population; 1-ml samples were withdrawn at 15-min intervals for measurements of adenine nucleotides after rapid quenching and extraction in chloroform. (*a*) Cell numbers and synchrony index, *F*; (*b*) total adenylates; (*c*) ATP; (*d*) ADP; (*e*) AMP; (*f*) ATP/ADP ratio; (*g*) adenylate charge. Adenylate concentrations are expressed as nmol/ml culture. From Edwards and Lloyd (1978).

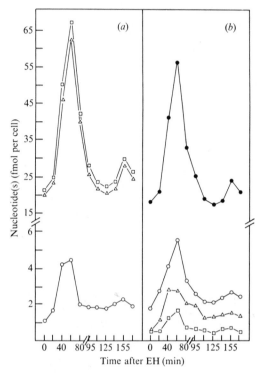

Figure 3. Amounts of ribonucleoside triphosphates in *Tetrahymena pyriformis* during the first and second synchronous division cycles. Cells were grown exponentially in PPY medium containing [^{32}P] orthophosphate (2–5 μCi mol^{-1}) for over six generations to achieve isotopic equilibrium, then induced for synchrony of divisions. At intervals after the end of the synchronizing heat treatment (EH), samples were taken, chromatographed, and the nucleotides were quantified. Results shown are the average from five experiments. (*a*) Sums of ribonucleoside triphosphates: □-, ATP + GTP + CTP + UTP; △, ATP + GTP; ○, UTP + CTP. (*b*) Individual ribonucleoside triphosphates: ●, ATP; △, UTP; ○, GTP; □, CTP. From Echetebu and Plesner (1977).

variation in energy charge appears to be at variance with previous reports for microorganisms (Chapman and Atkinson, 1977).

The unique role of the adenine nucleotides is in the stoichiometric coupling and energy transduction between metabolic sequences. All metabolic pathways are stoichiometrically related through this system, and similarly all such sequences must be kinetically regulated by it also. What is remarkable is the turnover time of ATP, which is probably less than 1 sec, and it is this short turnover time, the preservation of energy charge, and the coupling with biosynthetic pathways that regulate not only all biological homeostasis but also determine the rate of cell growth (Chapman and Atkinson, 1977). Differences, therefore, between the adenylate metabolism of amebae and ciliates point to a fundamental biochemical difference, which qualifies all generalities about the ecological role of protozoa. For the two species discussed, these differences are most significantly related, ecologically, to the cell division cycle and consequently the rate of reproduction and population growth.

2.1.2. Single Trophic Level

There is an extensive literature of the nutrition of axenic cultures of protozoa covered in numerous reviews (Lwoff, 1951; Hutner and Lwoff, 1955; Hutner,

1964; Kidder, 1967). Although the nutritional cycles of the organisms are implicit in nutrient and energy cycling, axenic conditions do not normally prevail in nature; where there are protozoa, there are normally other microorganisms—bacteria, yeasts and fungi—that mediate the food cycle. For this reason, cycling in axenic culture is not reviewed here.

2.1.3. Prey-Predator Relations

Curds and Bazin (1977) have recently reviewed protozoan predation in batch and continuous cultures and outlined the theory of prey-predator interactions, both the kinetics of predation and prey-predator dynamics. In the kinetics of predation, they discuss the effects of a limiting nutrient on the prey organism, which in the field is of decisive significance in determining the rate of cycling, but the theory proposed does not accommodate other environmental constraints which may be important in particular ecosystems, e.g., where there is variation, persistence, or patchiness of prey-predator systems (Chesson, 1978; Ziegler, 1977), which is a frequent characteristic of natural populations, or the temporal behavior of interacting populations in changing environments (Villarreal et al., 1977). Again, their theory assumes a prey population growing under favorable nutrient conditions to be immortal (the specific death rate being equated to zero), but that the predator population always dies in the absence of its prey. Neither of these assumptions would obtain in most natural ecosystems. The capacity of predators to resist starvation and the carrying capacity of prey may both prove to be important (Heller, 1978). A critical evaluation of prey-predator models is given by Wangersky (1978), who stresses the significance of time lag in population dynamics.

Further, in nature, ecosystems have an effective size, reflecting both their absolute size and their physical structure, which limits the mobility of species. Because of this, it is possible to show that those of larger size persist longer than smaller ones, because oscillations of population tend to be displaced further from extinction thresholds (Crowley, 1978). Crowley reexamined Gause's original data on *Paramecium* and *Didinium* and the later studies of Luckinbill (1973, 1974), who repeated and elaborated Gause's experiments varying culture volume, viscosity of the medium, and bacterial concentration (carrying capacity). Luckinbill found that larger culture volumes and higher viscosities yielded longer persistence times at low carrying capacities. Crowley therefore concluded that protozoan populations attained stable limit cycles if the system's effective size exceeded the densities at which only single individuals remained. He also pointed out that Luckinbill's data show that resource limitation can interact with predation to permit persistence. When the time scale of the predator's numerical response is about the same or longer than the time scale of prey increase, then the carrying capacity largely determines the amplitude of oscillations about the equilibria. A low carrying capacity slows or halts prey increase, so that predators can initiate the prey decline sooner, and with lower maximum population densities. A high carrying capacity implies relatively high density maxima and low minima, with consequent risk of extinction. The distinction is important, for in most natural ecosystems carrying capacity tends to be low, whereas in laboratory systems it is often artificially high. Theoretically, however, no matter how high the carrying capacity for the prey may be, a system can be sufficiently extensive to allow persistence through many oscillations. In fact, this is the case in most natural ecosystems, allowing for the persistence of natural diversity. Sardeshpande *et al.* (1977) found that the decline of bacterial populations in soil due to protozoan predation was much less than in broth culture.

From the theoretical kinetic assumptions have been derived a number of dynamic models that have been used for batch and continuous cultures. Simulation models may then be developed to compare experimental data with the dynamic models. In such a way it is possible to analyze the relations of substrate, bacterial prey, and protozoan concentration and the specific growth rates of the two populations, and the differences between a simple axenic culture of prey population and one in which there is also a predator (Fig. 4). Such data indicate the way in which an intermediate trophic stage can affect nutrient cycling.

More directly relevant are experiments measuring predator yield from a known prey. The data were summarized by Curds and Bazin (1977) and are shown in Table I. This information indicates a very high rate of conversion, relative to metazoan populations, the yield coefficient ranging from 0.37 to 0.78. This relationship is, however, closely dependent on the specific prey and its state of nutrition, and there is also a relationship with the population densities of both prey and predator. Greater energy is expended by the predator of a sparsely distributed prey or moving prey than that of a dense and stationary prey population.

Drake and Tsuchiya (1977) found that with *Colpoda steinii* as predator, fed on *Escherichia coli*, the yield of protozoan biomass per unit of prey consumed was constant at all growth rates and that predator growth could be equated with prey density. Laybourn and Stewart (1975) studied consumption, growth, and gross growth efficiency of *Colpidium campylum* fed on a species of *Moraxella*. Temperature influenced the level of growth with a Q_{10} of 3.4 over the range 10-20° but had little effect on bacterial consumption. They found that growth was influenced by food concentration at low levels and that the growth efficiencies ranged from 3 to 11%.

The most successful application of model studies has been to activated sludge systems, where large prey-predator populations occur and the continuous enrichment of nutrients approximates closely laboratory conditions. Curds (1973) was able to conclude from his batch and continuous cultures that the protozoa were able to remove suspended bacteria by predation alone. He was also able to predict the fate of a mixed bacterial population containing flocculating and nonflocculating species. However, Pirt and Bazin (1972) showed that where predation reduced the number of bacteria, the rate of nutrient consumption decreased, reducing the efficiency of nutrient removal.

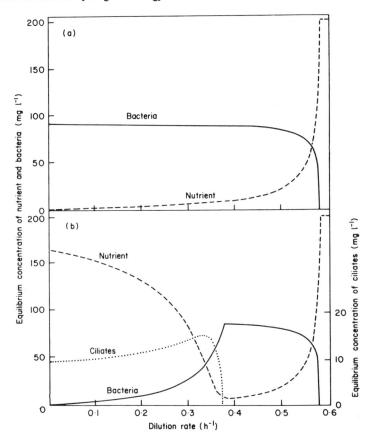

Figure 4. Effect of dilution rate upon theoretical steady-state values of limiting nutrient and bacteria (a) in the absence of a predator and (b) when a predatory ciliated protozoan is present. Adapted from Curds and Bazin (1977).

Table I. Yield Coefficients (W) in Dry Weight, Maximum Specific Growth Rates (λ_m), and Saturation Constants (L) of Protozoa in Batch Culture

Predator	Prey	Yield (W)	λ_m (per hr)	L (mg/liter)
Acanthamoeba sp.	*Saccharomyces cerevisiae*	0.37	0.07	—
Colpoda steinii	*Escherichia coli*	0.78	0.23	6.0
Entodinium caudatum	*Escherichia coli*	0.50	—	—
Tetrahymena pyriformis	*Klebsiella aerogenes*	0.50	0.22	11.6
Tetrahymena pyriformis	*Klebsiella aerogenes*	0.73	0.10	6.1
Uronema sp.	*Serratia marinorubra*	—	0.15	0.49 (as C)

In ecosystems the diversity of potential prey species makes selection and preference of critical importance. Where the prey perform quite different roles, the differential effected by selective predation could be important in nutrient cycling. This was the origin of the partial sterilization theory proposed by Russell and Hutchinson (1909) to explain "sick" soils. They argued that protozoa preyed on agronomically beneficial bacteria and that partial sterilization removed the protozoa but not the beneficial bacteria and was beneficial for this reason. There are two major questions posed by this thesis. First, is there in fact a significant population flux due to micropredation in the soil and is this micropredation selective; and second, does micropredation have any significant effect on the mineralization and turnover of nutrients? Neither of these questions has proved as easy to answer as to ask. The early studies at Rothamsted undertaken to examine Russell and Hutchinson's thesis established not only the existence of large numbers of trophic protozoa in the soil but an inverse relationship between the protozoan and bacterial numbers (Crump, 1920; Cutler, 1923; Cutler and Crump, 1935). They observed a daily fluctuation in the number of protozoa, particularly flagellates, but no obvious relationship with climatic parameters (Cutler and Crump, 1920). Their experiments suggested that amebae were more effective than flagellates in reducing the population (Cutler, 1927). Experimentally, they established a relationship between the reproductive rate of a soil ameba and the kind and quantity of available food supply (Cutler and Crump, 1927) and between the numbers of bacteria and the rate of ciliate reproduction (Cutler, 1927). When they measured the carbon dioxide production of sand and soil cultures with bacteria and amebae, they found conflicting results: in sands containing peptone, the amebae caused a decrease in carbon dioxide production, whereas with glucose or soil extract, they caused an increase (Cutler and Crump, 1929). These early studies were greatly extended by Singh, who improved counting techniques and investigated food preferences of the protozoa (Singh, 1946, 1960). In 1964, he summarized the findings of the Rothamsted workers on the probable role of soil protozoa in soil fertility, citing (1) the stimulation of nitrogen fixation by *Azotobacter* in the presence of protozoa (Nasir, 1923; Cutler and Bal, 1926), including dead protozoa (Harvey and Greaves, 1941), in this case clearly not due to micropredation; (2) the increase in ammonification in the presence of protozoa, reported by Meiklejohn (1930, 1932); and (3) the effect on cellulose and chitin and on carbon dioxide production (Singh, 1964). More recently, these observations have been confirmed by Alexander and his students at Cornell. They confirmed the regulatory effect of protozoa on bacterial populations (Habte and Alexander, 1975; Danso *et al.*, 1975; Habte and Alexander, 1977) and showed that the predator did not reduce the bacterial prey numbers below a certain point, and they found that the ability of the bacteria to survive was governed by their capacity to reproduce and replace the cells consumed by predation (Danso and Alexander, 1975; Habte and Alexander, 1978) (Fig. 5). These studies emphasize significant differences between the theoretical approach to batch or continuous culture and the more complex soil situation.

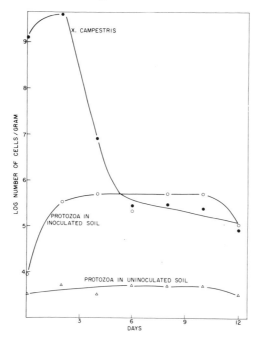

Figure 5. Numbers of protozoa in soil inoculated with *Xanthomonas* and in uninoculated soil. From Habte and Alexander (1978).

An attempt to develop a simulation model for the effect of predation on bacteria in continuous culture with greater reference to the complexities of the natural environment has recently been presented by Hunt *et al.* (1977). Their model for the C, N, and P content of bacteria and their chemostat medium distinguished structural components, synthetic machinery, building blocks and intermediates, C reserves, NH_4, orthophosphate, and polyphosphate (Fig. 6). Growth, incorporation of substrates, and production of wastes were related to physiological status as indicated by the various cell components. Chemical composition of bacteria growing in a chemostat in media limiting C, N, and P was used to explore the consequences of predation on bacterial populations. In C-limiting media, predation increased NH_4 uptake, despite a decrease in bacterial biomass, whereas in N-limiting media, both bacterial biomass and N uptake were decreased by predation. The model suggests that by lowering bacterial biomass, the predation increases the level of a limiting nutrient and thereby increases growth rate of the bacteria and increases uptake of the non-limiting nutrient. The effect of growth rate on the chemical composition and nutrient requirements of the bacteria is as important as the composition of the medium. This important paper does much to reconcile inconsistencies in earlier work and demonstrates the relationship of bacterial predation to nutrient cycling. These ideas have been further developed in a series of papers by the Fort

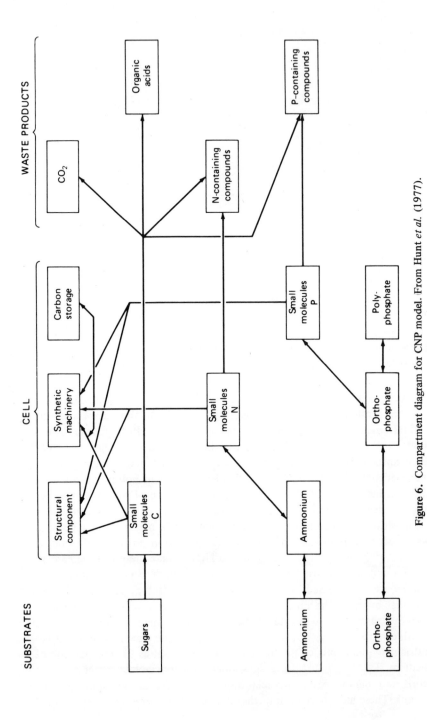

Figure 6. Compartment diagram for CNP model. From Hunt *et al.* (1977).

Collins group (Coleman *et al.*, 1978a,b; Herzberg *et al.*, 1978; Anderson *et al.*, 1978; Cole *et al.*, 1978) on microcosm studies of trophic interactions as they affect energy and nutrient dynamics. They studied the physiological response of selected rhizosphere bacteria (Herzberg *et al.*, 1978), the interactions of bacteria, amebae, and nematodes (Anderson *et al.*, 1978), the flows of metabolic and biomass carbon (Coleman *et al.*, 1978b), and phosphorus transformations (Cole *et al.*, 1978). In this series of studies, they drew together many of the points made separately by earlier workers. They showed the contrast between a typical autochthonous bacterium (*Arthrobacter*) and a typical zymogenous bacterium (*Pseudomonas*) and suggested that the former was dominant in the rhizosphere during low nutrient flux and the latter during high nutrient flux. They showed that amebae grazing on bacteria significantly reduced the *Pseudomonas* population, as did nematodes. The amebae have less effect on respiratory CO_2 loss than the nematodes, but both increased the CO_2 loss compared with the ungrazed control. The amebae, however, accumulated a relatively higher proportion of biomass C than the nematodes, indicating an assimilation efficiency of about 40%. Moreover, the effect of the amebae on mineralization of phosphate was much more pronounced than that of the nematodes. Most of the bacterial phosphate was mineralized and returned to the inorganic phosphorus pool in the microcosms with amebae, although the experiments did not show whether this was due to direct cycling or the effect of predation on the dynamics of the bacterial population.

2.1.4. Multiple Trophic Levels

Most of the field data quantifying prey–predator relations and most of the experimental data deal with a simple two-stage trophic system. However, in nature three or more trophic stages may involve protozoa, e.g., bacteriovore, carnivore, and parasite. The primary prey–predator relation would be of greatest significance with regard to nutrient flux and energetics, but if substantial populations of higher trophic stages were present, they could affect the rate of cycling in two ways: one, in preying on the primary predator, and two, in retaining some part of nutrients cycled in their biomass, i.e., providing a further nutrient pool or sink.

Some data on the activity of carnivores is given by Laybourn and her associates, who showed that there was considerable intraspecific variation in the parameters and significant differences in the temperature tolerance of different species (Laybourn and Finlay, 1976). The very active ciliate predator *Didinium nasutum* has a much higher O_2 uptake than the sedentary omnivore *Stentor coeruleus*, when cells of the same mass are compared at the same temperature (Laybourn, 1975; 1976a, 1977). *Stentor*, fed on *Tetrahymena*, showed a remarkably high assimilation rate of about 72%, and net growth efficiencies were very high (about 96%). *Podophrya fixa*, a sessile suctorian with an entirely different feeding mechanism, appears to be similar to *Stentor* in the rate of O_2 uptake, but its assimilation rate has not been measured (Laybourn, 1976b). Unlike the giant carnivorous ameba, *Chaos chaos*, the rate of metabolism diminishes following starvation (Holter and Zeuthen, 1947).

Data for a marine benthic ciliate is provided by Vernberg and Coull (1974) for the large psammobic ciliate, *Tracheloraphis*, which has a high O_2 uptake, though no data are available on assimilation efficiency. They pointed out that not only did the ciliate have a high O_2 consumption, but it also had a short generation time, so that the total energy flow through the ciliate population would tend to be higher than through a fauna with a much slower turnover rate.

All these data are for ciliates. Little comparable data are available for large rhizopods, such as the large amebae (Holter and Zeuthen, 1947), but in their case it is possible to demonstrate a definite growth cycle; smaller cells grow faster than larger cells, and the generation time is proportional to the weight or surface area (Satir and Zeuthen, 1961). Replication of reproduction rates is therefore difficult, doubling time for *Chaos chaos* being reported from 1.7 days to 4.1 days (Satir and Zeuthen, 1961). However, it appears that the relationship between growth rate and weight or surface area may hold generally for the large amebae. The much more complex mycetozoa and acrasiales, also strictly protozoa, are not dealt with in this review, since they merit separate treatment. The data suggest that the higher trophic level protozoa may be more efficient in conserving nutrients and energy and therefore less efficient in cycling.

3. Ecological Constraints of the Microhabitat

3.1. Physical Constraints

3.1.1. Interfaces and Living Space

Because of their size, protozoa live in microhabitats, and the most characteristic physical feature of microhabitats is interfaces. These may be liquid–liquid, gas–liquid, solid–liquid, or solid–gas interfaces, and such interfaces provide boundaries between two phases of a heterogenous system (Marshall, 1976). These boundaries possess physical and chemical properties which differ from either phase and are commonly the site of nutrient enrichment and population growth (Theng, 1974; 1979). In particular, unique physical forces operative at interfaces are effective only over short distances but have a significant effect on the distribution of ions, macromolecules, and colloids in the vicinity of interfaces (Marshall, 1976). More immediately, an interface presents the problem of adsorption or adhesion of the cell to the surface. In protozoa, this is associated with the development of highly specialized organelles, such as the stalks and loricae of ciliates (Plachter, 1979), and the evolution of ameboid movement (Preston and King, 1978) or specialized cirri and feeding disks (Kloetzel, 1974). One of the most common ecological classifications of protozoa distinguishes them by their habit—sessile or free swimming—or the form of their movement—gliding, rotatory, or saltatory. The physical alignment of the interfaces also determines nutrient status, pH, gas tensions, and the interaction with other organisms, colloids, or free enzymes (Burns, 1978). It also determines limitations of size and shape, particularly evident in psammobic protozoa, in which instance the size of the sand granules and consequent pore size limit the morphology to long, narrow species or very small species (Fenchel, 1967, 1968a,b, 1969; Dragesco, 1960, 1965a,b; Chardez, 1972; Golemansky, 1978). Similar physical constraints, however, operate in forest litter and soil or in sphagnum (Heal, 1962; Meisterfeld, 1977; Schonborn, 1964). The attraction of microbial substrates to interfaces arises from interaction between the physical, chemical, and biological components. The buildup of large colloidal molecules provides a nutrient source stimulating the growth of bacterial colonies, which in turn provide a food source for saprophytes and micropredators. Therefore, for both physical and biological reasons, there is a concentration of nutrients and energy at interfaces, and because many protozoa are able to colonize such interfaces, they may be more effective in accelerating turnover than larger animals. Few micropredators, for example, can move into the smallest pore spaces or move over air–liquid interfaces, and those few include a high proportion of protozoa. Protozoa are found on living leaf surfaces as well as in benthic detritus. They may be associated with lichens and are present in trickling filters. The interfaces provide a nexus of nutrient concentration and consequently of nutrient turnover, but they also are a physical constraint on nutrient dynamics, and the adaptation of protozoa to this constraint provides one reason for the importance of protozoa in nutrient cycling.

The interactions of the substrate are not simply concerned with nutrient cycling. A wide spectrum of physical and chemical reactions, operating over a time scale independent of nutrient cycling but interacting with it, determines availability of nutrients, their form, and their rate of interchange. Ion exchange, equilibria reactions, mineral restructuring, temperature changes, and moisture fluctuations are typical examples. These reactions provide the major contrasts between the different ecosystems and explain the conflicting opinions on the role of microbial populations in nutrient cycling.

3.1.2. Temperature and Energy

Temperature is normally the principal parameter determining rates of reactions, and there is a wide difference both in the temperature tolerance and the relationship of metabolic activity at different temperatures (regression factor, b) for different species (Laybourn and Finlay, 1976). In prey–predator dynamics, the differential reaction to temperature regime is important. Normally, the prey and the predator have different growth curves, and the temperature will determine whether growth of the prey will sustain the predator population or whether, and when, it may be wholly consumed. This has a direct implication not only for population dynamics but also for nutrient turnover, since it affects the size, composition, and rate of turnover of the biomass.

Energy is also limiting where food concentration is limiting to growth, whether this is reflected by the movement of a predator (e.g., Karpenko *et al.*, 1977; Laybourn, 1977; Seravin and Orlovskaya, 1977) or in the activity of energy-dependent phagocytosis (Skriver and Nilsson, 1978). These considerations are important for they materially affect the efficiency of assimilation and the rate of nutrient cycling in protozoan populations in nature (Capo *et al.*, 1974). Errors of extrapolation from laboratory experiments to field conditions can lead to misleading conclusions (Laybourn and Finlay, 1976).

3.1.3. Water and Ionic Balance

In aquatic environments, salinity and ionic balance may seriously affect metabolic activity (Organ *et al.*, 1978). In terrestrial environments, the effects of high water tensions vary with the composition of the microbial populations. The respiration of a mixed population is sustained at a higher moisture tension than a purely bacterial population (Wilson and Griffin, 1975). This could reflect the continual respiration of a larger biomass. Darbyshire (1976) studied the growth of *Azotobacter* and the ciliate *Colpoda steinii* in soil at different moisture tensions. He found no growth of the bacterium at pF 2.7 but growth at all lower tensions. The ciliate grew best and also reduced the *Azotobacter* population most dramatically in the saturated soil. Similarly, Laminger (1978) and others (Heal, 1962, 1964a) found the distribution of testacea related to the moisture regime.

3.1.4. Major Elements

Although carbon is not normally considered a limiting factor in terrestrial ecosystems, it may be significant in many aquatic systems since CO_2 concentrations affect the character of the algal flora and the pattern of photosynthesis (King, 1972). In soil, assessment of carbon dynamics closely reflects soil structure, and modification can bring significant alterations to the cycle (Ausmus and O'Neill, 1978). Part of this variation can reflect the distribution of organic materials in the pore spaces of soil aggregates, which will affect their availability to all microorganisms, including the protozoa (Adu and Oades, 1978a,b). Limitations of N and P commonly control the succession of phytoplankton in aquatic habitats and are therefore an expression of major nutrient constraint (Feierabend, 1978; de Noyelles and O'Brien, 1978; Stoermer *et al.*, 1978; Welch *et al.*, 1978). Although energy and carbon set ultimate constraints on nutrient cycling, in nature growth rate and growth yield are commonly limited by the conserved substrates of N and P. The kinetics of interaction of growth with levels of available N and P has been most carefully studied in algae, and recent work has shown two aspects of importance: first, the phenomenon of "cooperative" metabolism (where the plot of enzyme reaction against substrate concentration gives a sigmoid curve), and second, that this effect is more marked with N than with P (Panikov and Pirt, 1978).

3.2. Nutrient Transport and Turnover

Nutrient turnover is a function of nutrient concentration and nutrient transport. In general, the rate of turnover is dependent upon nutrients being available in stoichiometric balance (Parnas, 1976) and upon the availability of growth factors (Provasoli and Pinter, 1953; Droop, 1970, 1973). The assimilation of N and P is dependent, in heterotrophs, upon an available energy source, but the provision of N and P without an energy source can stimulate increased respiratory activity and loss of biomass. In general, metabolic activity is dependent upon nutrient concentration within the cell, and external concentrations of nutrients are not important directly (Cunningham and Maas, 1978; Droop, 1977). The transport of external nutrients into the cell may follow various pathways and various mechanisms. In algae and bacteria, nutrients may simply diffuse into the cell, i.e., passive transport, or be actively transported, in this case commonly following Michaelis–Menten kinetics. Of major significance in protozoan biology are the alternative active transport mechanisms—pinocytosis or phagocytosis (ingestion)—by which they can take up nutrients (Chapman-Andresen and Holter, 1964; Chambers and Thompson, 1976). Even in the saprobic species, this assures a rapid assimilation of available nutrients, the uptake of dissolved nutrients being accelerated by the ingestion of inert particles (Bowers and Olszewski, 1972; Ricketts, 1972; Rasmussen, 1973, 1976, Rasmussen and Modeweg-Hansen, 1973; Hoffman *et al.*, 1974; Orias and Rasmussen, 1977). In predatory species, the direct ingestion of the prey also provides an extremely efficient means of nutrient assimilation. In polymorphic species, such as *Tetrahymena vorax*, respiration rates vary with the morphology of the ciliate (Buhse and Hamburger, 1974). One reason for the high rate of metabolism and great efficiency of nutrient cycling by protozoa is their means of assimilation, and similarly the difference between a saprobic ciliate such as *Tetrahymena pyriformis* and a small ameba such as *Acanthamoeba* or *Hartmanella* can be ascribed in part to their differences in feeding habit. The role of protozoa in nutrient cycling is to facilitate nutrient availability and so the rate of nutrient turnover, rather than effect the major role themselves. This they can do by the efficiency with which they concentrate nutrients available in the habitat and make them available, in a stoichiometric ratio, for other heterotrophs, for it is known that microorganisms metabolize microbial cells more efficiently than plant residues (Kaszubiak *et al.*, 1976).

3.3. Energetics

The rates and pathways of nutrient cycling ultimately depend upon the energetics of the biological systems. In particular, the balance of assimilated nutrients and energy that is channeled into maintenance or cell synthesis varies widely from cell to cell and under different conditions. If the efficiency of metabolism is measured by the molar growth yield for ATP, then this can vary widely not only between different organisms but for the same organism grown on the same carbon substrate in different media (Stouthamer, 1977), the yield generally being higher the more complex the medium. More important is the relation of the growth rate and the maintenance energy to the molar growth yield for ATP. With many protozoa, the requirements of maintenance energy are relatively small, so that a very high proportion of a carbon substrate may be assimilated into new cell tissue. This has important implications in nutrient cycling for it suggests that a higher proportion of the available carbon source is available for future cycling in a protozoan system than in a nonprotozoan system.

4. Population Structure and Dynamics

Although the constraints of the microhabitat set the limits within which nutrient cycling can take place, it is the biological parameters that are fundamental. Negligible cycling can take place in the absence of living organisms; even cell-free enzymes have only a limited capacity for nutrient cycling and can effect only single-step transformations.

4.1. Population Density, Trophic Structure, and Available Resource

Three aspects of the structure and dynamics of the population are important: the size and number of organisms and their physical distribution; the structure of the community, the food web, and trophic patterns of the populations; and the overall interactions of the populations and the substrate, their metabolic activity, and the rate of nutrient turnover. The classical model of trophic structure and energy flow is that of Lindemann (1942), but a more recent model by Wiegert and Owen (1971) separates those feeding on living organisms from those feeding on decaying organic matter and so distinguishes between ecosystems dominated by detritus, particularly terrestrial ecosystems (Schlesinger, 1977), and those of small biomass and rapid turnover, such as pelagic ecosystems (Fig. 7). These structural differences are reflected in the role and relative importance of prey–predator relations, particularly at the microbial level. Since the rate of turnover is largely a function of the interaction of the two groups, this is a convenient model for comparing the role of protozoa in very different ecosystems.

4.1.1. Available Resource

It is the paradox of nutrient cycling that the greatest activity is in the smallest component—the biomass, and this may bear little relation to the magnitude of available resources. The resources of the major ecosystems— marine, freshwater, and terrestrial—vary enormously. The energy resources of light, day length, and

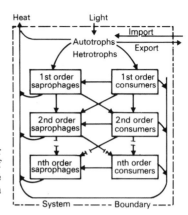

Figure 7. The proposed model of trophic transfer in an ecosystem. Transfers at levels below that of autotrophs are defined on the basis of whether the material is living or dead. After Wiegert and Owen (1971).

temperature divide the major ecosystems into zones of high productivity, and therefore high rate of turnover, and zones of low productivity, and low rate of turnover, whether by altitude, depth, or latitude. Nutrient resources show a similar gradation of concentration and availability, but nutrient limitations may be set by mineral imbalance or by the operation of Liebig's law of the minimum. The availability of mineral or energy resource not only limits the magnitude and flow of cycling but also the pattern of the population (Legner, 1975); for example, the climatic or edaphic limitations to tree growth or the effect of availability of silica as a structural component for plant growth. For all these reasons, it is less the absolute quantities of resource available than their global cycle that is significant. There is now an extensive literature on the global cycles of the main nutrient elements, particularly carbon (Woodwell and Pecan, 1973; Olson *et al.*, 1978), especially in relation to energy, climate, and carbon dioxide concentration (National Research Council, 1977). These studies provide estimates of the steady-state reservoirs of carbon and the carbon fluxes. Ultimately, it is against the background of these estimates that the contribution of any particular ecosystem or any particular component must be measured. Selected carbon fluxes estimated by Olson *et al.* (1978) are shown in Table II. The data indicate the importance of the terrestrial ecosystems in carbon and, therefore, energy flux. Similar data are available for the global nitrogen cycle (Delwiche, 1977) and also imply greater flux in the terrestrial ecosystems. Within these global cycles, the only flux to which protozoa are likely to contribute a significant biomass is the sedimentation of marine plankton to the abyssal zone, and this is a region of minimum cycling. In relation to available resource and biomass, therefore, protozoa commonly constitute a trivial component. In many cases, however, their populations are largest where flux is greatest.

4.1.2. Population Density and Biomass

Estimates of the biomass of various ecosystems are given by Whittaker (1975) and show that both on a square meter basis and on a global basis, the

Table II. Global Carbon Fluxes of Various
Reservoirs at Steady State

Flux (Gt/y)	Process
180	Air–ocean interface
85	Surface and thermocline to deep ocean
0.4	Sedimentation
17.7	Oceans
46	Aquatic organisms
112	Terrestrial ecosystems

terrestrial biomass constitutes by far the greatest part of the global standing crop, just as it provides the greater part of net global primary production. The implications are, therefore, that protozoa will constitute a far greater proportion of the biomass in aquatic environments than in terrestrial environments proportionately, but not absolutely. Since the nutrient flux largely involves organic matter, the differences in the size and nature of the biomass have implications for turnover times and, therefore, the role of separate components in nutrient turnover. Woody materials include a high proportion of recalcitrant molecules, and consequently in forest turnover there tend to be two cycles, a slow cycle and a relatively fast cycle (Olson et al., 1978). Similarly, aquatic ecosystems have a relatively rapid turnover time, at least in surface waters, although the standing crop is small. Keeling and Bacastow (1977) developed a model suggesting that the short-lived land biota carbon pool contains 4.6% of the total biotic carbon but contributes 54% of the carbon exchange with the atmosphere, the remaining flux and mass being attributed to the long-cycled carbon pool. The turnover times of the respective pools were estimated to be 2.5 and 60 years, but in this model no distinction was made between living and dead materials. The main point is that so little carbon is stored in the short-cycled biota. There is a similar contrast between the turnover of the biota of the surface waters and of the deeper waters (Keeling and Bacastow, 1977) so that in fact nutrient cycling, whether on a global or ecosystem basis, presents fluxes of very different rates and consequently of very different significance. It is this fact that redresses the imbalance of the standing crop or biomass, for the microbial biomass may be small in relation to that of higher plants, but its rate of cycling is infinitely faster.

4.1.3. Trophic Structure

Population dynamics are a function of the trophic structure of a community, and relevant factors are the stability or instability of the community, the relative significance of consumers and decomposers, and the importance of predators and parasites. Wiegert and Owen (1971) point out that the life history of the primary autotrophs determines the density restrictions on the primary consumers and that this results in the typical three-level trophic communities of terrestrial ecosystems and the four-level communities of aquatic systems. Probably more important in microbial ecology, including the ecology of the protozoa, is the relative contribution of biomass and detritus to energy and nutrient cycling. Where detritus constitutes a major component, decomposers will predominate over consumers, and microorganisms will tend to provide the major nutrient and energy flux. This is the case in many terrestrial ecosystems, in benthic habitats, and in polluted waters. Where living autotrophs prevail and where there is a relatively slow rate of turnover, the rate of nutrient cycling may be primarily due to large consumers, but where such consumers are active, microbial decomposers in the gut and in the feces play a significant role and these include protozoa (Stout, 1974). With short-lived autotrophs, such as algae or bacteria, protozoa can play a direct role as consumers as well as in the decomposition network. Indeed, if the photosynthetic protozoa are considered as primary producers, protozoa can be considered playing an autotrophic role. However, it is their role as part of the decomposer network that presents the still unsolved ecological riddle.

4.2. Marine Ecosystems

There is little information available on the nutrition and hence nutrient cycling of planktonic or pelagic protozoa, particularly of phagotrophic marine protozoa, which are commonly fastidious both with respect to their food and their feeding conditions. Hutner and Provosoli (1955) reviewed the nutrition of planktonic phytoflagellates, particularly dinoflagellates, chrysomonads, cryptomonads, and coccolithophorids. The photosynthetic dinoflagellates do not have the heterotrophic capabilities of freshwater flagellates and require both vitamin B_{12} and thiamine. No substrate appears to stimulate their growth in light, and efforts to grow them in the dark have failed. The nonphotosynthetic *Oxyrrhis* appears to need live food, such as phytoflagellates or yeasts, but does not require a solid surface for growth. Phagotrophic and planktonic chrysomonads have similar nutritional requirements to the planktonic dinoflagellates. Gold (1973) cultured tintinnids in continuous culture using a diet of four species of algae that were known to be growth promoting. Food input and flow rate were controlled to give a continuous yield of ciliates; in particular, the vitamin supply of the cultures was restricted to prevent blooming of the algae, which would have been detrimental to ciliate growth. Such cultures gave a steady yield of some 200–300 cells/ml over a period of some months, and the cultures were rapidly responsive to any changes in food levels. Because of multiplication of the algae in the culture vessels, it was not possible to calculate the exact food consumption and growth ratio of the ciliates, but populations of 200–250 cells/ml were maintained by a daily input of $1.2–1.3 \times 10^8$ algal cells. The ecological significance of vitamin requirements was discussed by Provasoli and Pinter (1953) and has been further investigated by Droop (1953, 1968, 1970). Droop has also investigated nutrient limitation in osmotrophic protozoa. He developed an empirical relationship relating specific growth rate in steady-state systems to nutrient status, with respect to more than one nutrient. He postulated (1) that the specific uptake of each substrate depends on its external concentration and is independent of the internal concentration; (2) that uptake and the cell quota are related linearly; and (3) that the steady-state uptake is the product of the cell quota and specific growth rate. From this he concluded that there could be "luxury consumption" where there were small numbers of rate-limited cells in an otherwise rich medium, and this would explain why natural waters become nutritionally impoverished during planktonic blooms. Blooms are not confined to algae and flagellated protozoa; ciliates may also cause "red water" (Bary and Stuckey, 1950), but the nutritional significance of these blooms is not known. Although amebae may be common in the surface microlayer (5 μm) and subsurface of pelagic water (Davis *et al.*, 1978) and their numbers correlate with particulate ATP and inversely with O_2 tensions, no quantitative data are available on their role in nutrient cycling. Little quantitative information is available on the trophic relations of radiolaria and foraminifera, but the implications of such data as are available on the planktonic ecosystem are of relatively slow and small flux, basically subject to nutrient limitation and specifically limited by vitamin concentration. Within these limits, however, the planktonic protozoa may comprise a major part of the nutrient cycle. With a very small biomass (mean 3 g/m^2) and an estimated 125 g dry matter/m^2/year mean productivity (Whittaker, 1975) largely subject to nutrient limitation, the planktonic ecosystem is relatively small and inactive, and measurement of the exact role of the different components is difficult because of diurnal and seasonal change.

Very different is the salt-marsh ecosystem. With a mean biomass of 15 kg/m^2 and a mean productivity of 2 kg/m^2/year, this is a highly productive system. Studies on highly productive psammolittoral and epiphytic communities indicate rapid changes in community structure with successive blooms of diatoms, foraminifera, and nematodes throughout the summer (Lee and Muller, 1973). Lee and Muller (1973) found that the foraminifera they studied were approximately as productive as the nematodes and crabs previously studied in such ecosystems and had a growth efficiency of approximately 18% and, in view of their large standing crops, turnover of species, and high growth efficiency, concluded that they could play a significant role in the carbon budget and nutrient cycling within the detritus microbial assemblages of salt marshes. Contrasting with the data for planktonic protozoa, this evidence suggests that the foraminiferan species studied are very opportunistic and well adapted to play an important trophic role in the rapidly changing food webs of the community. Similarly, folliculinids, which are commonly found attached to submerged surfaces and marine animals, are omnivorous in diet and tolerant of a wide range of environmental conditions (Das, 1947), contrasting strongly with the tintinnids.

Fenchel (1967, 1968a,b, 1969) studied the ecology of the marine microbenthos, and in particular the ciliate fauna. In the Scandinavian waters that he studied, ciliates numbered $10^6 - 10^7/m^2$, representing up to 2.3 g wet weight. Highest numbers were found in fine sand and in localities with a rich growth of sulfur bacteria. In these sediments, their biomass is of the same order, and sometimes larger, as that of the other microfauna (nematodes, turbellarians, gastrotrichs, etc.). In the coarser sands, ciliates are fewer, and the micrometazoans play a greater role. This distribution Fenchel explained by the small size of the ciliates and their ability to endure reducing and anaerobic conditions. He found ciliates less numerous in detritus layers covering clay sediments. Fenchel found large numbers of dinoflagellates ($10^5 - 10^7/m^2$) and fewer euglenoids (less than $5 \times 10^5/m^2$). Naked amebae played a small role, seldom numbering more than $10^5/m^2$. Based on their respiratory activity, Fenchel (1969) concluded that the microfauna was more important than the macrofauna and that the ciliates contributed the greatest single microfaunal component. He recognized three microfaunal communities, the sublittoral sand microbiocenosis, the estuarine sand microbiocoenosis, and the sulphuretum, and he illustrated the variety and diversity of the fauna and its trophic structure in the three communities. Like Lee and Muller (1973), his study indicates a diverse and adaptable population. He cited a total biomass of some $2-300$ g/m^2 at one site, a figure somewhat larger than Whittaker's (1975) figure for the continental shelf, but close to his estuarine figure. Fenchel studied the feeding habits, the community succession, and the reproductive potential of the ciliates and concluded that they play an important role, particularly in the decomposition of large amounts of organic detritus, by diminishing the number of bacteria and protophytes in the sediments. A pioneer student of the ecology of these psammobic and sulfuretum ciliates was Fauré-Fremiet (1950a,b, 1951a,b,c), who similarly remarked on the morphological adaptations, diversity, and trophic succession of these protozoa. His work was followed by Dragesco (1960, 1965a,b) who worked in both Europe and Africa, Raikov (1962) who worked in Russia, and Borror (1963a,b, 1965, 1968) and Elliott and Bamforth (1975) in North America. Their work was with ciliates and flagellates. Chardez (1972), in Belgium, and Golemansky (1976, 1978), in Rumania, worked on the amebae peculiar to this habitat. The psammobic fauna thus generally contrasts with the planktonic and salt-marsh faunas in its composition, with a predominance of ciliates, fewer flagellates and amebae, and specialized testacea, with the tendency for morphological adaptation to the interstitial character of the habitat (Corliss and Hartwig, 1977). Fenchel's (1969) data for a 10-m-deep site with sulfur bacteria indicated that bacteria form the greater part of the ciliate diet with flagellates, diatoms, and other ciliates comprising the rest. In the northernmost site, in midsummer, bacteria were less important, and flagellates, diatoms, and ciliates comprised the greater part of the diet. At a shallow beach locality, diatoms comprised the major part of the diet. Fenchel's data indicated both the vertical and seasonal distribution of the ciliate species, confirming the changing character of the fauna with site, season, and depth.

One marine ecosystem for which there is less evidence of protozoan ecology is that of coral reefs and related habitats, which Whittaker (1975) estimates have the highest mean biomass and highest net primary productivity not only of all marine but of all ecosystems. Although considerable work has been done on ciliate commensals of coelenterates, no estimate of their total contribution to the biomass or energetics of these systems appears to be available. Populations of bacteriovores, commensal in the sea urchin, appear to be correlated with the ingestion of suitable bacterial prey by the urchin (Beers, 1963), and the carnivores feed exclusively on the smaller endocommensal ciliates within the echinoid large intestine (Berger, 1967). A better understanding of the nutrient flow within this commensal system might permit a better evaluation of the role of protozoa in these highly productive ecosystems.

In summary, marine ecosystems so far studied display a wide variation in the size and distribution of the biomass, in community structure, and in population interaction and metabolism. The largest biomass appears to be concentrated in littoral areas, such as salt marshes, shallow benthic communities, or coral reefs, where light and nutrient supply favor a high net productivity and a fairly rapid turnover. In such systems, protozoa appear to form a significant component, though the structure of the communities, their metabolism, and their interaction show striking contrasts both in the protozoan taxa involved, in their physical relationships, and in their trophic behavior. In the salt marsh, there appears to be a greater dependence upon algae as the primary food source, whereas in the benthic communities, detritus and sulfur bacteria provide the major food source. In the sands studied by Golemansky and in the pelagic communities, the fauna is more specialized, more fastidious in its food requirements, less capable of change and fluctuation in numbers, and therefore more restrictive in its nutrient cycling potential. But in these communities, net primary production is very low, nutrients are typically limiting, and total and protozoan biomass is confined to a very small figure.

Nutrient cycling in pelagic surface communities is limited primarily by the availability of major nutrients, particularly P, that are commonly replenished in the convergences by the upwelling of cold nutrient-rich waters. Nutrient cycling could also be restricted by Droop's "luxury consumption" since this would tend to accentuate nutrient lack in a normally depleted environment. The long mean residence time of oceanic waters [50,000 years, (Whittaker, 1975)] indicates the importance of vertical stratification and mixing. The most obvious protozoan contribution to such nutrient cycling must then be the sedimentation of foraminiferan ooze, estimated to be between 0.005 and 0.01 mm/year (Whittaker, 1975). Relative to the total surface turnover, however, contributions by the flagellate and ciliate faunas could still be important, in particular where evidence of ciliates forming a significant part of the food chain is available. Berk et al. (1977) found that ciliates comprised about 20% of the total planktonic biomass in estuarine waters off Chesapeake Bay and that they were consumed by copepods. They concluded, therefore, that ciliates could form an important source of food for the copepods. Previously, copepod food had been considered to be exclusively phytoplankton, but several reports have indicated that the phytoplankton biomass appeared to be inadequate to feed the filter-feeding copepods. Thus, both flagellates and ciliates may be more important in planktonic communities than hitherto thought. Evaluations of this food chain will require more detailed measurements of nutrient turnover.

The work on salt-marsh communities appears to confirm the importance of foraminifera in the consumption of the algal biomass, and their short life and relatively rapid turnover imply a major role in nutrient cycling. A similar position appears to hold with the benthic ciliates and flagellates. In their case, the situation appears to be largely one of detritus decomposition, but in this case, as with the pelagic growth, N and P are important limiting nutrients. Within the psammobic habitat, the physical penetration of the ciliate fauna into the interstitial spaces ensures the retrieval and recycling of detrital nutrients where comminution by larger microfauna is no longer possible. However, in the detrital habitat and in the sulfuretum, bacteria play the key role in nutrient cycling. Johannes (1965, 1968) studied nutrient regeneration in experiments with just bacterial decomposers and in experiments with a bacteriophagous ciliate. His experiments showed the liberation of phosphate only when the ciliate predator was present. Fenchel, however, argued that because the bacterium had been previously grown on nutrient-rich medium, which was not present during the experiment, the situation was not a steady state and that without a carbon source the reduced bacteria in the grazed experiment were unable to grow and take up the released phosphate (Fenchel, 1978; Fenchel and Harrison, 1976; Fenchel and Jorgensen, 1977). It is clear that micropredation may increase nutrient availability, but whether this increases the rate of nutrient cycling in the long term will depend on many factors. Increased nutrient availability will lead to increased biomass, and depending upon the character of this biomass, the rate of nutrient cycling will be accelerated or diminished.

4.3. Freshwater and Polluted Ecosystems

Although comprising a far smaller part of the earth's surface and a far smaller part of the biomass than that of the oceans, freshwater ecosystems have been more intensively studied and in particular their protozoan ecology. Numerous surveys (Noland, 1925; Wang, 1928; Picken, 1937; Lackey, 1938) have revealed the diversity and extent of the fauna and provided information on their feeding habits, and these surveys have been supplemented with extensive experimental studies detailing the nutrition and growth of the fauna. Like the marine ecosystems, the fauna is smallest and the biomass least in open waters, such as lakes, or fast flowing streams (Lackey, 1938). The nannoplankton of mesotrophic and eutrophic lakes, like that of the oceans, has a high production rate but also a high turnover rate, probably through predation and lysis (Kalff and Knoechel, 1978). No quantitative data are available, however, on the role of the protozoa in this turnover. In most freshwater ecosystems, the protozoa are most numerous in association with plants or microbial populations and where there is a substrate to which they can adhere or over which they can move and feed. These occur in relatively still or stagnant waters. The protozoa may be associated with submerged plants or algal blooms, where there is an adequate O_2 supply, with detritus or sediments, or with fungal or bacterial colonies. Where the substrate is initially living and passes through a cycle of senescence and decay, the protozoan population will change, and there have been numerous studies of such successions, such as those of Picken (1937) and Fauré-Fremiet (1950c). This is shown by the autotrophic and osmotrophic protozoa but most conspicuously by the algivorous and bacteriophagous species. Normally the algivorous species give way to bacteriophagous species as decomposition takes place. Such a change is shown even more dramatically where a freshwater stream is polluted by effluent and has been studied in an English chalk stream by Gray (1951, 1952) and an American stream by Small (1973). Apart from such typical freshwater habitats, there are also specialized faunas associated with more restricted habitats, such as tree or stump holes, temporary ponds, the reservoirs of the pitcher plant, and anaerobic sediments (Lauterborn, 1916). But although the fauna and their feeding habits have been well characterized, few measurements have been made of their role in nutrient cycling or even of their relative biomass in the ecosystem. Schönborn (1977) studied protozoan productivity at two German river sites. One site was only moderately polluted and developed a fine algal mat on which testacea grew freely; the second, more strongly polluted site had a more varied microflora and fewer shell-bearing protozoa, but often with a massive production of diatoms and vorticellae. The investigations of biomass were restricted to a very small part of the biomass, the shell-bearing protozoa, including both the testacea and loricate ciliates, and the river is not an optimum habitat for testacea. Nevertheless, it was interesting that Schönborn found greater production and abundance at the less polluted site but a low turnover rate, whereas at the polluted site, although the biomass was smaller the turnover rate was greater because of the higher incidence of loricate ciliates. The estimated annual production figures were 1.0 and 0.35 g/m^2, respectively; Laminger (1973) estimated 2.48 g/m^2 for the testacean fauna of an alpine lake sediment. Such figures do not indicate a major role in nutrient cycling by these organisms in such habitats.

The increasing pollution of fresh waters by industrial, agricultural, and domestic wastes led Kolkwitz and Marsson (1909) to introduce a classification of waters which has been widely used and further developed (Bick, 1963). Four main categories were initially recognized: katharobic—water essentially devoid of organic matter, well oxygenated, and with very few bacteria, such as is found in some freshwater springs; oligosaprobic— well-oxygenated, relatively clean water in which mineralization of organic matter is largely completed and chlorophyll-bearing protozoa may occur in large numbers; mesosaprobic—water with organic matter and bacteria but having a good O_2 supply and normally the habitat with the richest protozoan fauna; polysaprobic—water with decomposable organic matter of high molecular weight sustaining very high populations of bacteria and with consequently reduced O_2 tensions but high sulfide and carbon dioxide tensions. The polysaprobic water contains a protozoan fauna of predominantly bacteriophagous ciliates.

The correlation between fauna and water quality has been extensively documented (Bick, 1963, 1968; Bick and Kunze, 1971) and widely used to assess the degree of pollution. However, although there is an obvious relation to nutrient cycling and relevant parameters, such as O_2 tensions and carbon dioxide, are commonly measured, there has been little quantitative study of the role of protozoa in nutrient cycling in these waters.

Much of the difficulty in evaluating the role of protozoa in natural populations lies in monitoring the small and rapid changes in nutrient balance in which they play their role. For this reason, microcosms have been developed which enable both closer and more complete monitoring of nutrient cycling than is possible in the field and also permit more exact measurement of the role of the different components. Initially, aquatic microcosms tended to have short lives because of the rapid and irreversible changes in the systems, but more recent developments have shown that steady states can be preserved for as long as 3 years, permitting a convincing study of nutrient cycling and the participation of the different biota (Ringelberg and Kersting, 1978).

The most extensive series of laboratory experiments have been carried out by Bick and his associates. Many of their studies relate the distribution of ciliates to specific environmental parameters, such as O_2 tension, carbon dioxide tension, salinity, ammonia concentration and pH, and numbers of bacteria, and comprise therefore autecological characterization of a wide range of species normally found in different habitats or different stages of a faunal succession (Bick and Kunze, 1971). However, this work was extended to study the environmental conditions and succession of ciliates associated with the decomposition of peptone and cellulose and related the number of species and the number of individuals to the rate of breakdown of these nutrients (Bick, 1967; Bick and Schmerenbeck, 1971; Schmerenbeck, 1975). Although these studies related population changes to fluctuations of mineral nitrogen, for example, and in some cases measured both the bacterial and protozoan biomass, it was not possible to quantify the role of the protozoa, although they clearly were of major importance in these laboratory systems. What these experiments do show, however, is the importance of faunal diversity in sustaining decomposition rates.

The applicability of the laboratory microcosms to natural waters is still uncertain, but in one case, that of sewage treatment plants, such as trickling filters, activated sludge, or Imhoff tanks, which are naturally partly closed and controlled systems, the relation of the laboratory experiments to the field system is much more convincing. Ciliates are the dominant protozoa in activated sludge, and those which either attach themselves to or crawl over the surface of the sludge are the most common. There is a close relationship between the composition and activity of the bacterial and protozoan populations (Reid, 1969). Experimental work has demonstrated that ciliated protozoa are essential for the production of good-quality effluent (Curds, 1973). This is possibly the clearest case of protozoa as a highly significant component in the nutrient cycling of a near-natural ecosystem and is partly due to the restricted fauna of an activated sludge in which the ciliates are such a conspicuous component. The experimental evidence indicates that the ciliates function to disperse bacterial growth by predation and reduce the number of coliform bacteria present, thus facilitating substrate diffusion (Maxham and Hickman, 1974). They can also use sugars known to occur in the activated sludge (Brown, 1967). Because of the relative simplicity of this ecosystem, it has proved particularly susceptible to simulation modeling, and mathematical models have been devised which relate both the bacterial and protozoan populations (Curds, 1973). The experimental data had shown that the presence of the protozoa reduced the half-life of *E. coli* from 16.1 hr to 1.8 hr, indicating a dramatic effect on nutrient turnover. However, the presence of carnivorous protozoa feeding on the bacteriophagous populations caused fluctuations in their numbers and consequently affected the rate of sewage clarification (Curds, 1973). More recently, there has been further development of the modeling of protozoan predation in batch and continuous culture, providing a detailed analysis of the kinetics of predation relating the growth and mortality of the prey and predator and the specific rate of predation (Curds and Bazin, 1977). It has been possible to study food preferences and food selection and predator yield. In continuous culture, it is possible to establish a steady-state relationship, but what is of interest is that the predator may take time to adapt to the steady state (Curds and Bazin, 1977). These experiments, although not directly related to field situations nor generally related to nutrient turnover, indicate that quantification of nutrient turnover is possible in terms of population dynamics and as such provide promise of further developments.

One of the most interesting and successful attempts to do this has been a study of the phosphorus cycle in a model tundra ecosystem by Barsdate *et al.* (1974). Using ^{32}P and a simple fractionation technique, they investigated phosphorus pools in simple laboratory microcosms consisting of *Carex* detritus from a tundra pond seeded with the natural bacterial microflora, with or without protozoan predators. Sterile microcosms were used as controls. Their conceptual model (Fig. 8) allowed for five pools: free dissolved phosphate, adsorbed phosphate, surplus bacterial phosphate, bound bacterial phosphate, and protozoan phosphate. From their measurements over periods of 4–10 hr on cultures considered to have reached steady state, they estimated transfer rates between the pools. They found that transfer rates varied greatly depending upon the initial concentration of dissolved phosphate, but that generally the transfer rate from

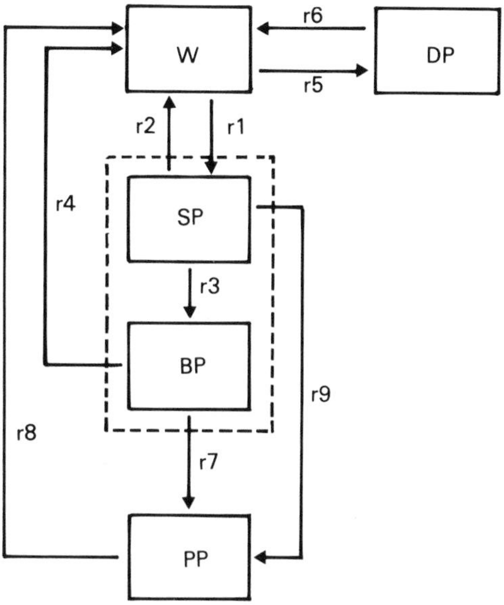

Figure 8. Conceptual model of the phosphorous cycle of the systems. W, DP, SP, BP, and PP are pool sizes of phosphorus of the water, sorbed to detritus, surplus phosphorus, and biologically bound bacterial phosphorus and protozoan phosphorus, respectively. rl–9 are transfer rates between the pools believed to be important. From Barsdate, Prentki, and Fenchel, 1974.

the free dissolved phosphate to the organisms was greater when the protozoan predators were present than when they were absent, and that the uptake by the bacterial cells when the predator was present was invariably much greater. They concluded that the fourfold increase of uptake rate by the bacteria in the presence of the protozoan predators which they recorded was due to the high proportion of young, rapidly dividing bacterial cells present in the grazed microcosm, compared with the static population in the ungrazed microcosm. Their data, however, indicated that only a small fraction of the phosphate (4.3%) was cycled by the protozoa and that therefore the activity of the P cycle was independent of the activity of the C cycle in which the protozoa were presumed to play a larger role. Further, the rapid cycling of P was shown to be largely independent of the release of P from the detritus. They suggested that the turnover time of P in the pool of free dissolved P could be about 2 min, but they noted that the highest uptake velocities occurred in the systems with a low free phosphate concentration and were much slower with high initial free phosphate concentrations. In developing their simplified concepts, Barsdate *et al.* (1974) made two major assumptions: first, that the cycling of one major nutrient, such as P, could be treated independently of others, e.g., N or C, and second, that differences in transfer rates were a function of cell turnover rates. Neither of these assumptions can be accepted uncritically. As they themselves suggest, differences in bacterial uptake rates of P can reflect the operation of distinct uptake systems. These are quite well known, and the low-affinity system that operates at low P concentrations may have a higher rate of uptake than the high-affinity system that operates at high P concentrations (Medvecsky and Rosenberg, 1971; Beever and Burns, 1978). The high cell-turnover rates are commonly attributed to the continual breakdown and resynthesis of cell protein. An alternative interpretation is that phosphate uptake by the cell is governed by a feedback system which relates cell uptake to cell demand (Beever and Burns, 1978). On this hypothesis, the differences recorded by Barsdate *et al.* (1974) are a direct function of the physiological state of the bacterial population, as they themselves suggest, but the differences in uptake rate are not a function of turnover but of sustained cell synthesis in a grazed population. Whatever may be the mechanism of phosphate exchange between the free dissolved pool and the bacteria, their experiments clearly demonstrated the role of protozoa in accelerating the uptake of P. They also demonstrated a highly significant increase in the release of P from the detritus in the presence of the protozoa, compared with a simple bacterial microcosm (Table III). This is probably the most convincing data available demonstrating the role of protozoa in mineral cycling, but it is important to stress that the protozoa themselves play an indirect, rather than a direct, part in the cycling of the P.

Barsdate *et al.* (1974) included in their study computer simulation of their experiments using the conceptual model already described and changing the values of pool size or transfer rate to provide various alternatives. In particular, they illustrated the difference between a microcosm with a relatively large initial free pool size and one with a low initial free pool size in the subsequent transfer to the bacteria-bound and bacteria-surplus pools (Fig. 8). They assumed, however, that the total transfer rate would be increased by the addition of grazers, though they pointed out that there would be a change in the bacterial population so that a steady-state model was not applicable. However, the assumption that rates would increase when the food chain is lengthened appears to be untenable, and their equations appear to overlook the fractionation of the initial P flux that would follow the addition of further steps in the cycle. What is important, however, is that for the first time an attempt has been made to link an

Table III. Uptake Rates of Phosphate by Bacteria in Grazed and Ungrazed Systems and the Regeneration of Phosphate from *Carex* Litter

Transfer velocity	Bacteria alone	Bacteria and ciliates
Water to organism		
μg P/liter/hr	18.9	147.3
μg P/bacterium/hr $\times$ 10^{-7}	16.7	191.3
Carex litter to water and resin bags in water		
μg P/liter/hr	0.028	0.84
μg P/bacterium/hr $\times$ 10^{-7}	0.025	1.09

experimental laboratory technique to a real ecosystem both by modeling, which provides the opportunity of devising test measurements, and by the use of radio-tracers which facilitate such measurements. The development of this approach offers a very fruitful field for future study.

In summary, freshwater, polluted, and sewage ecosystems provide a range comparable to that of marine ecosystems, although there is little quantitative data available of the total biomass, protozoan biomass, or rate of nutrient cycling in the field. Probably the greatest protozoan biomass is to be found in activated sludge, where Curds (1973) calculated the protozoa to form 5% of the suspended dry matter. Far more is known of the composition of the population, its food preferences, and its ecological tolerance. In the least organically enriched waters, autotrophic flagellates predominate. With increasing organic content, the diversity of the population increases, and sequences dominated by flagellates, bacteriophagous or algivorous ciliates, and predators are commonly associated with nutritional strata. In polysaprobic conditions, such as heavily polluted waters or activated sludge, ciliates tend to become dominant, and often population fluctuations are induced by the activity of protozoan carnivores. Where anaerobic conditions prevail, in lake sediments or Imhoff tanks, a specialized fauna capable of reproducing under anaerobic conditions is found, and again this is predominantly a ciliate fauna.

The role of the protozoa in the energetics of such ecosystems and in nutrient cycling is far less understood. For a very long time, the association of polysaprobic ciliates with dense bacterial clumps has been well known, and it is generally accepted that they play a major role in dispersing the clumps and clearing the liquor, particularly in activated sludge. This implies accelerating the mineralization of N, P, and S. But quantitative data are not yet available. The possibility of extrapolating from laboratory cultures to field conditions has been greatly enhanced, however, by the development of simulation modeling, which now only requires the integration of population dynamics and nutrient cycling.

4.4. Terrestrial Ecosystems

Terrestrial ecosystems differ from aquatic ecosystems in two important particulars. Although water still comprises a substantial part of the system, its distribution is discontinuous, both in time and space. This has important implications on the bionomics of terrestrial organisms and on the circulation of nutrients: life histories tend to have intermittent periods of activity and inactivity—for the protozoa this often means encystment and excystment; and isolation of small communities is the rule rather than the exception. Although there is an enormous range of biomass in terrestrial ecosystems, the smallest tends to be as great as the oceanic biomass, whereas grasslands may be 2-3 orders of magnitude greater and forests 3-4 orders of magnitude greater (Whittaker, 1975). Productivity is generally as great, and in forests an order of magnitude or more greater. Terres-

trial organisms are dominated, therefore, by plant biomass and plant productivity and the enormous mass of detritus associated with it (Wiegart and Owen,
1971, Schlesinger, 1977). These provide the principal parameters of nutrient
cycling. Because of their small size and distribution, protozoa colonize epiphytic
habitats readily, and quite complex communities may be found on the surface
of living leaves or even bark (Bonnet, 1973). However, they are particularly well
represented in association with mosses and lichens, which have proved one of
the richest sources of terrestrial protozoa (Fantham and Porter, 1945; Wenzel,
1953; Gellert, 1955, 1956). Their distribution is related to the moisture regime
of the habitat, the type of moss association, and the exposure to light (Meisterfeld, 1977), and the size and form of the protozoa, as with the psammobic fauna,
reflect the limitations of their habitat (Bonnet, 1975, Chardez, 1968). Many of
these associations are oligotrophic, being in strongly acid sites where nutrients
have tended to be leached from the soil and where soil processes tend to be
degrading. Because of this, such communities are often living in a nutrient situation where little is added from rainfall or plant exudate and where there are only
small pools of nutrients to cycle in a diverse microfaunal population. Although
vastly different in structure, the nutrient dynamics of the moss ecosystems are
not too different from those of the open ocean.

Because they are the most obvious and easily determined populations, most
quantitative data refer to the testacea. This is unfortunate since large populations of ciliate protozoa, often with quite large species, are present in most of
the wetter moss habitats, and since their growth and rate of turnover are typically greater than that of the testacea, estimates of protozoan activity based only
on the latter can be misleading. Meisterfeld (1977), who studied 189 samples of
sphagnum from 15 bog and fen sites in Germany, estimated a mean biomass of
living testacea of 0.7 g/m^2 and a maximum value of about 4 g/m^2. Schönborn
(1977), who studied moss from the drier habitat of a beech forest floor in North
Germany, estimated the biomass of the testacea to be only 0.015 g/m^2 and
showed that it fluctuated widely with rainfall. The annual production was only
0.11 mg over 16–17 generations. However, there is in this case no overall measure of metabolic activity or nutrient cycling against which such a contribution
can be judged, although the testacea are certainly a major component of this
community.

In fens and bogs, the soil or peat underlying such vegetation or that which
has developed under it appears equally inert. It is often very old, comprising recalcitrant organic materials not readily decomposed, often under anaerobic conditions, and has very low metabolic activity. Nevertheless, there may be a large
protozoan fauna associated with it, comprising a wide range of flagellates, amebae, and ciliates (Stout, 1971). This evidence again suggests that, although the
protozoa may play an appreciable role in such nutrient cycling as takes place,
the activity of the whole ecosystem tends to be depressed. Much the same situation obtains in moorland soils, in tundra, and in alpine and polar communities

(Latter *et al.*, 1967; Smith, 1973a,b). Smith (1973b) studied the population ecology of the testacean *Corythion dubium* in Signy Island (Antarctica), where there is a marked seasonal contrast between the winter, when the peat is frozen, and the summer, initiated by a spring thaw. There was quite a large protozoan fauna present, though ciliates were very rare and amebae entirely absent, the flagellates and testacea constituting almost the entire fauna. In this case, therefore, the population dynamics of a dominant testate species are a good indication of total protozoan activity and of their possible role in nutrient cycling. Smith found that the dominant species, *Corythion dubium*, grew rapidly following the spring thaw and that the population peaked again in the autumn, and it appeared that this species reacted more rapidly than others to changes in the environment. The testacea formed a significant part of the microbial biomass, apparently substantially greater than the bacteria (Collins *et al.*, 1975). During the one spring month, bacterial populations were estimated to have 2.2 generations in the surface peat and to produce 1.8 mg dry weight/m^2, and *Corythion dubium* was estimated to have 2.65 generations in the same time. While the protozoan population remained relatively stable during the summer, the bacteria declined sharply, and it was concluded that this was due to protozoan predation (Collins *et al.*, 1975). If this were so, it provides a dramatic example of protozoan predation limiting bacterial growth. Conversely, it suggests that predation accelerates nutrient turnover in a nutrient-deficient environment where competition would be intense between the primary producers and the decomposers.

Contrasting strongly with such habitats are those associated with the normal decomposition cycle of temperate and tropical forests. These ecosystems have a very large biomass (6–200 kg/m^2, mean 35 kg/m^2) and a very high net productivity (600–3500 g/m^2/year) (Whittaker, 1975) as well as a high rate of turnover. They provide a wealth of niches for protozoa (Bamforth, 1973), which are found from the leaf surface, through the senescent and decaying leaf, in the litter, and in the mineral soil horizons, in fact with all stages of decomposition and nutrient cycling. Numbers, and hence biomass, of the protozoa vary greatly, both with the climate and type of vegetation (Fig. 9, after Bamforth, 1973). Readily decomposed broadleaf forests tend to harbor the largest populations and conifers the smallest, but there are significant population differences within the same type of vegetation depending on soil and climatic conditions (Stout, 1974), particularly the contrasting soil patterns of mull and mor (Stout, 1968). Two recent studies have provided some quantitative data of population dynamics: those of Coûteaux (1975, 1976), who studied three mor-type soils, an oak wood and a spruce wood in Belgium and an oak wood near Paris, and Lousier (1974a,b, 1976), who studied an aspen woodland in the Rocky Mountains of Canada (Lousier and Parkinson, 1976). Both confined their studies to the testacean populations.

Coûteaux found that populations were greatest, but diversity least, in the spruce wood and that the Belgian oak wood had a larger population but a less

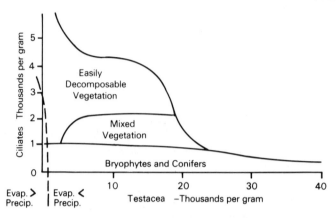

Figure 9. The relationship of the number of ciliates and testacea in litters to the type of vegetation. Region to the left of the broken line denotes arid habitats. From Bamforth (1973).

diverse fauna than the Paris oak wood. Her data, based on weekly counts over 15 months, were expressed as the number of trophic individuals, encysted individuals, and those with a closed epiphragm, expressed on a dry weight basis. There are contrasting seasonal patterns, the numbers dropping in the winter in the Belgian oak wood more markedly than at the other two sites. Moisture was considered to be the prime cause of population fluctuation, but there was a seasonal cycle of 6 months for the Belgian sites and 40 weeks for the Paris wood, with individual species showing some variation from the basic pattern. Many species, for example, had their main growth in the spring. The population dynamics of other species showed no obvious correlation with seasonal or other factors. This study presented a great deal of detailed information on population dynamics, but it is unfortunate that it was not related more closely to nutrient dynamics; in particular, population density could have been expressed on a square meter basis, and sufficient data were available to have attempted an estimation of biomass and particularly changes in biomass during the seasons. This could well have provided data on the energetics, and possibly nutrient turnover, in the testacean population.

Although Lousier's study was confined to one site with soil and vegetation very different from those of Coûteaux, his findings supplement and extend her conclusions. Like Coûteaux, he found that moisture controlled the fluctuations of the testacean population, and the experiments he carried out showed a significant increase in populations of trophic cells in his well-drained soil when the moisture content was increased during a seasonal dry period (Lousier, 1974b) and a decline in the number of encysted forms. Both the numbers of trophic cells and the total populations showed positive correlations with the moisture status of the soil. Lousier also estimated the biomass of his population, its gener-

Table IV. Generation Times and Secondary Production of Testacea
in Watered and Unwatered Plots

Horizon	Watered plots		Unwatered plots	
	A_0F	A_0H	A_0F	A_0H
Production numbers	15,363	13,746	6367	4709
Intrinsic rate of increase (r)				
(individual animal per day)	0.102	0.114	0.064	0.097
Generation time (T, days)	6.8	6.1	10.9	7.2
Secondary production (g/m^2)	1.56	2.97	0.65	1.02
Standing crop numbers	1089	533	370	288
Standing crop biomass				
(g/m^2, wet wt)	0.11	0.12	0.04	0.06

ation time, and production. These were significantly greater in the irrigated soil than in the control soil. Although there was little difference in the standing crop between the two horizons studied (A_0F and A_0H) or in generation time, production was greater in the A_0H horizon (Table IV). Lousier developed a model to represent population processes (Fig. 10) which, given the biomass, turnover, and production figures, would allow some estimate of the energetics and nutrient turnover of the population. It is obvious, however, from the figures for the whole ecosystem (Lousier and Parkinson, 1976) that the direct role of the testacea in nutrient cycling must be small. One point of interest is that Lousier's figures of 6-11 days for generation time are similar to those obtained by Heal (1964b) from culture studies, suggesting that Lousier's assumptions were reasonably valid.

Far less data are available on the flagellate, amebae, and ciliate dynamics in forest ecosystems than for the testacea. Counts (Stout, 1962; Nielsen, 1968) indicate that very large numbers of ciliates can be attained and that, like the testacea, the counts vary sharply with the moisture status of the soil. No attempt has been made to assess the generation time or productivity of the ciliates, but it would undoubtedly be greater than that of the testacea, probably by an order

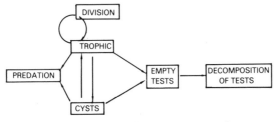

Figure 10. Population processes that can affect the turnover of testacean numbers. After Lousier (1974b).

of magnitude, since laboratory data on generation times indicate that reproduction can take place in a few hours rather than over several days.

Grassland ecosystems show a diversity comparable to forest ecosystems. Tall tussock grasslands have a large aboveground biomass and a slow cycling of nutrients, similar to woodlands. Semiarid prairies may have a small aboveground biomass and nutrient cycling limited by severe seasonal climatic fluctuations. Highly productive grazed pasture may have a productivity and rate of nutrient turnover comparable to that of the tropical forest. The protozoan populations are equally diverse, but there have been few studies of their dynamics. In the main, studies have concentrated on the influence of the rhizosphere (Newman, 1978) on the size, composition, and presumably the activity of the protozoan population, but little quantitative work is available on their role in nutrient cycling (Biczók, 1954; Nikolyuk, 1968, 1969; Darbyshire, 1969; Darbyshire and Greaves, 1967, 1973). Such studies showed the predominant protozoa to be flagellates and small amebae, both of which tended to show higher numbers in the rhizosphere than in the unplanted soil. However, grassland soils do possess a significant testacean and ciliate fauna, though the numbers never attain those of the smaller flagellates and amebae. In soils not subject to extreme climate conditions, a significant part of the protozoan biomass will consist of these groups as the individuals are commonly much larger than the small flagellates or amebae.

Elliott and Coleman (1977) studied the population dynamics of the protozoa in a short-grass prairie in Colorado for which considerable data were available on nutrient cycling. They found the dominant protozoa to be small amebae (about 10 μm in diameter), with small flagellates being much less frequent and few ciliate species occurring in any number at all. Curiously, cysts did not form a large part of the population despite the aridity of the environment. Like Lousier, Elliott and Colemann (1977) found that population dynamics responded principally to the moisture regime (in the prairie soil, a wetting and drying cycle), with populations increasing in response to irrigation. They showed that the amebae also responded to increased nitrogen availability. From their data, they calculated a mean generation time for the amebae under laboratory conditions of 2.4 days, but under field conditions they estimated it would be about 10 times longer. They estimated the protozoan biomass to be about 0.3 g/m^2 and the production to be about 1.3 g/m^2/year implying a lower turnover than in Lousier's woodland. However, their laboratory experiments demonstrated the significance of the wetting and drying cycle in this arid environment, and probably the population fluctuations could have been greater, and therefore the turnover, than their figures indicated.

Agricultural soils have been examined most closely for the relationship between protozoan populations and nutrient cycling, but in the main, such studies have compared the composition of the fauna and the size of the population under different manurial treatments or cultural practices (Singh, 1949). The most intensive study of population dynamics in a field of known history has been carried out at Rothamsted in England. Here, Cutler *et al*. (1922) followed daily population changes of bacteria and protozoa from July 5, 1920 to July 4, 1921 in a plot that annually received farmyard manure, and they also collected a great deal of meterological and soil data. Population data for six species (two small amebae and four small flagellates) were susceptible to statistical treatment, and they were able to show, like Coûteaux, that over and beyond daily fluctuations, there were seasonal patterns, in general the bacteria and protozoa being most numerous at the end of November and fewest during February. For one flagellate species, there was a 2-day periodicity. They found an inverse relationship between the numbers of bacteria and the numbers of active amebae in 86% of their total observations. From these data, Stout and Heal (1967) calculated the standing crop and annual production of the protozoa and estimated that the generation time was from 1 to 3 days. They concluded that the turnover was 50–300 times the standing crop. Singh (1946) also studied Rothamsted soils and, like Cutler *et al*. (1922), found the predominant species to be flagellates and amebae, with ciliates being two orders of magnitude less numerous. Even allowing for their greater size, this still suggests that in cultivated soils the ciliates are not as significant as the smaller protozoan species. Singh (1949) also studied numbers of amebae in two cropped soils under three different agricultural treatments: with minerals, farmyard manure, or unfertilized. The ratio of total counts in these three treatments was roughly 3:4:1 and the ratio of trophic cells 4:5:1. Although there was substantially more organic matter in the plots treated with farmyard manure, there was no difference between the other two. There were comparatively small differences in the bacterial and fungal populations, and Russell (1961) observed that only the protozoa tended to reflect differences in soil fertility. The data imply that there is a far greater biomass of amebae in the fertilized plots, implying a much greater pool of nutrients, and that there is probably a greater turnover since there is a higher proportion of trophic to encysted cells.

In summary, terrestrial ecosystems sustain sharply contrasting protozoan populations, from antarctic mosses with no amebae and numerous testacea to grasslands dominated by small amebae and flagellates. The richest fauna is found in the ecosystems with the largest standing crop, forests, and it is also the most diverse. The data available indicate that the protozoan biomass is in all cases minimal, not more than a few grams per square meter, but that in some habitats they may still form a significant part of the microbial population. Turnover rates are generally slow, either because of an inherently slow metabolic rate, as with the testacea, or because of prolonged inactive or encysted stages in their life history, as with ciliates. The most significant populations appear to be those of the small amebae, which are primarily predatory on bacteria. They appear to combine the most effective response to increased turnover with the largest biomass in systems in which there is rapid nutrient cycling. Where there is a greater proportion of the nutrients locked up in detritus, testacea and ciliates can form a larger part of the protozoan biomass, but their metabolic activity is probably less than that of the amebae of agricultural and grassland soils.

5. Conclusions

To determine the role of protozoa in nutrient cycling, it is necessary to have information on the magnitude and rate of flux; pools in which the nutrients are held and their residence times; and the differentials, both of cycling and residence time. Only in the artificial and restricted conditions of a laboratory microcosm are we likely to approximate this information. Nevertheless, it is worth examining analyses of natural ecosystems, since they should both provide orders of magnitude and suggest the areas where more critical data may provide valuable information. Of the major ecosystems, soil is the one for which there has been the most intensive analysis of nutrient cycling, particularly for agricultural, grassland, and forest soils.

Jenkinson and Rayner (1977) developed a model of organic turnover in the long-term classical field experiments at Rothamsted, England, based on five main sources of information: (a) long-term (10–100 years) changes in the organic matter of the soils; (b) incubation experiments with ^{14}C-labeled plant material over a 1 to 10-year period; (c) radiocarbon dating over a 1000-year period; (d) the pulse of radiocarbon generated by thermonuclear testing ("bomb" carbon); and (e) from information of the soil biomass. They recognized five compartments in their model: decomposable plant material, resistant plant material, soil biomass, physically stabilized soil organic matter, and chemically stabilized soil organic matter. The half-life of these pools were calculated to be 0.165, 2.31, 1.69, 49.5, and 1980 years, respectively. The age of the biomass component was estimated to be 25.9 years, although its half-life was only 1.69 years due to the age of part of the organic matter entering the biomass. Similarly, there was a wide range of ages of the soil organic matter, ranging from virtually zero to 2565 years. They predicted that the biomass derived from the annual application of 1 ton of fresh plant C/ha after 10,000 years would be 0.28 ton and the other components would be 0.1 ton of decomposable plant material, 0.47 ton of resistant plant material, 1.3 tons in physically bound organic matter, and 12.2 tons in chemically bound organic matter. This model was based wholly on the C cycle, and it is known that the N cycle follows a different pattern (Paul and van Veen, 1978). Their estimates of soil biomass, essentially microbial biomass, based on the biocidal technique and optical measurement, divided the size of organisms into a series of logarithmic volume classes, and they found that, although the pattern varied with different soils, each logarithmic volume class contained the same biovolume. Thus, a class of rare large organisms contains as much biovolume as a class of numerous small organisms. If the larger organisms comprise a proportion of protozoa, then such an estimate sets the limits both for the total microbial biomass and the possible protozoan component. In this way, within the time scale and pool size of the C cycle, constraints are placed on estimating the role of protozoa in soil nutrient cycling. The alternative approach is to estimate the biomass of protozoa directly, and from such laboratory information as is available to estimate the limits of activity of which such a biomass could be possible. Unfortunately, such estimates are liable to errors of at least one magnitude (Clark, 1967). For the sites discussed by Jenkinson and Rayner (1977), we know that amebae are the most important protozoa, and we have a good deal of information of their population dynamics and of their possible feeding habits. From such data, Stout and Heal (1967) estimated that the amebae may have contributed from 3 to 8% of the total respiratory activity of the soil and metabolized a substrate of about 150-900 g/m^2/year. As this would be largely derived from bacterial cells, such activity presumes an initial bacterial production of 15-85 times the standing crop of acceptable bacterial prey. From the difference in the ratios of C:N:P:S of bacteria and protozoa, the N, P, and S in excess of their needs that the amebae assimilate with the bacterial substrate will be excreted, accepting the data of Cole et al. (1978) and Coleman et al. (1978b) quoted previously. They calculated a C assimilation by the amebae of 40% and claimed that most of the bacterial P was returned to the inorganic P pool by the amebae. However, despite the differences in stoichiometric ratios between bacterial and protozoan cells, the P released probably approximates to the same proportion as that of the respiration, i.e., 3-8%. This is of the same order as Fenchel and Harrison (1976) cite for a laboratory system of bacteria and the ciliate *Tetrahymena*, an organism known to have a higher reproduction rate and a higher P turnover than the amebae (Section 2.1.1c). Fenchel and Harrison argue that, because the turnover of P in bacteria is much more rapid than in protozoan cells, the principal regeneration of inorganic P in these systems is by bacteria rather than by protozoa or by other micropredators. This conclusion appears to be substantiated by the biomass data of Jenkinson et al. (1976), for clearly only a fraction of the total microbial biomass is protozoa and similarly only a fraction of the biomass P pool is in protozoan cells.

What is perhaps of greater interest is the differential activity of the different protozoan communities and populations. It is here that the development of new and more sophisticated techniques, better monitoring of overall nutrient cycling, and the development of models which can accommodate both field situations and laboratory experiments offer hopes of advance. The real ecological problem is accommodating estimates of activity of very different orders of magnitude within a single matrix and of ensuring that fluctuations inherent in any field situation are not ignored, either in the modeling or monitoring of the field situation or in devising laboratory experiments. Therefore, the present challenge to protozoan ecologists wishing to evaluate the role of protozoa in natural situations is: first, to devise a convenient, accurate, and reproducible technique of population census, which will not be confined simply to a dominant species or a single taxonomic group; second, to identify and measure the parameters with which protozoan activity interacts; and third to analyze the data in a way that makes them applicable not simply to other measurements made on that particular ecosystem but also for comparison with other ecosystems.

The unique properties of protozoa, which have for centuries attracted diverse studies, are as challenging today for the microbial ecologists as they have been in the past to morphologists, biochemists, and geneticists. May they prove equally rewarding!

ACKNOWLEDGMENTS. The author is grateful to Mrs. J. Davin, librarian, for help with the references and to Mrs. H. Kinloch and Mr. Q. Christie for preparing the figures.

References

Adu, J. K., and Oades, J. M., 1978a, Physical factors influencing decomposition of organic materials in soil aggregates, *Soil. Biol. Biochem.* **10**:109-115.

Adu, J. K., and Oades, J. M., 1978b, Utilization of organic materials in soil aggregates by bacteria and fungi, *Soil Biol. Biochem.* **10**:117-122.

Anderson, R. V., Elliott, E. T., McClellan, J. F., Coleman, D. C., and Cole, C. V., 1978, Trophic interactions in soil as they affect energy and nutrient dynamics. III. Biotic interactions of bacteria, amoebae, and nematodes, *Microb. Ecol.* **4**:361-371.

Ausmus, B. S., and O'Neill, E. G., 1978, Comparison of carbon dynamics of three microcosm substrates, *Soil Biol. Biochem.* **10**:425-429.

Bamforth, S. S., 1973, Population dynamics of soil and vegetation protozoa, *Amer. Zool.* **13**:171-176.

Barsdate, R. J., Prentki, R. T., and Fenchel, T., 1974, Phosphorus cycle of model ecosystems: Significance for decomposer food chains and effect on bacterial grazers, *Oikos* **25**:239-251.

Bary, B. M., and Stuckey, R. G., 1950, An occurrence in Wellington Harbour of *Cyclotrichium meuneri* Powers, a ciliate causing red water, with some additions to its morphology, *Trans. R. Soc. N.Z.* **78**:86-92.

Beers, C. D., 1963, Relation of feeding in the sea urchin, *Strongylocentrotus droebachiensis*, to division in some of its endocommensal ciliates, *Biol. Bull.* **124**:1-8.

Beever, R. E., and Burns, D. J. W., 1978, Does cycloheximide-induced loss of phosphate uptake in *Neurospora crassa* reflect rapid turnover?, *J. Bacteriol.* **134**:1176-1178.

Berger, J., 1967, The morphology and biology of a carnivorous commensal ciliate from echinoids, *J. Protozool.* **14**(Suppl.):24-25.

Berk, S. G., Brownlee, D. C., Heinle, D. R., Kling, H. J., and Colwell, R. R., 1977, Ciliates as a food source for marine planktonic copepods, *Microb. Ecol.* **4**:27-40.

Bick, H., 1963, A review of central European methods for the biological estimation of water pollution levels, *Bull. W.H.O.* **29**:401-413.

Bick, H., 1967, Vergleichende Untersuchungen der Ciliatensukzession beim Abbau von Pepton und Cellulose (Modellversuche), *Hydrobiologia* **30**:353-373.

Bick, H., 1968, Autokologische und saprobiologische Untersuchungen an Süsswasserciliaten, *Hydrobiologia* **31**:17-36.

Bick, H., and Kunze, S., 1971, Eine Zusammenstellung von autokologischen und saprobiologischen Befunden an Süsswasserciliaten, *Int. Rev. Gesamten Hydrobiol.* **56**:337-384.

Bick, H., and Schmerenbeck, W., 1971, Vergleichende Untersuchungen des Peptonabbaus und der damit verknüpften Ciliaten Besiedlung in strömenden und stagnierenden Modellgewässern, *Hydrobiologia* 37:409-446.

Biczók, F., 1954, Testaceae in the rhizosphere, *Ann. Biol. Univ. Hung.* 2:385-394.

Bonnet, L., 1973, Aspects généraux du peuplement thécamoebien des mousses corticoles, in: *Progress in Protozoology* (P. de Puytorac and J. Grain, eds.), p. 51, Université de Clermont, Clermont-Ferrand.

Bonnet, L., 1975, Types morphologique, écologie et evolution de la thèque chez les Thécamoebiens, *Protistologica* 11:363-378.

Borror, A. C., 1963a, Morphology and ecology of the benthic ciliated Protozoa of Alligator Harbor, Florida, *Arch. Protistenkd.* 106:464-534.

Borror, A. C., 1963b, Morphology and ecology of some uncommon ciliates from Alligator Harbor, Florida, *Trans. Am. Microsc. Soc.* 82:125-131.

Borror, A. C., 1965, New and little-known tidal marsh ciliates, *Trans. Am. Microsc. Soc.* 84:550-565.

Borror, A. C., 1968, Ecology of interstitial ciliates, *Trans. Am. Microsc. Soc.* 87:233-243.

Bowers, B., and Olszewski, T. E., 1972, Pinocytosis in *Acanthamoeba castellanii:* kinetics and morphology, *J. Cell Biol.* 53:681-694.

Brown, T. J., 1967, Utilization of 4 sugars known to occur in activated sludge by *Aspidisca cicada* (Ciliata, Hypotrichida); effect on division rate, *J. Protozool.* 14:340-344.

Buhse, H. E., Jr., and Hamburger, K., 1974, Induced macrostome formation in *Tetrahymena vorax* strain V_2: Patterns of respiration, *C.R. Trav. Lab. Carlsberg* 40:77-89.

Burns, R. G., 1978, Enzyme activity in soil: Some theoretical and practical considerations, in *Soil Enzymes* (R. G. Burns, ed.), pp. 295-340, Academic Press, London.

Capo, C., Bongrand, P., Benoliel, A. M., and Depieds, R., 1974, Phagocytosis, *J. Theor. Biol.* 47:177-188.

Chambers, J., and Thompson, J., 1976, Phagocytosis and pinocytosis in *Acanthamoeba castellanii, J. Gen. Microbiol.* 92:246-250.

Chapman, A. G., and Atkinson, D. E., 1977, Adenine nucleotide concentrations and turn-over rates, their correlation with biological activity in bacteria and yeast, *Adv. Microb. Physiol.* 15:254-306.

Chapman-Andresen, C., and Holter, H., 1964, Differential uptake of protein and glucose by pinocytosis in *Amoeba proteus, C.R. Trav. Lab. Carlsberg* 34:211-226.

Chardez, D., 1968, Études statistique sur l'écologie et la morphologie des Thécamoebiens (Protozoa, Rhizopoda Testacea), *Hydrobiologia* 32:271-287.

Chardez, D., 1972, Étude sur les thécamoebiens des biotypes interstitiels, psammons littoraux et zones marginales souterraines des eaux douces, *Bull. Rech. Agron. Gembloux* 6: 257-268.

Chesson, P., 1978, Predator-prey theory and variability, *Annu. Rev. Ecol. Syst.* 9:323-347.

Clark, F. E., 1967, Bacteria in soil, in *Soil Biology* (A. Burges and F. Raw, eds.), pp. 15-49, Academic Press, London.

Cole, C. V., Elliott, E. T., Hunt, H. W., and Coleman, D. C., 1978, Trophic interactions in soils as they affect energy and nutrient dynamics. V. Phosphorus transformations, *Microb. Ecol.* 4:381-387.

Coleman, D. C., Cole, C. V., Hunt, H. W., and Klein, D. A., 1978a, Trophic interactions in soils as they affect energy and nutrient dynamics. I. Introduction, *Microb. Ecol.* 4: 345-349.

Coleman, D. C., Anderson, R. V., Cole, C. V., Elliott, E. T., Woods, L., and Campion, M. K., 1978b, Trophic interactions in soils as they affect energy and nutrient dynamics. IV. Flows of metabolic and biomass carbon, *Microb. Ecol.* 4:373-380.

Collins, N. J., Baker, J. H., and Tilbrook, P. J., 1975, Signy Island, maritime antarctic, in *Structure and Functions of Tundra Ecosystems* (T. Rosswall and O. W. Heal, eds.), pp. 345-374, Ecological Bulletin 20, Swedish Natural Research Council, Stockholm.

Corliss, J. P., and Hartwig, E., 1977, The 'primitive' interstitial ciliates: Their ecology, nuclear uniqueness, and postulated place in the evolution and systematics of the phylum Ciliophora, *Mikrofauna Meeresboden* 61:65–88.

Coûteaux, M.-M., 1975, Écologie des Thécamoebiens de quelques humus bruts forestiers: l'espéce dans la dynamique de l'équilibre, *Rev. Ecol. Biol. Sol* 12:421–447.

Coûteaux, M.-M., 1976, Dynamisme de l'équilibre des thécamoebiens dans quelques sols climaciques, *Mem. Mus. Natl. Hist. Nat. Ser. A Zool.* 96:1–183.

Crowley, P. H., 1978, Effective size and the persistence of ecosystems, *Oecologia* 35:185–195.

Crump, L. M., 1920, Numbers of protozoa in certain Rothamsted soils, *J. Agric. Sci.* 10:182–198.

Cunningham, A., and Maas, P., 1978, Time lag and nutrient storage effects in the transient growth response of *Chlamydomonas reinhardii* in nitrogen-limited batch and continuous culture, *J. Gen. Microbiol.* 104:227–231.

Curds, C. R., 1973, The role of Protozoa in the activated-sludge process, *Am. Zool.* 13:161–169.

Curds, C. R., 1977, Microbial interactions involving protozoa, in *Aquatic Microbiology* (F. A. Skinner and J. M. Shewan, eds.), pp. 69–105, Society of Applied Bacteriology, Symposium Series No. 6, Academic Press, London.

Curds, C. R., and Bazin, M. J., 1977, Protozoan predation in batch and continuous culture, in *Advances in Aquatic Microbiology* (M. R. Droop and H. W. Jannasch, eds.), pp. 115–176, Academic Press, London.

Cutler, D. W., 1923, The action of protozoa on bacteria when inoculated into sterile soil, *Ann. Appl. Biol.* 10:137–141.

Cutler, D. W., 1927, Soil protozoa and bacteria in relation to their environment, *J. Quekett Micros. Club* 15:309–330.

Cutler, D. W., and Bal, D. V., 1926, Influence of protozoa on the process of nitrogen fixation by *Azotobacter chroococcum, Ann. Appl. Biol.* 13:516–534.

Cutler, D. W., and Crump, L. M., 1920, Daily periodicity in the numbers of active soil flagellates: With a brief note on the relation of trophic amoebae and bacterial numbers, *Ann. Appl. Biol.* 7:11–24.

Cutler, D. W., and Crump, L. M., 1927, The qualitative and quantitative effects of food on the growth of a soil amoeba (*Hartmanella hyalina*), *Br. J. Exp. Biol.* 5:155–165.

Cutler, D. W., and Crump, L. M., 1929, Carbon dioxide production in sands and soils in the presence and absence of amoebae, *Ann. Appl. Biol.* 16:472–482.

Cutler, D. W., and Crump, L. M., 1935, *Problems in Soil Microbiology*, Longmans, Green and Co., London.

Cutler, D. W., Crump, L. M., and Sandon, H., 1922, A quantitative investigation of the bacterial and protozoan population of the soil, with an account of the protozoan fauna, *Philos. Trans. R. Soc. London Ser. B* 211:317–350.

Danso, S. K. A., and Alexander, M., 1975, Regulation of predation by prey-density: The protozoan-*Rhizobium* relationship, *Appl. Microbiol.* 29:515–521.

Danso, S. K. A., Keya, S. O., and Alexander, M., 1975, Protozoa and the decline of *Rhizobium* populations added to the soil, *Can. J. Microbiol.* 21:884–895.

Darbyshire, J. F., 1969, Protozoa in the rhizosphere of *Lolium perenne* L., *Can. J. Microbiol.* 12:1287–1289.

Darbyshire, J. F., 1976, Effect of water suctions on the growth in soil of the ciliate *Colpoda steini* and the bacterium *Azotobacter chroococcum, J. Soil Sci.* 27:369–376.

Darbyshire, J. F., and Greaves, M. P., 1967, Protozoa and bacteria in the rhizosphere of *Sinapsis alba* L., *Trifolium repens* L., and *Lolium perenne* L., *Can. J. Microbiol.* 13:1057–1068.

Darbyshire, J. F., and Greaves, M. P., 1973, Bacteria and protozoa in the rhizosphere, *Pestic. Sci.* **4**:349–360.

Das, S. M., 1947, The biology of two species of Folliculinidae (Ciliata, Heterotricha) found at Cullercoats, with a note on the British species of the family, *Proc. Zool. Soc.* **117**: 441–456.

Davis, P., Caron, D., and Sieburth, J., 1978, Oceanic amebae from the North Atlantic: Culture, distribution and taxonomy, *Trans. Am. Microsc. Soc.* **97**:73–88.

Delwiche, C. C., 1977, Energy relations in the global nitrogen cycle, *Ambio* **6**:106–111.

de Noyelles, F., and O'Brien, W. J., 1978, Phytoplankton succession in nutrient enriched experimental ponds as related to changing carbon, nitrogen and phosphorus conditions, *Arch. Hydrobiol.* **84**:137–165.

Doyle, W. L., and Harding, J. F., 1937, Quantitative studies on the ciliate *Glaucoma*. Excretion of ammonia, *J. Exp. Biol.* **14**:462–469.

Dragesco, J., 1960, Ciliés Mésopsammique Littoraux, *Trav. Sta. Biol. Roscoff* **12**, n.s., 1–36.

Dragesco, J., 1965a, Étude cytologique de quelques Flagelles mésopsammiques, *Cah. Biol. Mar.* **6**:83–115.

Dragesco, J., 1965b, Ciliés mésopsammique d'Afrique Noire, *Cah. Biol. Mar.* **6**:357–399.

Drake, J. F., and Tsuchiya, H. M., 1977, Growth kinetics of *Colpoda steinii* on *Escherichia coli*, *Appl. Environ. Microbiol.* **34**:18–22.

Droop, M. R., 1953, Phagotrophy in *Oxyrrhis marina* Dujardin, *Nature (London)* **172**:250–251.

Droop, M. R., 1968, Vitamin B_{12} and marine ecology. IV. The kinetics of uptake, growth and inhibition in *Monochrysis lutheri*, *J. Mar. Biol. Assoc. U.K.* **48**:689–733.

Droop, M. R., 1970, Vitamin B_{12} and marine ecology, *Helgol. Wiss. Meeresunters* **20**: 629–636.

Droop, M. R., 1973, Some thoughts on nutrient limitation in algae, *J. Phycol.* **9**:264–272.

Droop, M. R., 1977, An approach to quantitative nutrition of phytoplankton, *J. Protozool.* **24**:528–532.

Echetebu, C. O., and Plesner, P., 1977, The pool of ribonucleoside triphosphates in synchronized *Tetrahymena pyriformis*, *J. Gen. Microbiol.* **103**:389–392.

Edwards, S. W., and Lloyd, D., 1977a, Changes in oxygen uptake rates, enzyme activities, cytochrome amounts and adenine nucleotide pool levels during growth of *Acanthamoeba castellanii* in batch culture, *J. Gen. Microbiol.* **102**:135–144.

Edwards, S. W., and Lloyd, D., 1977b, Cyanide-insensitive respiration in *Acanthamoeba castellanii*: Changes in sensitivity of whole cell respiration during exponential growth, *J. Gen. Microbiol.* **103**:207–213.

Edwards, S. W., and Lloyd, D., 1978, Oscillations of respiration and adenine nucleotides in synchronous cultures of *Acanthamoeba castellanii*: Mitochondrial respiratory control *in vivo*, *J. Gen. Microbiol.* **108**:197–204.

Elliott, P. B., and Bamforth, S. S., 1975, Interstitial protozoa and algae of Louisiana salt marshes, *J. Protozool.* **22**:514–519.

Elliott, E. T., and Coleman, D. C., 1977, Soil protozoan dynamics in a shortgrass prairie, *Soil Biol. Biochem.* **9**:113–118.

Fantham, H. B., and Porter, A., 1945, The microfauna, especially the protozoa, found in some Canadian mosses, *Proc. Zool. Soc. London* **115**:97–174.

Fauré-Fremiet, E., 1950a, Écologie des ciliés psammophiles littoraux, *Bull. Biol. Fr. Belg.* **84**:35–75.

Fauré-Fremiet, E., 1950b, Caulobactéries épizoiques associés aux *Centrophorella* (Ciliés holotriches), *Bull. Soc. Zool. Fr.* **75**:154–157.

Fauré-Fremiet, E., 1950c, Ecology of ciliate infusoria, *Endeavour* **9**:183–187.

Fauré-Fremiet, E., 1951a, Associations infusoriennes à *Beggiatoa*, *Hydrobiologia* **3**:65–71.

Fauré-Fremiet, E., 1951b, The marine sand-dwelling ciliates of Cape Cod, *Biol. Bull. Mar. Biol. Lab. Woods Hole Mass.* **100**:59–70.

Fauré-Fremiet, E., 1951c, The tidal rhythm of the diatom *Hantzschia amphioxys*, *Biol. Bull. Mar. Biol. Lab. Woods Hold Mass.* **100**:173–177.

Feierabend, R., 1978, Untersuchungen über freie und gelöste Aminosäuren in natürlichen Gewässern, *Arch. Hydrobiol.* **84**:454–479.

Fenchel, T., 1967, The ecology of marine microbenthos. I. The quantitative importance of ciliates as compared with metazoans in various types of sediments, *Ophelia* **4**:121–138.

Fenchel, T., 1968a, The ecology of marine microbenthos. II. The food of marine benthic ciliates, *Ophelia* **5**:73–121.

Fenchel, T., 1968b, The ecology of marine microbenthos. III. The reproductive potential of ciliates, *Ophelia* **5**:123–136.

Fenchel, T., 1969, The ecology of marine microbenthos. IV. Structure and function of the benthic ecosystem, its chemical and physical factors and the microfauna communities with special reference to the ciliated protozoa, *Ophelia* **6**:1–182.

Fenchel, T., 1974, Intrinsic rate of natural increase: The relationship with body size, *Oecologia (Berlin)* **14**:317–326.

Fenchel, T. M., 1978, The ecology of micro- and meiobenthos, *Annu. Rev. Ecol. Syst.* **9**:99–121.

Fenchel, T., and Harrison, P., 1976, The significance of bacterial grazing and mineral cycling for the decomposition of particulate detritus, in: *The Role of Terrestrial and Aquatic Organisms in Decomposition* (J. M. Anderson and A. MacFadyen, eds.), pp. 285–299, 17th Symposium of the British Ecological Society, Blackwell Scientific Publications, Oxford.

Fenchel, T., and Jørgensen, B. B., 1977, Detritus food chains of aquatic ecosystems: the role of bacteria, in: *Advances in Microbial Ecology*, Vol. 1 (M. Alexander, ed.), pp. 1–58, Plenum Press, New York.

Findlay, B. J., 1977, The dependence of reproductive rate on cell size and temperature in freshwater ciliated protozoa, *Oecologia (Berlin)* **30**:75–81.

Garland, P. B., 1977, Energy transduction and transmission in microbial systems, in: *Microbial Energetics* (B. A. Haddock and W. A. Hamilton, eds.), pp. 1–21, 27th Symposium for General Microbiology, Cambridge University Press, Cambridge.

Gellért, J., 1955, Die Ciliaten des sich unter der Flechte *Parmelia saxatilis* Mass. gebildeten Humus, *Acta Biol. Acad. Sci. Hung.* **6**:77–111.

Gellért, J., 1956, Ciliaten des sich unter dem Moosrasen auf Felsen gebildeten Humus, *Acta Biol. Acad. Sci. Hung.* **6**:337–359.

Gessat, M., and Jantzen, H., 1974, Die Bedeutung von Adenosin-3′, 5′-monophosphate für die Entwicklung von *Acanthamoeba castellanii*, *Arch. Microbiol.* **99**:155–166.

Gold, K., 1973, Methods for growing Tintinnida in continuous culture, *Am. Zool.* **13**:203–208.

Golemansky, V., 1976, Rhizopodes psammobiontes (Protozoa, Rhizopoda) du psammal supralittoral des côtes guinéenes de l'Atlantique, *Acta Zool. Bulg.* **4**:23–29.

Golemansky, V., 1978, Adaptations morphologique des thécamoebiens psammobiontes du psammal supralittoral des mers, *Acta Protozool.* **17**:141–152.

Gray, E., 1951, The ecology of the bacteria of Hobson's Brook, a Cambridgeshire chalk stream, *J. Gen. Microbiol.* **5**:840–859.

Gray, E., 1952, The ecology of the ciliate fauna of Hobson's Brook, a Cambridgeshire chalk stream, *J. Gen. Microbiol.* **6**:108–122.

Habte, M., and Alexander, M., 1975, Protozoa as agents responsible for the decline of *Xanthomonas campestris* in soil, *Appl. Microbiol.* **29**:159–164.

Habte, M., and Alexander, M., 1977, Further evidence for the regulation of bacterial populations in soil by protozoa, *Arch. Microbiol.* **113**:181–183.

Habte, M., and Alexander, M., 1978, Mechanisms of persistence of low numbers of bacteria preyed upon by protozoa, *Soil Biol. Biochem.* **10**:1-6.

Harding, J. P., 1937a, Quantitative studies on the ciliate *Glaucoma*. I. The regulation of the size and the fission rate by the bacterial food supply, *J. Exp. Biol.* **14**:422-430.

Harding, J. P., 1937b, Quantitative studies on the ciliate *Glaucoma*. II. The effects of starvation, *J. Exp. Biol.* **14**:431-439.

Harvey, R. J., and Greaves, J. E., 1941, Nitrogen fixation by *Azotobacter chroococcum* in the presence of soil protozoa, *Soil Sci.* **51**:85-100.

Heal, O. W., 1962, The abundance and micro-distribution of testate amoebae (Rhizopoda, Testacea) in *Sphagum, Oikos* **13**:35-47.

Heal, O. W., 1964a, Observations on the seasonal and spatial distribution of testacea (Protozoa, Rhizopoda) in *Sphagnum, J. Anim. Ecol.* **33**:395-412.

Heal, O. W., 1964b, The use of cultures for studying Testacea (Protozoa, Rhizopoda) in soil, *Pedobiologia* **4**:1-7.

Heller, R., 1978, Two predator-prey difference equations considering population growth and starvation, *J. Theor. Biol.* **70**:401-413.

Herzberg, M. A., Klein, D. A., and Coleman, D. C., 1978, Trophic interactions in soils as they affect energy and nutrient dynamics. II. Physiological responses of selected rhizosphere bacteria, *Microb. Ecol.* **4**:351-359.

Hoffman, E. K., Rasmussen, L., and Zeuthen, E., 1974, Cytochalasin B: Aspects of phagocytosis in nutrient uptake in Tetrahymena, *J. Cell Sci.* **15**:403-406.

Holter, H., and Zeuthen, E., 1947, Metabolism and reduced weight in starving *Chaos chaos, C. R. Trav. Lab. Carlsberg* **26**:277-296.

Hunt, H. W., Cole, C. V., Klein, D. A., and Coleman, D. C., 1977, A simulation model for the effect of predation on bacteria on continuous culture, *Microb. Ecol.* **3**:259-278.

Kandatsu, M., and Horiguchi, M., 1962, Ocurrence of ciliatin (2-Aminoethylphosphonic acid) in *Tetrahymena, Agric. Biol. Chem. (Tokyo)* **26**:721-722.

Hutner, S. H. (ed.), 1964, *Biochemistry and Physiology of Protozoa*, Vol. 3, Academic Press, New York.

Hutner, S. H., and Lwoff, A. (eds.), 1955, *Biochemistry and Physiology of Protozoa*, Vol. 2, Academic Press, New York.

Hunter, S. H., and Provasoli, L., 1955, The Phytoflagellates, in: *Biochemistry and Physiology of Protozoa*, Vol. 1 (A. Lwoff, ed.), pp. 27-128, Academic Press, New York.

Jantzen, H., 1974, Das Adenosinphosphat-System während Wachstum und Entwicklung von *Acanthamoeba castellanii, Arch. Microbiol.* **101**:391-399.

Jenkinson, D. A., and Rayner, J. H., 1977, The turnover of soil organic matter in some of the Rothamsted classical experiments, *Soil Sci.* **123**:298-305.

Jenkinson, D. S., Powlson, D. S., and Wedderburn, R. W. W., 1976, The effects of biocidal treatments on metabolism in soil. III. The relationship between soil biovolume, measured by optical microscopy, and the flush of decomposition caused by fumigation, *Soil Biol. Biochem.* **8**:189-202.

Johannes, R. E., 1965, Influence of marine protozoa on nutrient regeneration, *Limnol. Oceanogr.* **10**:434-442.

Johannes, R. E., 1968, Nutrient regeneration in lakes and oceans, in: *Advances in Microbiology of the Sea*, Vol. 1 (M. R. Droop and E. J. Ferguson Wood, eds.), pp. 203-213, Academic Press, London.

Kalff, J., and Knoechel, R., 1978, Phytoplankton and their dynamics in oligotrophic and eutrophic lakes, *Annu. Rev. Ecol. Syst.* **9**:475-495.

Karpenko, A., Railkin, A. I., and Seravin, L. N., 1977, Feeding behaviour of unicellular animals. II. The role of prey mobility in the feeding behaviour of protozoa, *Acta Protozool.* **16**:333-344.

Kaszubiak, H., Kaczmarek, W., and Durska, G., 1976, Feeding of soil microbial community on organic matter from its dead cells, *Ekol. Pol.* **24**:391–397.

Keeling, C. D., and Bacastow, R. B., 1977, Impact of industrial gases on climate, in: *Energy and Climate*, National Research Council, pp. 72–95, National Academy of Sciences, Washington, D. C.

Kidder, G. W. (ed.), 1967, Protozoa, in: *Chemical Zoology*, Vol. 1 (M. Florkin and B. T. Scheer, eds.), Academic Press, New York.

King, D. L., 1972, Carbon as a limiting factor in lake ecology, in: *Trace Substances in Environmental Health*, Vol. 5 (D. D. Hemphill, ed.), pp. 109–115, University of Missouri, Columbia.

Kloetzel, J. A., 1974, Feeding in ciliated protozoa. 1. Pharyngeal disks in *Euplotes*: a source of membrane for food vacuole formation? *J. Cell. Sci.* **15**:379–401.

Kolkwitz, R., and Marsson, M., 1909, Oekologie der tierischen Saproben, *Int. Rev. Gesamten Hydrobiol. Hydrogr.* **2**:126–152.

Lackey, J. B., 1938, A study of some ecologic factors affecting protozoa, *Ecol. Monogr.* **8**: 501–527.

Laminger, H., 1973, Untersuchungen über Abundanz and Biomasse der sedimentbewohnenden Testaceen (Protozoa, Rhizopoda) in einem Hochgebirgssee (Vorderer Finstertaler See, Kühtal, Tirol), *Int. Rev. Gesamten Hydrobiol.* **58**:543–568.

Laminger, H., 1978, The effects of soil moisture fluctuations on the testacean species *Trinema enchelys* (Ehrenberg) Leidy in a high mountain brown-earth podzol and its feeding behaviour, *Arch. Protistenkd.* **120**:446–454.

Latter, P. M., Cragg, J. B., and Heal, O. W., 1967, Comparative studies on the microbiology of four moorland soils in the Northern Pennines, *J. Ecol.* **55**:445–464.

Lauterborn, R., 1916, Die sapropelische Lebewelt. Ein Beitrag zur Biologie des Faulschlammes natürlicher Gewässer, *Verh. Naturforsch. Med. Ver. Heidelberg, n.s.* **13**: 395–481.

Laybourn, J., 1975, Respiratory energy losses in *Stentor coeruleus* Ehrenberg (Ciliophora), *Oecologia (Berlin)* **21**:273–278.

Laybourn, J., 1976a, Energy budgets for *Stentor coeruleus* Ehrenberg (Ciliophora), *Oecologia (Berlin)* **22**:431–437.

Laybourn, J., 1976b, Respiratory energy losses in *Podophrya fixa* Müller in relation to temperature and nutritional status, *J. Gen. Microbiol.* **96**:203–208.

Laybourn, J., 1977, Respiratory energy losses in the protozoan predator *Didinium nasutum* Müller (Ciliophora), *Oecologia (Berlin)* **27**:305–309.

Laybourn, J., and Finlay, B. J., 1976, Respiratory energy losses related to cell weight and temperature in ciliated protozoa, *Oecologia (Berlin)* **24**:349–355.

Laybourn, J. E. M., and Stewart, J. M., 1975, Studies on consumption and growth in the ciliate *Colpidium campylum* Stokes, *J. Anim. Ecol.* **44**:165–174.

Lee, J. J., and Muller, W. A., 1973, Trophic dynamics and niches of salt marsh Foraminifera, *Am. Zool.* **13**:215–223.

Legner, M., 1975, Concentration of organic substances in water as a factor controlling the occurrence of some ciliate species, *Int. Rev. Gesamten Hydrobiol.* **60**:639–654.

Lindemann, R. L., 1942, The trophic-dynamic aspect of ecology, *Ecology* **23**:399–418.

Lloyd, D., Phillips, G. A., and Statham, M., 1978, Oscillations of respiration, adenine nucleotide levels and heat evolution in synchronous cultures of *Tetrahymena pyriformis* St prepared by continuous-flow selection, *J. Gen. Microbiol.* **106**:19–26.

Lousier, J. D., 1974a, Effects of experimental soil moisture fluctuations on turnover rates of testacea, *Soil Biol. Biochem.* **6**:19–26.

Lousier, J. D., 1974b, Response of soil testacea to soil moisture fluctuations, *Soil Biol. Biochem.* **6**:235–239.

Lousier, J. D., 1976, Testate amebae (Rhizopoda, Testacea) in some Canadian Rocky Mountain soils, *Arch. Protistenkd.* **118**:191–201.

Lousier, J. D., and Parkinson, D., 1976, Litter decomposition in a cool temperate deciduous forest, *Can. J. Bot.* **54**:419–436.

Luckinbill, L. S., 1973, Coexistence in laboratory populations of *Paramecium aurelia* and its predator *Didinium nasutum, Ecology* **54**:1320–1327.

Luckinbill, L. S., 1974, The effects of space and enrichment on a predator–prey system, *Ecology* **55**:1142–1147.

Lwoff, A. (ed.), 1951, *Biochemistry and Physiology of Protozoa*, Vol. 1, Academic Press, New York.

Marshall, K. C., 1976, *Interfaces in Microbial Ecology*, Harvard University Press, Cambridge, Massachusetts.

Matheja, J., and Degens, E. T., 1971, *Structural Molecular Biology of Phosphates*, p. 180, G. Fisher Verlag, Stuttgart.

Maxham, J. V., and Hickman, H. J., 1974, Substrate diffusion and uptake within bacterial flocs, *J. Theor. Biol.* **43**:229–239.

Medvecsky, N., and Rosenberg, H., 1971, Phosphate transport in *Escherichia coli, Biochim. Biophys. Acta* **241**:494–506.

Meiklejohn, J., 1930, The relationship between the numbers of soil bacterium and the ammonia produced by it in peptone solution, with some reference to the effect of this process on the presence of amoebae, *Ann. Appl. Biol.* **19**:584–608.

Meiklejohn, J., 1932, The effect of *Colpidium* on ammonia produced by soil bacteria, *Ann. Appl. Biol.* **19**:584–608.

Meisterfeld, R., 1977, Die horizontale und vertikale Verteilung der Testaceen (Rhizopoda, Testacea) in *Sphagnum, Arch. Hydrobiol.* **79**:319–356.

Nasir, S. A., 1923, Some preliminary investigations on the relationship of protozoa to soil fertility, with special reference to nitrogen fixation, *Ann. Appl. Biol.* **10**:122–133.

National Research Council, 1977, *Energy and Climate*, p. 158, National Academy of Sciences, Washington, D. C.

Needham, D. M., Robertson, M., Needham, J., and Baldwin, E., 1932, Phosphagen and protozoa, *J. Exp. Biol.* **9**:332–335.

Newman, E. I., 1978, Root microorganisms: Their significance in the ecosystem, *Biol. Rev. Cambridge Philos. Soc.* **53**:511–554.

Nichols, G. L., and Syrett, P. J., 1978, Nitrate reductase deficient mutants of *Chlamydomonas reinhardii*. Isolation and genetics, *J. Gen. Microbiol.* **108**:71–77.

Nielsen, L. Brunberg, 1968, Investigations on the microfauna of leaf litter in a Danish beech forest, *Nat. Jutl.* **14**:79–87.

Nikolyuk, V. F., 1968, The effect of root systems of wild growing and cultivated plants on Protozoa in the soils of Uzbekistan, in: *Methods of Productivity Studies in Root Systems and Rhizosphere Organisms: International Symposium*, pp. 126–129, Nauka, Leningrad.

Nikolyuk, V. F., 1969, Some aspects of the study of soil protozoa, *Acta Protozool.* **7**: 99–109.

Noland, L. E., 1925, Factors influencing the distribution of fresh water ciliates, *Ecology* **6**: 437–452.

Olson, J. S., Pfuderer, H. A., and Chan, Y.-H., 1978, Changes in the Global Carbon Cycle and the Biosphere, Environmental Sciences Division Publication No. 1050, Oak Ridge National Technical Information Service, U. S. Department of Commerce, Springfield, Virginia, p. 169.

Organ, A. E., Bovee, E. C., and Jahn, T. L., 1978, Effects of ionic ratios vs. osmotic pressure on the rate of the water expelling vesicle of *Tetrahymena pyriformis, Acta Protozool.* **17**:177–190.

Orias, E., and Rasmussen, L., 1977, Dual capacity for nutrient uptake in *Tetrahymena*. II. Role of the two systems in vitamin uptake, *J. Protozool.* **24**:507–511.

Panikov, N., and Pirt, S. J., 1978, The effects of cooperativity and growth yield variation on the kinetics of nitrogen or phosphate limited growth of *Chlorella* in a chemostat culture, *J. Gen. Microbiol.* **108**:295–303.

Parnas, H., 1976, A theoretical explanation of the priming effect based on microbial growth with two limiting substrates, *Soil Biol. Biochem,* **8**:139–144.

Paul, E. A., and van Veen, J. A., 1978, The use of tracers to determine the dynamic nature of organic matter, in: *Symposia Papers,* Vol. 3, pp. 61–102, Transactions of the 11th Congress International Society of Soil Science, Edmonton.

Pauls, K. P., and Thompson, J. E., 1978, Growth and differentiation-related enzyme changes in cytoplasmic membranes of *Acanthamoeba castellanii, J. Gen. Microbiol.* **107**: 147–153.

Payne, W. J., and Wiebe, W. J., 1978, Growth yield and efficiency in chemosynthetic microorganisms, *Annu. Rev. Microbiol.* **32**:155–183.

Peak, J. G., and Peak, M. J., 1977, Regulation by ammonium of variation in heterotrophic CO_2 fixation by *Euglena gracilis* during growth cycles, *J. Protozool.* **24**:441–444.

Picken, L. E. R., 1937, The structure of some protozoan communities, *J. Ecol.* **25**:368–384.

Pirt, S. J., and Bazin, M. J., 1972, Possible adverse effect of Protozoa on effluent purification systems, *Nature (London)* **239**:290.

Plachter, H., 1979, Untersuchungen zur Feinstruktur der Stiele und Gehäuse einiger symphorionter Ciliaten, *Arch. Protistenkd.* **121**:193–210.

Preston, T. M., and King, C. A., 1978, Cell-substrate associations during the amoeboid locomotion of *Naegleria, J. Gen. Microbiol.* **104**:347–351.

Provasoli, L., and Pinter, I. J., 1953, Ecological implications of *in vitro* nutritional requirements of algal flagellates, *Ann. N.Y. Acad. Sci.* **56**:839–851.

Raikov, I. B., 1962, Les ciliés mésopsammique du litoral de la Mer Blanche (URSS) avec une description de quelques espèces nouvelles ou peu connues, *Cah. Biol. Mar.* **3**:325–361.

Rasmussen, L., 1973, On the role of food vacuole formation in the uptake of dissolved nutrients by *Tetrahymena, Exp. Cell. Res.* **82**:192–196.

Rasmussen, L., 1976, Nutrient uptake in *Tetrahymena pyriformis, Carlsberg Res. Commun.* **41**:143–167.

Rasmussen, L., and Modeweg-Hansen, L., 1973, Cell multiplication in Tetrahymena cultures after addition of particulate material, *J. Cell Sci.* **12**:275–286.

Reid, R., 1969, Fluctuations in populations of 3 *Vorticella* species from an activated-sludge sewage plant, *J. Protozool.* **16**:111–120.

Ricketts, T. R., 1972, The interaction of particulate material and dissolved foodstuffs in food uptake by *Tetrahymena pyriformis, Arch. Mikrobiol.* **81**:344–349.

Ringelberg, J., and Kersting, K., 1978, Properties of an aquatic microecosystem: I. General introduction to the prototypes, *Arch. Hydrobiol.* **83**:47–68.

Roti, L. W., and Stevens, A. R., 1975, DNA synthesis in growth and encystment of *Acanthamoeba castellanii, J. Cell. Sci,* **17**:503–515.

Russell, E. J., and Hutchinson, H. B., 1909, On the effect of partial sterilization of soil on the production of plant food, *J. Agric. Sci.* **3**:111–144.

Russell, E. W., 1961, *Soil Conditions and Plant Growth*, 9th ed., Longmans, Green and Co., London.

Sardeshpande, J. S., Balasubramanya, R. H., Kulkarni, J. H., and Bagyaraj, D. J., 1977, Protozoa in relation to Rhizobium S-12 and *Azotobacter chroococcum, Plant Soil* **47**: 75–80.

Satir, P., and Zeuthen, E., 1961, Cell cycle and the relationship of growth to reduced weight (RW) in the giant ameba *Chaos chaos* L., *C.R. Trav. Lab. Carlsberg,* **32**:241–264.

Schlesinger, W. H., 1977, Carbon balance in terrestrial detritus, *Annu. Rev. Ecol. Syst.* **8**: 51–81.

Schmerenbeck, W., 1975, Experimentelle Untersuchungen an strömenden Modellgewässern zur Frage der Beziehung Zwischen dem Abbau organischer Substanz und der Ciliaten-besiedlung, Thesis, Institut für Landwirtschaftliche Zoologie und Bienenkunde, No. 2, Bonn, p. 95.

Schönborn, W., 1964, Lebensformtypen und Lebensraumwechsel der Testaceen, *Limnologica* 2:321-335.

Schönborn, W., 1977, Production studies on Protozoa, *Oecologia* 27:171-184.

Seravin, L. N., and Orlovskaya, E. E., 1977, Feeding behaviour of unicellular animals. I. The main role of chemoreception in the food choice of carnivorous protozoa, *Acta Protozool.* 16:309-332.

Singh, B. N., 1946, A method of estimating the numbers of soil protozoa, especially amoebae based on their differential feeding on bacteria, *Ann. Appl. Biol.* 33:112-119.

Singh, B. N., 1949, The effect of artificial fertilizers and dung on the numbers of amoebae in Rothamsted soils, *J. Gen. Microbiol.* 3:204-210.

Singh, B. N., 1960, Inter-relationship between micropredators and bacteria in soil, Proceedings 47th Indian Science Congress, Part II, Agricultural Sciences, pp. 1-14.

Singh, B. N., 1964, Soil protozoa and their probable role in soil fertility, *Bull. Natl. Inst. Sci. India* 26:238-244.

Skriver, L., and Nilsson, J. R., 1978, The relationship between energy-dependent phagocytosis and the rate of oxygen consumption in *Tetrahymena*, *J. Gen. Microbiol.* 109: 359-366.

Small, E. B., 1973, A study of ciliate protozoa from a small polluted stream in East Central Illinois, *Am. Zool.* 13:223-230.

Smith, H. G., 1973a, The Signy Island Terrestrial Reference Sites II. Protozoa, *Br. Antarctic Survey Bull.* 33, 34:83-87.

Smith, H. G., 1973b, The Signy Island Terrestrial Reference Sites. III. Population Ecology of *Corythion dubium* (Rhizopoda, Testacida) in Site I, *Br. Antarctic Survey Bull.* 33 34:123-135.

Stoermer, E. F., Ladewski, B. G., and Schelske, C. L., 1978, Population responses of Lake Michigan phytoplankton to nitrogen and phosphorus enrichment, *Hydrobiologia* 57: 249-265.

Stout, J. D., 1962, An estimation of microfaunal populations in soils and forest litter, *J. Soil Sci.* 13:314-320.

Stout, J. D., 1968, The significance of the protozoan fauna in distinguishing mull and mor of beech (*Fagus silvatica*), *Pedobiologia* 8:387-400.

Stout, J. D., 1971, Aspects of the microbiology and oxidation of Wicken Fen soil, *Soil Biol. Biochem.* 3:9-25.

Stout, J. D., 1974, Protozoa, in: *Biology of Plant Litter Decomposition*, Vol. 11 (H. Dickinson and G. J. F. Pugh, eds.), pp. 383-420, Academic Press, New York.

Stout, J. D., and Heal, O. W., 1967, Protozoa, in: *Soil Biology* (A. Burges and F. Raw, eds.), pp. 149-195, Academic Press, New York.

Stouthamer, A. H., 1977, Energetic aspects of growth of micro-organisms, in: *Microbial Energetics* (B. A. Haddock and W. A. Hamilton, eds.), pp. 285-316, Symposium 27, Society for General Microbiology, Cambridge.

Theng, B. K. G., 1974, *The Chemistry of Clay–Organic Reactions*, Hilger, London.

Theng, B. K. G., 1979, *Formation and Properties of Clay–Polymer Complexes*, Elsevier, Amsterdam.

van Niel, C. B., Thomas, J. P., Ruben, S., and Kamen, M. D., 1942, Radioactive carbon as an indicator of carbon dioxide utilization. IX. The assimilation of carbon dioxide by protozoa, *Proc. Natl. Acad. Sci. U.S.A.* 28:157-161.

Vernberg, W. B., and Coull, B. C., 1974, Respiration of an interstitial ciliate and benthic energy relationships, *Oecologia (Berlin)* 16:259-264.

Villarreal, E., Canale, R., and Arcasu, Z., 1977, Transport equations for a microbial predator–prey community, *Microb. Ecol.* 3:131-142.

Wang, C. C., 1928, Ecological studies of the seasonal distribution of Protozoa in a fresh water pond, *J. Morphol.* **46**:431–478.

Wangersky, P. J., 1978, Lotka-Volterra population models, *Annu. Rev. Ecol. Syst.* **9**: 189–218.

Welch, E. B., Sturtevant, P., and Perkins, M. A., 1978, Dominance of phosphorus over nitrogen as the limiter to phytoplankton growth rate, *Hydrobiologia* **57**:209–215.

Wenzel, F., 1953, Die der Moosrasen trockner Standorte, *Arch. Protistenkd.* **99**:70–141.

Whittaker, R. H., 1975, *Communities and Ecosystems*, 2nd ed., Macmillan, New York.

Wiegert, R. G., and Owen, D. F., 1971, Trophic structure, available resources and population density in terrestrial vs aquatic ecosystems, *J. Theor. Biol.* **30**:69–81.

Wilson, J. M., and Griffin, D. M., 1975, Water potential and the respiration of microorganisms in the soil, *Soil Biol. Biochem.* **7**:199–204.

Woodwell, G. M., and Pecan, E. V. (eds.), 1973, *Carbon and the Biosphere*, U. S. Atomic Energy Commission, U.S. Department of Commerce, Springfield, Virginia.

Ziegler, B., 1977, Persistence and patchiness of predator–prey systems induced by discrete event population exchange mechanisms, *J. Theor. Biol.* **67**:687–713.

2

Surface Microlayers in Aquatic Environments

BIRGITTA NORKRANS

1. Introduction

> The role of the infinitely small in Nature is infinitely great.
>
> —Louis Pasteur

The sea–air interface covers three-quarters of the world's surface. The microlayers formed at the interface by chemical and microbial constituents, however, represent only an infinitesimal part of the water body, easily housed within, what MacIntyre (1974a) calls, the "top millimeter of the ocean." Probably due to this meager dimension, the real significance of microlayers was long disregarded. Not until their presence was highlighted by frequent oil spills was increasing attention paid to them. Their profound influence—even under unpolluted conditions—on several interrelated processes was revealed. The processes are summarized in Garrett's (1972) classic diagram (Fig. 1). They affect the gaseous exchanges (Quinn and Otto, 1971; Liss, 1977; Quickenden and Barnes, 1978) and transport mechanisms from the water column to the atmosphere and *vice versa*. Dissolved substances, particles, and microorganisms are brought to the interface by simple diffusion, rising bubbles (Garrett, 1967; Jarvis, 1967), convection, and upwelling from sediments and subsurface water, and at the same time, the microlayer is a sink for fallout from the atmosphere (Duce *et al.*, 1976). Thus it becomes an accumulation layer, where the concentration of various chemical compounds and microorganisms exceeds that of the subsurface

---

BIRGITTA NORKRANS • Department of Marine Microbiology, Botanical Institute, University of Göteborg, Carl Skottsbergs Gata 22, S-413 19 Göteborg, Sweden.

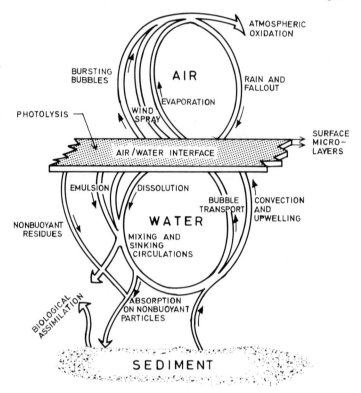

Figure 1. Interrelated processes in aquatic environments. After Garrett (1972).

water by orders of magnitude (Parker and Barsom, 1970; Hatcher and Parker, 1974a; MacIntyre, 1974b).

The purpose of this review is firstly to present and discuss current knowledge of the surface microlayers at the air–water interface, its chemistry, and certain physical phenomena characteristic of the interface and then to proceed to the relatively scarce data on its structure and the function of its biotic components, termed neuston (Naumann, 1917), focusing on bacteria.

2. The Surface Microlayer(s)

Natural bodies of water contain dissolved organic matter. In seawater, this exists within the concentration range 1–5 mg/liter, but it may exceed 40 mg/liter in some inshore waters. For lakes, values of 1–50 mg/liter are usually reported (Fenchel and Jørgensen, 1977). The dissolved organic matter originates from autolysis of dead plant material, from animals, from active bacteria, and from growing primary producers, mainly phytoplankton, and its composition may reflect the whole spectrum of biogenic compounds. The same authors refer to data showing that as much as 10-30% of marine phytoplankton production is lost as dissolved organic matter; whether or not this is from healthy or dead phytoplankton has been discussed (Sharp, 1977; Fogg, 1977). Wangersky (1978) also discusses this production in different water bodies and the contribution of organic matter from land, air, and human activity. Among all the organic substances, those with high surface activity, low water solubility, and specific gravity below that of the aqueous medium tend to accumulate at the air/water interface to which organisms and particles also adhere to form the surface microlayers(s).

Surface microlayer(s) is a a loosely defined term often implying no particular thickness, chemical composition, or concentration, but simply referring physically to the thin surface layer collected by the particular sampling device used. As a prelude to a discussion of the microlayers, it is therefore appropriate to describe the main types of sampling devices.

2.1. Sampling Devices for Collection of the Surface Microlayers

The various sampling devices used differ in a number of ways as regards the thickness of the water surface layer sampled (Table I), the mechanisms by which they collect surface material, the ability to collect specific strata, the sample volume which can be obtained with a reasonable effort, and the water conditions under which they can be used.

Table I. Various Devices for Sampling Surface Microlayers

Sampler	Sample thickness (μm)	Samples from	References
Freezing probe	1000	Seawater Model system	Hamilton and Clifton (1979)
Mesh screen	250-150	Seawater	Garrett (1965) Sieburth (1965) Duce et al. (1972)
Rotating drum	60-100	Seawater	Harvey (1966)
Glass plate	60-100	Seawater	Harvey and Burzell (1972)
Hydrophilic Nuclepore membrane	40-4	Seawater estuarine	Crow et al. (1975)
Perforated Teflon plate	40-5	Seawater Model system	Larsson et al. (1974) Norkrans and Sörensson (1977)
Teflon sheet	7	Seawater Model system	Kjelleberg et al. (1979)
Hydrophobic Nuclepore membrane	1	Seawater Model system	Kjelleberg et al. (1979)
Bubble microtome	1	Seawater	MacIntyre (1968)
Germanium prism	10^{-2}	Seawater	Baier (1972)

Attempts to compare the efficiency of different conventional methods of collection have been made (Hatcher and Parker, 1974b; Daumas *et al.*, 1976). The drum and glass plate, for instance, tended to indicate greater degrees of surface enrichment than the screen. This seems to indicate that most surface enrichment occurs in microlayers thinner than that collected by the screen, although each device may also be selective with respect to its ability to recover different chemical constituents of the surface microlayers (Odham *et al.*, 1978).

Whereas most of the bacteria adhere to the uppermost layer which is collected by a hydrophobic "few-micron layer" sampler, the size of the phytoplankton necessitates the use of a "hundreds-micron layer" sampler such as screens or drums to ensure their adequate sampling. A useful sampler has to collect efficiently both chemicals and microorganisms.

2.1.1. *"Few-Micron Layer" Sampler*

Since a lipid film is considered to be the uppermost layer of the surface microlayers, a pronounced hydrophobic material such as Teflon applied parallel to the surface was expected to be suitable as a sampler for it. With the requirements of ease and simplicity in mind, Miget *et al.* (1974) developed a sampler consisting of a 2-mm Teflon disk attached to an aluminum backing by means of bolts. Larsson *et al.* (1974), simultaneously and independently, designed a Teflon plate densely perforated with conical holes. No contaminants are leached out from the Teflon during the extraction. It was also found that the Teflon plate served satisfactorily in the collection of microorganisms associated with the lipid microlayer. The lipid-collection ability and the extent of water withdrawal for the Teflon plate are given in Table II. The amount of water retained was determined gravimetrically using a model system of aged sea water covered with a monomolecular film of stearic acid. The lipid-collecting ability was tested in model systems prepared from various lipids applied dropwise by syringe as light petroleum–ethanol $(9:1 \text{ v/v})$ solutions to the air–water interface in different amounts corresponding to gaseous monolayers or condensed mono-, tri-, or decalayers.

Table II also summarizes other pronounced hydrophobic samplers, samplers of membraneous type, which can float on water and be retrieved with sterile forceps or which can be fitted into special holders before use. These remove as nearly as possible only those organic molecules situated at the air–water interface, leaving behind subsurface waters whose chemical constituents and biological inhabitants do not contribute to the modification of that interface. A Teflon sheet and hydrophobic Nuclepore membrane were tested by Kjelleberg *et al.* (1979) in both field and model systems. There seems to be no tendency to favor any thickness of the condensed film since equally good yields are obtained in all cases. This is important since the lipid-layer thickness varies on natural waters. These samplers were also compared with the hydrophilic Nuclepore membrane first used by Crow *et al.* (1975).

Table II. Lipid Collecting Ability and Water Withdrawal of Nuclepore
Membranes and Teflon Samplers Determined in Model Systems

Sampler	Amount of water withdrawn[b] (μl/dm^2)	Calculated sample thickness (μm)	% Recovery of lipid film[c]				
			Oleic acid		Olive oil (mainly triglycerides)		Triglycerides of isotridecanoic acid (gaseous film)
			(Monolayer)	(Decalayer)	(Monolayer)	(Trilayer)	
Hydrophobic Nuclepore membrane[d]	8 ± 5	0.8	n.d.[f]	93	59	94	45
Hydrophilic Nuclepore membrane[d]	294 ± 62	29	n.d.	n.d.	n.d.	n.d.	n.d.
Teflon sheet[e]	65 ± 29	7	80	99	72	62	45
Teflon plate[e]	356 ± 90	36	90	83	83	124	52

[a]Data from Kjelleberg et al. (1979).
[b]Based on gravimetrical determinations, 20 weighings.
[c]Gas–liquid chromatographic analysis used.
[d]n-Hexane was used to remove the lipids.
[e]Chloroform was used to remove the lipids. The smallest possible volume of solvent was used in both cases.
[f]n.d., Not determined.

Bubbles are produced by living organisms and by breaking waves. Model experiments have shown that during the slow rise of bubbles through the water column, organic molecules and particulate matter are adsorbed on their surfaces. At the water surface, the bubbles eventually burst. When the surface free energy of the bubble is converted to kinetic energy, aerosol droplets are formed and ejected into the air in millisecond reactions. This formation of droplets has been described as a bubble microtome (MacIntyre, 1968) since the ejected droplets carry away material only from the bubble surface and the top micrometers of the surface microlayers.

2.1.2. "Hundreds-Micron Layer" Sampler

Garrett (1965) was the first to develop a reasonably simple technique for the collection of the aquatic surface microlayer. He used a 16-mesh, 60 X 75 cm, aluminum-framed Monel screen to collect the upper 150 μm of sea water. The screen is lowered vertically into the water and then oriented horizontally through the liquid surface. This procedure introduces vertical contact surfaces before the surface layer is reached. The main mechanism for film removal is the entrapment between the 0.14-mm diameter Monel wires. Garrett reports a recovery of approximately 75% after the first surface contact, the yield also obtained for the removal of oleic and stearic acid and oleyl alcohol monolayers in model systems. In field studies, 150-250 dippings were found necessary for collecting satisfactory amounts of material. Sieburth (1965) used a stainless-steel version, reduced in size to fit a portable autoclave, to study microorganisms in the top 250-μm layer, whereas Duce *et al.* (1972) used a 20-mesh polyethylene screen. The latter screen material, however, changes the surface potential of a clean water surface so much, even after repeated cleanings, that it cannot be recommended (MacIntyre, 1974b).

Harvey (1966) used a rotating hydrophilic drum device mounted in front of a slow-moving boat. This device acts as a skimmer. The sampler is buoyant and moves through water as the drum rotates. Before the drum surface reenters the water, a stationary neoprene blade removes much of the adhering layer, which then flows into a plastic cup. Harvey himself mentions the disadvantages associated with the use of a hydrophilic ceramic cylinder. Other authors point to the difficulties involved in using neoprene, plastic cups, and polyethylene bottles. They may all add some substances to the sample. The thickness of the layer collected is temperature dependent. It varies between 60 and 100 μm, being approximately 60 μm at 20°C. Samples of 20 liters can easily be collected. Brockmann *et al.* (1976) used a modified skimmer to obtain a large-volume (18-liter) sample of a 250-μm layer.

Harvey and Burzell (1972) used a similar technique for smaller samples, consisting of a glass plate withdrawn vertically from the surface at about 20 cm/sec and wiped off with a neoprene windshield wiper blade. A liter of microlayer can be collected in about 45 min by this method.

Screen samplers have been frequently used. However, aqueous boundary layers cannot be kept intact on these or on the skimmer. A recently reported technique using a different approach (Hamilton and Clifton, 1979) seems to offer this advantage even though the sample thickness is about 1000 μm. This is achieved by a freezing process that takes place in <1 sec. A probe, consisting of a disk of 0.25-m diam polymethyl metacrylate, encased on its lower side by a thin (<0.1 mm) polyvinyl chloride (PVC) membrane attached with adhesive tape to the upper surface of the disk, is precooled in liquid N_2 and attached to a plastic rod with freezing water. The freezing probe is brought into contact with the water surface, which immediately freezes to a depth of about 1000 μm onto the lower surface of the PVC membrane. By removing the fixing tape, the PVC membrane with the frozen surface microlayers can be detached and stored in a conventional freezer until sectioned with a sledge microtome for further analysis. The freezing probe, because of the speed of sampling, is said to be useful when the surface of the sea is rough and to sample surface foam and bubble-burst debris. The method seems convincingly useful for chemical examination of the distribution of elements and compounds, even if the recoveries of ^{14}C-labeled fatty acid films in some cases were low. The reported recoveries of living *Phaeodactylum tricornutum* and *Artemia salina* from frozen microlayers of a stirred laboratory system seem to demonstrate the applicability of the method not only for counting and microscopical recognition of biological specimens in different strata but also for physiological and biochemical studies.

Another unorthodox collecting method devised by Hamilton and Clifton (1979) does not offer the same possibilities. It merely involves the entrapment in a PVC film of nonaqueous and particulate material floating on the surface of the water body. The film is formed after spraying a solution of PVC in a suitable solvent and subsequent evaporation of the solvent.

During all sampling, care has to be taken to ensure that the samples are not contaminated by material from the collection craft.

2.2. The Lipid Film

Most workers (e.g., Harvey, 1966; Garrett, 1967; Jarvis, 1967; Larsson *et al.*, 1974; Kattner and Brockmann, 1978) have considered lipids to be the major constituents of the uppermost surface microlayer on all natural waters—marine or limnic. The amphiphilic molecules lower the free energy of the system. They have a preferred orientation with respect to the water surface, with the long-chained hydrophobic parts extending into the air, thus enabling them to form an ordered film. The film consists mainly of even-numbered free fatty acids, both saturated and unsaturated, with chain lengths of from 12 to 22 carbon atoms. The most frequently occurring one is palmitic acid (C 16:0). (In the convention used, the value before the colon designates the number of carbons, and the value after the colon refers to the number of double bonds in the molecule.) Branched acids and odd-numbered acids, however, with C 15 and C 17 acids dominating, are also observed, suggesting at least a partial bacterial derivation (Blumer, 1970). Besides free fatty acids, glycerides are present of which triglycerides are the most common. In the surface film, there are characteristic deviations in the pattern of the fatty acid distribution compared to that of the subsurface water. The relative amounts of short-chain fatty acids and alcohols are higher in the subsurface water than in the film. These fatty acids and alcohols have been forced out of the film by longer-chain species when competitive adsorptive processes occur (Garrett, 1967). The latter are less water soluble and more surface active. The relative amount of unsaturated acids is lower in the surface film probably because of photooxidation and direct oxidation through air contact. The amount of glycerides relative to fatty acids is always higher in the surface multifilms than in the subsurface water (Larsson, 1973).

Besides free fatty acids and glycerides (40-65%), phospholipids are also present in the films, together with hydrocarbons (15-30%) and small amounts of steroids, etc. The hydrocarbons are largely derived from phytoplankton (Blumer *et al.*, 1971). They are few and of a simple structure compared to those of ancient sediments and fossil fuels (Bocard *et al.*, 1977). Normal paraffins accumulate in the water column, whereas branched and aromatic paraffins occur at the surface (Ledet and Laseter, 1974).

This generalized description of the lipid films probably gives the impression of a certain monotony in their composition. This is, however, not so. Plankton samples from different geographical and climatic regions show, for instance, pronounced differences in their lipid chemistry, often for the same species. Such chemical differences may be reflected in the molecular composition of the matter dissolved in the sea and in the film. This may provide distinguishing tags that are characteristic for specific water masses and persisting for long time periods. Larsson *et al.* (1974) compared surface films produced by cod, herring, grayback, and plaice, and the fatty acid patterns seem to be specific enough for discrimination.

The lipid film of so-called dry surfactants is a real surface microlayer; it is from 1 to 2 nm thick, corresponding to the length of fatty acids, and up to approximately 10 nm thick when multilayered. Natural monomolecular films, "slicks," often occur on the sea surface in coastal regions, but under certain conditions they also occur on the open oceans. These films have the ability to strongly dampen capillary waves. They have been observed to remain coherent at wind speeds of 4-7 m/sec. According to the observed chain-length distribution of the lipids, the minimum amount of lipid surface film giving rise to visible capillary wave damping was calculated to be $2 \, mg/m^2$ (Larsson *et al.*, 1974; Lange and Hühnerfuss, 1978) at a film pressure of less than 1 mN/m (Garrett, 1976). The absolute amounts of lipids reported for the air–water interface, even in unpolluted areas, differ markedly. This is partly due to the sampling methods used. Daumas *et al.* (1976) found the concentrations of fatty acids to be twice as high in drum samples as in screen samples from the same unpolluted area. There are, however, more important differences arising from variations in productivity between different waters and from seasonal and daily variations. Almost certainly, differences also originate in patchiness in the film. That sea slicks are not uniformly distributed was shown at an early stage by Langmuir. Wind and action can break up the film into streaks. The influence of wind and hydrographic factors has recently been studied by Lange and Hühnerfuss (1978) both in the field and in a laboratory wind-wave tunnel. In the case of gaseous films, the lipids are always unevenly distributed.

Table III gives the total amounts of free and bound fatty acids in samples collected on the Swedish west coast with a strict lipid film sampler, the Teflon plate. The distribution of fatty acids in surface and subsurface samples was determined, and their concentration in the surface film was found to be three orders of magnitude higher than that in the subsurface water. In this description of the natural surface microlayers, as in most others, the quoted chemical and biological quantities are implicitly assumed to be static. Investigations by Brockmann *et al.* (1976) and Kattner and Brockmann (1978), however, show how dynamic these quantities are. They were able to follow the formation of a big surface slick area off the island of Sylt over a period of 4 days, making observations at short intervals. For example, they estimated changes in surface pressure by measuring the surface tension according to Adam (1937) and analyzed the lipid contents. They found an increase in the surface pressure from 7 to 18 mN/m within a period of 1.5 hr and a variation in the lipid content by

Table III. Enrichment Factor for Bacteria $(E)^a$, Amounts of Total, Free, and Free and Bound Fatty Acids in Samples[b] from Surface, S (μg/ml), and Bulk, B (μg/1000 ml), at Different Times of the Year in a Nonpolluted Area (Gullmarsfjorden)[c]

Date	Number of bacteria/ml[d]	E	Total	Fatty acids Free and bound 12:0	14:0	16:0	18:0	18:1	Free 12:0	14:0	16:0	18:0	18:1
Aug. 23, 1974	S 12,600 ± 870	76	24	0.5	3.0	10.0	5.0	5.0	0.7	1.7	4.8	1.5	<0.2
	B 165 ± 26		22	<0.2	0.2	1.0	0.8	0.8	0.5	0.5	1.0	0.6	
Oct. 24, 1974	S 2,200 ± 360	11	11	0.8	2.0	5.0	2.0	1.0	0.5	1.0	2.4	1.4	0.4
	B 200 ± 28		6.6	<0.2	0.3	1.0	0.6	0.6	0.5	0.5	1.4	1.0	
Nov. 11, 1974	S 5,200 ± 560	15	13	0.8	2.0	7.3	3.1	9.0	0.7	2.0	7.0	4.0	0.6
	B 340 ± 37		1.9	<0.2	0.3	0.9	0.5	0.5	0.2	0.2	1.0	0.6	
Feb. 6, 1975	S 50,000 ± 5,200	10	16										
	B 5,000 ± 580		n.d.[e]										

[a] E defined as the number of bacteria/ml in the Teflon-plate-collected surface layer relative to that of the subsurface water (bulk).
[b] Ten Teflon plate samplings. Note the difference: S values in μg/ml, B values in μg/1000 ml.
[c] Data from Odham and Norkans (unpublished).
[d] Average value from generally six samplings.
[e] n.d., Not detected.

a factor of 4 within a few hours. Besides an increasing population of the dominating dinoflagellate, *Prorocentrum*, the calm conditions contributed to the accumulation. As the occurrence of such films is rather unpredictable, it is possible that satisfactory information about the events during slick formation can be obtained only by applying artificial films in natural environments, which is what the above authors intend to do.

2.3. Polysaccharide–Protein Complex Layer

Baier *et al.* (1974), however, consider all waters, except heavily polluted ones, to be coated with films of polysaccharide–protein complexes ranging in thickness from 10 to 30 nm. Their proposal is based on material collected on a germanium slide and analyzed with a "nondestructive, direct analytical technique," viz., multiple attenuated infrared reflectance (Baier, 1972, 1975). These "wet surfactants" can form a more irregular, submerged stratum. They are essentially hydrophilic but stick to the surface by virtue of their few hydrophobic chains. As pointed out by Wangersky (1976), it remains to be seen whether the infrared absorption technique overestimates the contribution of proteinaceous material. Even if significant amounts of this material are present, considerably smaller amounts of lipids are sufficient to dominate the interface by virtue of their higher surface activity. Furthermore, when the lipids are forced out of the film by wind, in foam, or by other means, they do not denature, as do the proteins in unfolding processes, but dissolve again and return to an interface position.

These two strata, the lipid film and the polysaccharide–protein layer taken together, have been called the *surface microlayer* by Sieburth *et al.* (1976), who quote a thickness of 0.1 μm. They reported a calculated mean value of 875 mg/ liter of total carbohydrates for this layer.

2.4. Bacterioneuston Layer

The surface microlayers have been reported to have a different microflora and a considerably greater number of microorganisms than the subsurface water (Blanchard and Syzdek, 1970; Sieburth, 1971; Marumo *et al.*, 1971; Bezdek and Carlucci, 1972; Crow *et al.*, 1975; Tsiban, 1975; Kjelleberg and Håkansson, 1977; Dutka and Kwan, 1978; Odham *et al.*, 1978; Young, 1978). When comparing different, pronouncedly hydrophobic samplers (Table II) along the Swedish west coast and in model systems, the hydrophobic Nuclepore membrane with a minimum of water withdrawal gave the highest enrichment factor (E) for bacteria in all experiments (Kjelleberg *et al.*, 1979). E varied over a range of 10^2–10^4 with corresponding bacterial concentrations in the subsurface water of 10^4–10^2/ml. The bacterial concentrations were calculated in numbers per unit volume, and E was defined as the bacterial concentration in the collected surface layer relative to that of the subsurface water. A calculation per unit volume, and not per area, is necessary when comparisons between subsurface and surface samples have to be made (see Wangersky, 1976). The absolute numbers of bacteria in the surface layer obtained at one and the same sampling station were fairly constant irrespective of the sample thickness, being approximately in the range 1-40 μm. This may indicate that in this case the operationally dependent dimension, 1 μm, coincides fairly well with that of the bacterioneuston layer, the biological stratum (Norkrans, 1979). Furthermore, only a strong interaction between the lipid film and the bacteria seems to render the bacterial collection possible.

From a taxonomical point of view, the bacterioneuston is poorly characterized. When the bacteria have been identified, the genus *Pseudomonas* has repeatedly been reported as predominant; e.g., by Sieburth (1971) in the Caribbean Sea and in the Atlantic and Pacific Oceans, and by Tsiban (1975) in the Black Sea, the Sea of Azov, the Caspian Sea, and Alaska Bay. The latter author also found the genera *Chromobacterium* and *Micrococcus* generally distributed. Carty and Colwell (1975) classified bacterial isolates from the air–sea interface during a Pacific Ocean expedition. Besides *Pseudomonas* species, they found *Vibrio*, *Aeromonas*, and *Spirillum* species to be predominant. The same

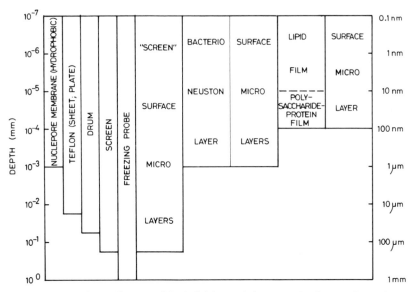

Figure 2. Illustrating various possible definitions of the strata in the top 1-mm layer of the air–water interface in aquatic environments. Defined, for example, in terms of (1) samplers; e.g., "screen" surface microlayers; (2) the organisms; the bacterioneuston layer, defined here also as the surface microlayers; (3) organic chemical composition; the lipid and polysaccharide-protein complex films together defined by Sieburth (1976) as the surface microlayer.

species were also found in the air above the sea surface, transferred to the air by bursting bubbles and foam. Normally, however, terrestrial *Bacillus* species dominated among the air-collected bacteria. These spore-forming species are more resistant to desiccation and solar radiation than the bubble-transported marine species are. This contributes to the dominance by *Bacillus*.

The term surface microlayer(s) has been used in very different ways in the literature, and generally the term has been operationally defined. However, as described, it is possible to recognize different strata in the top 1-mm layer of the air–water interface in aquatic environments (cf. MacIntyre, 1974a). Considering the biological and organic chemical characteristics of the microlayers, the following suggestions, illustrated in Fig. 2, can be made. The lipid film (1–10 nm thick) together with the polysaccharide–protein layer form the surface microlayer, 0.1 μm (Sieburth *et al.*, 1976). The bacterioneuston layer extends over 1 μm and forms the surface *microlayers*. Thicker surface microlayers have to be defined in terms of samplers, e.g., "screen" surface microlayers or "screen" microlayers. Sieburth *et al.* (1976) term their screen-collected layer (150 μm) simply "screen."

3. Some General Concepts for Lipid Films—Gaseous, Liquid, and Solid States

At low densities, the film behaves as a two-dimensional gas (Gaines, 1966) in which the molecules move almost independently of each other. Such a film is known to interact with larger molecules, like polymers and proteins, and with microorganisms. A large amount of organic matter can be accommodated within such an open structure (Fig. 3A).

Alternatively, the film can be in the liquid or solid state (Figs. 3B and C). The architecture of the liquid film is very much dependent on the magnitude of lateral cohesive forces between the film molecules. For straight-chain fatty acids, the bonding between the hydrophobic hydrocarbon chains extending into the air is strong enough to maintain the film molecules in small clusters or islands, a condensed film even at low surface density. For an unsaturated acid, such as oleic acid, the cohesion between the hydrocarbon chains is smaller. Due to the double bond in the hydrocarbon chains, the molecules are less densely packed at a certain lateral pressure, and the film remains liquid even at high compression due to steric hindrance (see Fig. 4).

With increased condensation, a solid monolayer is formed, and stratified multilayered films (Larsson, 1973) can be formed at higher concentrations. Normally, solid or liquid aggregates (microcrystals or microdrops) are built up. With increased condensation, the ability to form complexes with larger entities decreases. If additional nonpolar hydrocarbons are present, a continuous hydrocarbon phase of varying chain length is formed. It may become fairly thick

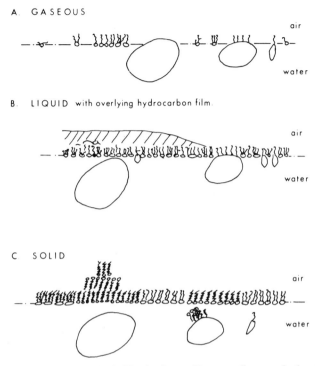

Figure 3. Aggregation states of a lipid microlayer. The wavy lines are hydrophobic chains and the open structures are polar groups. (A) Microlayer of low density behaving as a two-dimensional gas. (B) Liquid surface monolayer. It aids in forming extended sheets of oily films. (C) Solid monolayer. It usually contains small crystalline regions. Bulky molecules are excluded. Crystalline chains are represented as 〰〰〰. Large bodies represent particles. From Odham *et al.* (1978).

before it is broken up into islands or drops. If the film consists mainly of long-chain polar hydrocarbons, it is easily crystallized, and the film is transformed into particulate matter. These aggregates remain at the sea surface by buoyancy.

4. Microbial Interaction at the Surface Lipid Film

Model systems lack the multiplicity of a natural ecosystem, but they offer great advantages. Environmental factors may be kept constant, and individual variables can be studied under controlled conditions. In the case of interface studies, they seem especially justified since the initial events of interaction are governed by purely physicochemical forces, in which the microorganisms behave as colloid particles, their biological activity being manifested only in subsequent stages.

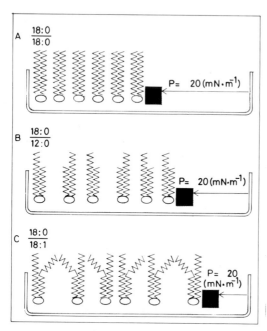

Figure 4. Models of monolayered lipid films illustrating the molecular area at a given surface pressure, *P*. (A) Molecules with two long saturated chains (18:0/18:0) occupying a minimum space; (B) with (18:0/12:0), the space required is greater because of lessened attraction between short chains; (C) with one unsaturated chain (18:0/18:1), the greatest space is required due to the kink at the double bond. After Bangham (1975).

Our own model systems for studies of the microbial interaction at the air-water interface were in principle all alike (Kjelleberg *et al.*, 1976; Norkrans and Sörensson, 1977; Hermansson *et al.*, 1979; Kjelleberg and Stenström, 1980). The film-forming substances were triglycerides, saturated or unsaturated acids, octadecylamine, or zwitterionic phospholipids. The acids were negatively charged and the octadecylamine positively charged at the experimental pH. They were applied to the interface of an aqueous saline subphase ("bulk") containing suspended bacteria by dissolving them in a light petroleum–ethanol mixture (9:1 vol) and spreading them dropwise with a micrometer syringe. The amounts of lipid were calculated to correspond to a mono- or "decalayer" under the static film conditions used in the bacterial accumulation experiments. In surface balance studies, an amount of lipid generally corresponding to a sixth of a monolayer was spread.

4.1. Bacterial Accumulation at the Surface Lipid Film

Bacterial accumulation has been found in natural surface microlayers. To get some information about the factors which contribute to the accumulation, model system studies were performed.

An accumulation of bacteria occurred at the lipid film in all experiments, measured after a 2-hr equilibration. The enrichment factor (*E*) for the bacteria was defined, as above, as the number of bacteria per milliliter in the surface microlayer relative to that in the subsurface water (about 2×10^6).

1. The accumulation varied with the test organisms used, the E values being in a range from 7 to 100. *Serratia marinorubra* showed an enrichment factor of about 100, about ten times higher than those obtained for *Aeromonas* and *Pseudomonas* species. *P. fluorescens* consistently gave the lowest value. Results of gradient centrifugation eliminated differences in specific cell density between the test bacteria as a possible reason for differences in accumulation tendency. Silll lower E values were obtained for the two gram-positive bacteria tested.

2. The accumulation varied with the thickness of the film. For a decalayer of oleic acid, it was about twice that for a monolayer. Field experiments show a positive correlation between the number of microorganisms and the amount of surface film material sampled (e.g., Table III and Brockmann *et al.*, 1976).

3. The accumulation varied with the cell density in the subsurface water. The E values of *S. marinorubra* decreased about 100-fold with increases in the number of subsurface cells from 2×10^6 to 50×10^6 cells/ml (Fig. 5). This might be due to a decrease in "film-bacteria sites" until a point of saturation, after which the further apparent change in accumulated bacteria lies within experimental error. Fletcher and Loeb (1979) observed a similar leveling out for the attachment of a marine pseudomonad to solid surfaces. They explained it tentatively as due to the conversion of these surfaces, initially hydrophobic and favorable for attachment, into hydrophilic, negatively charged, unfavorable ones. This occurred when a monolayer had formed of attached bacteria known to have an acidic, hydrophilic surface polymer (Fletcher and Floodgate, 1973). Under natural conditions, changes provoked by the pioneer species may govern the regular succession of bacterial types observed on immersed solid surfaces (Marshall, 1976). In bubble-bursting experiments (Blanchard and Syzdek, 1970), in which relatively high numbers of bacteria were used, the number of bacteria transferred into the air was also relatively constant irrespective of the number of

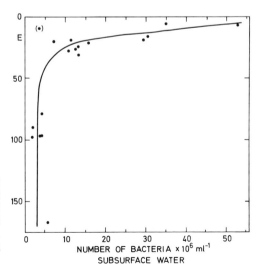

Figure 5. Enrichment factors (E) for *Serratia marinorubra* in model systems with a monolayered lipid film of oleic acid, at different cell densities in the subsurface water. After Norkrans and Sörensson (1977).

bacteria in the solution from which they were ejected. Here also a competition for film sites, namely around the bubble, can be assumed.

The E values of *S. marinorubra* decreased with an increasing number of sub-surface cells whether or not these consisted of *Serratia* cells alone or mixed with *Aeromonas* cells. Electrostatic repulsion between bacteria may block further accumulation. In these experiments (Norkrans and Sörensson, 1977), the lipid film systems were collected with the Teflon plate, and the enumeration of bacteria was made by plate count. E values of about 100 were also found for *S. marcescens* at bulk concentrations of *Serratia* cells of about 2×10^6/ml when the surface films were collected by Teflon sheet and the bacteria counted by epifluorescence microscopy after acridine-orange staining (Hermansson *et al.*, 1979).

The enrichment of bacteria at the air–water interface in the accumulation studies could be assumed to be due to bacterial reproduction in that layer. Such an assumption, however, was not supported by growth experiments with *Serratia* cells in which the conditions were kept as close as possible to those in the model systems. After a 2-hr incubation, no growth could be detected. Moreover, the apparently greater ability of *Serratia* cells to be enriched by the presence of a lipid surface layer does not seem to be attributable to the occurrence of fimbriae (see Ottow, 1975). Fimbriation certainly occurs within all the genera—*Serratia*, *Pseudomonas*, and even *Aeromonas* (Tweedy *et al.*, 1968)—represented in these studies. Due to our use of shake culture cells, however, fimbriae-like structures appeared only very sparsely. This was confirmed by electron microscopic examination of negatively stained cells of our strains.

Rising bubbles and water movements are undoubtedly decisive in nature for "coarse regulation" of the transport of bacteria to the water surface. Aerotaxis and chemotaxis may also play a part, albeit only for "fine regulation" of motile bacteria. Nevertheless, besides Brownian motion, only these processes can be active in the model systems.

4.2. Physicochemical Characterization of the Microbial Interaction at the Surface Lipid Film

The proposal of bacterial interaction at sites in or just beneath the surface lipid film is supported by surface balance studies in which the behavior of various monolayered films with and without bacteria present is compared (Kjelleberg *et al.*, 1976). The bacterial interaction is indicated by an increase in the surface pressure at a given area per molecule. Dissolution of the film, e.g., as a result of enzymatic activity, leads to a decrease in the film pressure. As could be expected from the E values obtained in the accumulation studies, *S. marinorubra* again showed the strongest interaction and the greatest strengthening of an oleic acid and a zwitterionic dipalmitoyl phosphatidyl choline film among the bacteria examined, except for the cell-wall-less *Acholeplasma laid-*

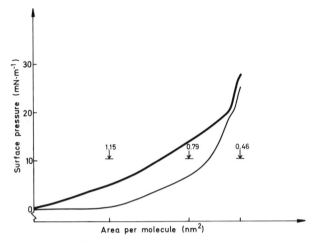

Figure 6. Surface pressure (decrease in surface tension) π (mN/m) as a function of allotted molecular area for a monolayer of dipalmitoylphosphatidyl choline on a saline solution with (—) and without (−) cells of *Acholeplasma laidlawii A*. Calculated molecular areas (nm^2) are indicated by arrows. After Kjelleberg *et al.* (1976).

lawii (Figs. 6 and 7) which has its lipid bilayered plasma membrane exposed. From the fact that part or all of the outer surface of some bacteria is hydrophobic, it can be deduced that such bacteria are rejected from the aqueous phase and attracted toward any nonaqueous phase (Marshall, 1976). More extensive surface balance studies (Kjelleberg and Stenström, 1980) revealed that hydro-

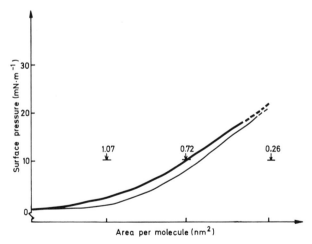

Figure 7. Surface pressure π (mN/m) as a function of allotted molecular area (nm^2) of oleic acid on saline solution with and without cells of *Serratia marinorubra*. Broken lines indicate viscous film. Same symbols as in Fig. 6. After Kjelleberg *et al.* (1976).

phobic interaction alone is not involved but that the bacterial interaction with the lipid film is a multifactor phenomenon, as reviewed by Marshall (1975, 1976) or discussed by Dexter *et al.* (1975) and Fletcher and Loeb (1979) as regards solid–liquid interfaces.

Factors besides the hydrophobic interaction shown to be of significance include the following.

1. As indicated above, the lateral cohesive forces between the hydrocarbon chains have a strong influence on the architecture of the film (Fig. 4) and thereby on the microbial interaction. If the surface molecules are sufficiently apart to prevent interchain reactions or if the lateral cohesion is strong enough to bring about and maintain island formations, as can be the case for straight-chain acids such as palmitic acid (C 16:0) or stearic acid (C 18:0), the possibility for microbial interaction is enhanced, as reflected in increased surface pressure.

2. The surface charge, both of the monolayered surface film and of the bacteria, affects the interaction. Most bacteria carry a characteristic negative charge (Harden and Harris, 1953; Longton *et al.*, 1975). The negative charge on the test bacteria used in the investigations discussed here was found by electrophoretic measurements. As expected, it was found that replacing negatively charged fatty acid films with octadecylamine films, which are positively charged at the experimental pH (7.4), provoked a large increase in surface pressure. Furthermore, *S. marinorubra,* which is characterized by a greater negative charge than *P. halocrenaea,* caused a larger surface pressure increase with octadecylamine films.

3. Reactions of specific surface localized groups may also contribute to the total interaction. The red, pyrrol-containing prodigiosins that are characteristic secondary metabolites for *Serratia* cells at late growth phases are, for instance, presumed to be codeterminants for the interaction. This is supported by the finding that pigmented cells exhibited a stronger interaction with lipid films than nonpigmented mutant cells both when the interaction was observed as bacterial accumulation by means of plate count or an epifluorescence technique and when studied by a surface balance technique (Hermansson *et al.*, 1979). Furthermore, when expressing as a concentration factor (C) the water-to-air transfer of *Serratia* cells by jet drops, Blanchard and Syzdek (1978) found C values of 200, 11, and 0.3 for red, pink, and nonpigmented cells, respectively. Hydrophobic interaction, however, may be the main force, as is suggested by Marshall (1976) and supported by a current study (Kjelleberg *et al.*, 1980). When determining the number of labeled fatty acid molecules bound to each bacterial cell, it was found that pigmented, late-logarithmic-phase cells of *S. marcescens* showed a higher number of surface-localized binding sites and a higher value of "total hydrophobicity" than mid-logarithmic-phase cells with less pigmentation. The hydrophobicity was here defined by the number of sites per cell times the binding affinity constant for each site.

4.3. Some Remarks about the Microbial Interaction at Gas-Liquid and Solid-Liquid Interfaces

The microbial interaction at gas-liquid interfaces in microbial ecology has been studied much less than that at solid-liquid interfaces in aquatic ecosystems and in general. At both interfaces, the microbial interaction is dependent on several similar forces. A solid substrate, however, partly fixes the solid-liquid system, whereas air-water interfaces consist of an organic layer on a dynamic water surface whose chemical composition is constantly being modified by physical, chemical, and biological processes. For this reason, simple generalizations to predict the interaction at natural air-water interfaces are very difficult. A gas-water interface system which has dealt with the mechanisms of the interaction more frequently is that consisting of transporting and bursting bubbles in water bodies, mostly studied in model systems, and summarized by Marshall (1976).

From our own model system studies and the literature, we draw the following conclusions about the probable forces taking part in the microbial interaction.

The microbes move by random movements or, possibly, by chemotaxis or aerotaxis until the distance from the surface film becomes small enough to allow the initial, reversible, physicochemical interaction forces to come into play. They are as follows:

1. *van der Waals forces*: a weak physical attraction between two polarizable groups. This has been suggested as part of an initial reversible stage of bacterial interaction (see Marshall, 1976).

2. *Electrostatic interactions*, in particular a charge-charge interaction: a stronger chemical adsorption in which the generally negatively charged bacteria interact with a positively charged microlayer material (Kjelleberg and Stenström, 1979) or counterions serve as bridges between the organisms and negatively charged groups on the microlayer material or solid substrate (Olsson *et al.*, 1976).

3. *Hydrophobic interaction*: This plays a dominant role in the stabilization of ordered systems in water, in which nonpolar groups tend to adhere strongly to each other. The attraction forces are considered to be the result of the avoidance of water rather than to be the forces of the nonpolar groups per se (Kushner, 1978) and ultimately due to the strong self-attraction of the water molecules (Tanford, 1979). Bacteria are known to be hydrophobic, although they vary in their degree of hydrophobicity, which seems to be one of the reasons for differences in interaction between pigmented and nonpigmented *Serratia marcescens* (Hermansson *et al.*, 1979). It remains to be seen whether a general positive correlation exists between the different degrees of hydrophobicity and the varying tendencies for interaction such as are observed, for example, in bubble-bursting experiments, for *Achromobacter* being $>S.$ *marinorubra* $>Micro-*

coccus infirmis > *Vibrio haloplanktis* (Carlucci and Williams, 1965) and in experiments on accumulation at lipid films (Norkrans and Sörensson, 1977).

Studies are now under way that compare the average degree of hydrophobicity of a great number of bacteria isolated from the surface film or the water beneath at different stations along the Swedish west coast. The hydrophobicity data are obtained by means of hydrophobic interaction chromatography (HIC). A positive correlation seems to exist between the bacterial accumulation (E) at the surface film and the average degree of hydrophobicity, being higher for surface than for subsurface bacteria.

4. *Hydrogen bonding interactions*: brought about by the sharing of a proton from the microbial surface with a pair of electrons from an interface material. This interaction might be stronger for solid–liquid systems.

5. *Irreversible adhesion* (possibly): as is found in solid–liquid systems, in which production of adhesive material has been shown (Hirsch and Pankratz, 1970; Marshall *et al.*, 1971; Hirsch, 1972), and the adhesive has been found to be formed of polysaccharides (e.g., Fletcher and Floodgate, 1973; Gibbons and van Houte, 1975) or proteinaceous material (Danielsson *et al.*, 1977).

5. Factors Affecting Microbial Structure and Function of the Surface Microlayers

5.1. Availability of Nutrients for Bacterial Growth

Natural sea water is an extremely dilute nutrient solution, too dilute for growth of many marine bacteria. However, accumulation of organic matter occurs at all interfaces—gas–liquid, liquid–liquid, or solid–liquid—and the accumulated substances have been considered as potential nutrients for heterotrophic bacteria (e.g., ZoBell, 1943). Aggregates in the surface microlayers as well as subsurface aggregates take part in the interchange of metabolites and reduce enzyme escape (Jones and Jannasch, 1959; Jannasch and Pritchard, 1972), although other factors may also be involved in these processes. The outer membrane of gram-negative bacteria functioning as a molecular sieve (Decad and Nikaido, 1976) and the periplasm acting as a "forecourt" outside the cytoplasmic membrane with periplasmic enzymes and binding proteins (see Costerton *et al.*, 1974) may also offer the gram-negative bacteria means for the utilization of very dilute substrates (Hagström, 1978). The concentration of dissolved organic carbon in subsurface water in the North Atlantic has been estimated as $\leqslant 1.3$ mg carbon/liter, to be compared with a mean concentration of 1427 mg/liter (2.9 g organic matter/liter) calculated for the 0.1-μm surface microlayer (Sieburth *et al.*, 1976). The latter corresponds to the amount of nutrients in media generally used to grow marine bacteria. Thus, an overall lack of nutrients need not limit reproduction. The surface microlayer may even be a too "rich soup" for some bacteria from the subsurface water.

Different natural surface microlayers differ with respect to chemical composition and physicochemical conditions. From the model system studies it was obvious that different bacteria differ in the degree of their initial interaction at the lipid film. These initial interactions, therefore, could result in a certain selection of bacteria from the water body, even though the interaction forces are weak. It is easy to predict that of these heterotrophs only those having the ability to utilize the energy-yielding substances present in the surface microlayers will be able to compete successfully.

The pseudomonads are characterized by an extremely high versatility with respect to both the kind and the number of organic compounds that they can use as energy and carbon sources. Furthermore, they are known to live in a broad spectrum of conditions, even harsh ones. However, are even these tough bacteria sufficiently hardy to be able to reproduce under such hostile conditions as those offered by the surface microlayers? (See Section 5.3.)

5.2. Biochemical Activity of Neustonic Bacteria

Studies have not been focused specifically on the activity of *Pseudomonas* spp. in the surface microlayers with one exception (Sieburth, 1971). In this case, a higher percentage of lipolytic and proteolytic isolates were found in the surface microlayers than in the subsurface. However, data now being accumulated on the biochemical activity of bacteria in general in the surface microlayers point to their relatively low activity. This has been found to be true for all water types investigated and with all methods used. Dietz *et al.* (1976) conclude from their data on material from the Vancouver area that ATP levels and heterotrophic activity were lower for bacterioneuston (from a 70 to 80-μm sampled layer) than for bacterioplankton. Calculated per colony-forming unit (CFU), heterotrophic activity was generally about one-tenth of that for bacterioplankton. Labeled glucose was used for these uptake studies, the method being described in greater detail in a later work (Dietz *et al.*, 1977). Since the additions of substrate (labeled glucose or amino acids) can be made at concentrations in the same range as those occurring in natural water, it is possible to diminish the problems involved in a drastic disturbance of the natural equilibrium between substrates and microbes.

There has been and remains some criticism of all techniques based on uptake of certain labeled organic compounds, especially with regard to determination of total heterotrophic activity (Zimmermann *et al.*, 1978). The labeled nutrients represent a minute proportion of the many natural nutrients present in the investigated aquatic environment and may act selectively on the bacteria during incubation. Furthermore, the incubation may induce activity of dormant bacteria. This is, however, the eternal problem in microbial ecology, viz., to design experimental conditions with a maximum of ecological relevance, allowing conclusions to be drawn about the dynamic processes in nature.

Hoppe (1977) applied his modified macroautoradiographic method (Hoppe, 1974) to determine some physiological groups of heterotrophic bacteria in the neuston, the subsurface water, and the sediment in a brackish water area of the Limfjord, Denmark. He used a series of ^{14}C-labeled substrates with different degrees of degradability (glucose, xylose, lactose, fats, and toxic compounds such as phenol and DDT). He showed that there was a lower percentage of glucose-utilizing bacteria in neuston than in the water column and sediment. This was tentatively explained as being due to a larger proportion of neustonic bacteria having specialized uptake mechanisms with an insignificant glucose uptake. Whether or not glucose is a suitable substrate in studies of the heterotrophic activity of neustonic bacteria, selected for utilization of lipids, was also discussed by Dietz et al. (1976). It has occurred to us that conventional substrates do not offer bacteria that are dependent on interfacial enzymes suitable conditions to display their biochemical activity (Norkrans and Stehn, 1978).

Kjelleberg and Håkansson (1977) found a very clear trend indicating a higher number of biochemically active bacteria in the bulk than in the surface microlayers (sampled depth 30 μm) when testing lipolytic, proteolytic, and amylolytic activity in samples from the Swedish west coast. Some introductory studies have also been made on membrane-collected material from the same area with regard to respiratory potential, according to the method by Gocke and Hoppe (1977) using INT (2-(p-iodophenyl)-3-(p-nitrophenyl)-5-phenyl tetrazolium chloride) as electron acceptor. The studies indicate an activity/CFU about 15 times lower for surface bacteria than for subsurface bacteria (B. Dahlbäck, personal communication). Similarly, Fehon and Oliver (1977) concluded that bacteria from subsurface samples were more active crude-oil degraders than neuston bacteria obtained using a membrane in a slightly polluted area. Marumo et al. (1971) proposed that the bacteria accumulate in the surface microlayers as a result of physical forces rather than by reproduction since an extremely small viable count relative to the total count was found in the surface region.

It is a well-known fact that only a fraction of bacteria taken directly from nature can be detected by plate count (Jannasch and Jones, 1959). This means that the activity/total count will be some orders of magnitude lower than activity/CFU. So far, very few studies of the total count in surface microlayers have been published, and until recently, very few methods for determination of bacteria in natural environments have existed. The increasing awareness of the importance of the bacterial biomass in the oceans and other aquatic environments, a topic previously largely neglected, has, however, encouraged new efforts to develop reliable methods for the enumeration of these small creatures. These efforts have evidently been successful. Several new methods have recently been published based on epifluorescence microscopy after staining with acridine orange (Francisco et al., 1973; Zimmermann and Meyer-Reil, 1974;

Daley and Hobbie, 1975). Bührer (1977) adapted the fluorescence technique to bacterial enumeration in sediments; Zimmermann (1977) combined the technique with scanning electron microscopy, a method also used by Bowden (1977). Larsson *et al.* (1978) made a comparison between epifluorescence microscopy, bright-field light microscopy, and electron microscopy (TEM, sectioned specimens) for determinations of the number of bacteria and algae in lake water. Electron microscopy was used by Watson *et al.* (1977) on replicas prepared from membranes with adhering bacteria. The same authors made quantitative determinations of lipopolysaccharide (LPS) in seawater samples. The biomass of gram-negative (LPS-containing) bacteria was shown to be related to the LPS content of the samples. A factor of 6.35 was determined for converting LPS to bacterial carbon. King and White (1977) analyzed muramic acid, a component specific to the cell walls of bacteria and blue-green algae. Hagström (1978) has tried to use the frequency of dividing cells (FDC) as a measure of bacterial growth rate in the phototrophic layer of the pelagic ecosystem. By combining epifluorescence microscopy with autoradiography (Meyer-Reil, 1978) or with INT as electron acceptor (Zimmermann *et al.*, 1978), a simultaneous determination can be made of the total number and the proportion of active bacteria. The two latter methods seem to us to be especially valuable for bacterioneuston studies since the dream of *in situ* determinations is unlikely to be realized.

5.3. Potentially Detrimental Factors for Survival and Growth in Surface Microlayers

Potentially detrimental factors such as intense solar radiation, temperature and salinity fluctuations, and the presence of toxic organic substances and heavy metals at high concentrations can make heavy demands upon the microorganisms in their struggle for survival and growth. All factors are selective and govern the microbial composition of the community. Furthermore, since hardly any natural environment is as constant as the marine water body with respect to temperature, pH, concentrations of salts and nutrients, etc., it can be assumed that microorganisms derived from such a milieu have a weak potential to meet fluctuations of different kinds. The development of mechanisms to cope with such fluctuations would hardly have had any general selective value.

5.3.1. Intense Solar Radiation

Gunkel (1970) deals with the harmful effects of sunlight upon marine bacteria, and in a recent comprehensive review, Nasim and James (1978) are concerned with the effect on microorganisms in general of biologically destructive irradiation, both nonionizing and ionizing. They discuss the nature of the damage and the cellular mechanisms that exist for protection and repair. Data given show the large variation in radiation sensitivity among microbes. The radiation dose of UV light for 37% survival goes from 50,000 to 500 ergs/mm^2 for the marine flagellate *Bodo marina* and *Escherichia coli*, respectively. In this respect, a recent study by Smith and Baker (1979) is of a certain interest. They studied the penetration of biologically active UV (UV-B radiation) and effective dose rates in natural waters. From their data referred to a "standard atmosphere," i.e., with the now prevailing ozone layer at noon San Diego, it is possible to calculate an energy value for the light at the surface. Taking into account diffusive scattering and the spectral contribution of natural light and furthermore approximating the biologically active part of the spectrum at wavelengths <310 nm by means of the Setlow DNA action spectrum (Setlow, 1974), it is possible to give an estimate of 10^2 J/m^2/day. This leads to an estimated approximate time for 60% killing of 50 and 0.5 days for *B. marina* and *E. coli*, respectively.

Microbes surviving and reproducing at the air–water interface must possess high UV resistance. Initial experiments have shown bacterial isolates from the subsurface water to be markedly more UV sensitive than those obtained from the bacterioneuston layer (B. Dahlbäck, personal communication).

To withstand the insolation in the surface microlayers, the bacteria may be characterized by a DNA with a high GC mole percent value and consequently a low thymine content, thus reducing the probability for thymine dimerization (cf. Marshall, 1976). This could be the case for pseudomonads which have GC values in the upper range for bacteria, i.e., between 57 and 70 mole%. The bacteria may also exhibit highly active and complete repair systems for photoreactivization, excision repair, and postreplication repair (see Nasim and James, 1978). The occurrence of mutations leading to selection for UV resistance is to be expected. An effective positioning away from the light as described for protozoa (Barcelo and Calkins, 1979) seems less possible in these microlayers. Dormant cells and physiologically inactive cells are generally less sensitive to environmental stress than active cells; we might find cells from surface microlayers less vulnerable for that reason also.

Photodynamic effects are also operative. These involve the sensitization of living tissue by light from near UV to light with wavelengths up to 700 nm, sensitization of endogenous or exogenous pigments, and O_2. Both their mutational effect and, more often, their destruction of cellular components, such as enzymes and membranes, have been described (see Krinsky, 1976). Carotenoids, with more than eight conjugated double bonds, can protect cells against potentially harmful or lethal photodynamic effects by quenching the excited molecules of oxygen, $'O_2$ or chemical products, thus decreasing the $'O_2$ available for photodynamic action. Mathews and Sistrom (1959) were the first to propose the ecological importance of colored carotenoid pigments in nonphotosynthetic bacteria. They concluded that organisms that normally exist in an aerobic environment with the possibility of exposure to light would be protected by the presence of carotenoid pigments, whereas organisms lacking these pigments would suffer lethal effects. Most bacteria reported to contain carotenoids are gram-positive (e.g., Weeks and Andrewes, 1972). This result could merely be due to a fortuitous choice of subjects for study. If it does reflect a true difference between gram-positive and gram-negative bacteria, then not many bacteria in the marine surface microlayers can depend on this form of protective mechanism since only about 5% of marine bacteria are gram-positive. Ninety-five percent of them, however, are colored, and at least some of the yellow ones may contain carotenoids (ZoBell, 1946). High contents in the membranes of components that are less sensitive to near-UV radiation, such as cyclopropane fatty acids, might also be of positive value (cf. Kates et al., 1964).

Hoober (1977) showed that, in an Arthrobacter species, photodynamic action resulted in a remarkable stimulation of the synthesis of a polypeptide, 21,000 daltons in mass, residing on the cell surface. This response of the cell may be related to protection against photooxidation. According to Hoober, the polypeptide does not seem to show similarities with membrane-, additional surface-, or flagella-protein subunits but with units of fimbriae found on Corynebacterium species. Furthermore, coryneform cells revealed numerous fimbriae after but not before the photooxidation treatment. Whether a fimbriae protein or not, such a dense covering could clearly provide light protection for bacteria in the surface microlayers either per se or indirectly by facilitating agglomeration of bacteria whereby some individuals would come into the "shadow."

5.3.2. Salinity Fluctuations

Fluctuations in salinity apply mainly to microbes in the surface microlayers of coastal waters such as estuaries, lagoons, and rock pools, although the uppermost surface layers in the open sea may also be subject to salinity variations following extensive rainfall or evaporation.

No studies on this subject directed specifically to surface microlayer bacteria have come to the author's attention. In general, however, marine gram-negative bacteria vary in their susceptibility to low salt concentrations, some lysing completely, others partially, when suspended in fresh water (MacLeod, 1965, 1971). Those lysing partly maintain more easily the integrity of their cell envelope at lowered salt concentrations, and they do not lose the cross-linkings within the mucopeptide layer. Such a loss weakens the cell sufficiently for the osmotic pressure within the cell to burst the remaining wall layers. Clearly, bacteria in this group, the euryhaline bacteria, are the ones most fitted for an environment with salinity fluctuations. It seems to be accepted that marine bacteria can easily adjust their internal salt concentration to counterbalance the external one without expending an appreciable amount of energy. Some gram-negative bacteria, however, such as Vibrio parahaemolyticus, Salmonella oranienburg, E. coli, and

Pseudomonas aeruginosa, have been shown to react to an increase in external salt concentration by accumulation of glutamic acid and proline (Tempest *et al.*, 1970; Christian and Hall, 1972; Measures, 1975, 1976).

5.3.3. Temperature Fluctuations

Microorganisms in the surface microlayers are exposed to heat both from the water and from direct absorption and hence to rapidly fluctuating temperatures due to cloud cover, etc. There seems to be no fluctuation data for these surfaces in the literature, but the phenomenon has been of interest to plant physiologists, who give data for leaf surfaces and other part of plants (e.g., Björkman *et al.*, 1972; Gates, 1973; Larcher, 1973). Oppenheimer (1970) gives an example of temperatures registered at the leaf surface of *Ficus*, being 49°C when the air temperature was 28°C and dropping to 30°C within 30 sec when a cloud passed. Dommergues *et al.* (1978) state that microorganisms are very sensitive to temperature changes. The membranes may be the most susceptible cell structure since microorganisms are known to manipulate their membrane in response to temperature changes. For example, the free fatty acid fraction from *E. coli* cells incubated at 15°C contained 10 times more unsaturated fatty acids than that of cells grown at 43°C, and the nature of the chemical modifications define the range of optimal function for the membranes (Cronan and Gelmann, 1975). Immediate "adjustments," such as a response to "upshocks" or "downshocks," do not seem very probable, but rather the creation of a disordered, nonfunctioning membrane with—in the worst case—lethal effects.

5.3.4. Organic Toxic Substances, Heavy Metals, and Their Transformation to Particles

Toxic substances originating from organisms in the surface microlayers have been reported (Sieburth *et al.*, 1976). Also, for many hydrophobic pollutants, such as polychlorinated biphenyls (PCBs), insecticides, and pesticides, markedly enhanced concentrations have been shown in the surface microlayers as compared to the levels in the subsurface water. Enrichment factors of 10^3-10^4 for PCBs have been reported from different waters (Duce *et al.*, 1972; Larsson *et al.*, 1974).

By compressing surface films in an apparatus with movable barriers, similar to a surface balance, Wheeler (1975) showed how films formed from filter-passing material (pore size < 1 μm) from lake water, coastal water, and Sargasso Sea water were brought to film collapse and particle formation at increased surface pressures. The large pore size certainly was a disadvantage in the preparation of the material.

Wind and wave action and the drag of rising bubbles can compress organic films to their collapse point and may contribute to the particle formation of organic material originally in the subsurface water. Concentrations of dieldrin and DDT-group pesticides were found to be consistently high in wind-generated foam from Lake Mendota, Wisconsin. PCBs were also detected (Eisenreich et al., 1978). The particles are of special importance if they act as conveyors into the atmosphere of such persistent substances that withstand most degradation processes except photodegradation.

Natural films are believed to influence the distributions of near-surface cations, and a particle formation can be assumed. Accumulation of heavy metals in the surface microlayers has been observed by several authors, mainly in marine systems (Liss, 1975). The surface concentrations were found to be higher by more than four orders of magnitude than those in mean ocean water, the latter being at micrograms per liter levels. The same enriched concentrations have also been found in wind-generated foam (Szekielda et al., 1972). Quite recently, the aqueous surface microlayers in a Delaware salt marsh (Pellenbarg and Church, 1979) and in Lakes Michigan, Ontario, and Mendota (Elzerman and Armstrong, 1979) have been examined for dissolved and particulate trace metals, and Zn, Cd, Cu, and especially Pb were found to be enriched in proportion to the amounts of accumulated film material. Wind-generated foam from Lake Mendota was also analyzed (Eisenreich et al., 1978) and exhibited chemical properties similar to surface microlayers with respect to organic and inorganic constituents. Blanchard and Parker (1977) give the same information in a review about the freshwater-to-air transfer of microorganisms and organic matter but also reveal the significance of aerosols as nuclei at the time of snow and rain formation and, under certain conditions, as dispersants of pathogenic microbes, bacteria, virus, amebae, etc., into the atmosphere. The literature dealing with these aspects has not been considered in the present review.

Fenchel and Jørgensen (1977) report the ratio of dissolved organic matter (DOM) to particulate dead organic matter and living biomass to be 100:10:2 in the oceans, and they give similar values for lakes. It is not possible to calculate how much of the particulate matter might derive from the above-mentioned processes. The importance of this contribution may be more qualitative than quantitative. Particle production by surface film collapse may be a mechanism whereby toxic substances enter the food webs via surface feeders and near-surface particle feeders in the subsurface water.

Generally, we consider bacteria as decomposers, important agents in biogeochemical cycling, but we may overlook their role as food for various consumers. Sorption to or uptake by bacteria of various persistent compounds may serve as a means of introducing these toxic compounds into aquatic food chains (Grimes and Morrison, 1975; Paris and Lewis, 1976). Within all major groups of protozoa (flagellates, rhizopods, and ciliates), single species and larger taxa are found that are totally specialized to a bacterial diet. There are also some heterotrophic microflagellates that can themselves be contaminated by the toxic substances and then serve as food for grazing phagotrophic protists and feed the special blue-pigmented copepods floating on the surface (David, 1965).

The behavior of surface oil slicks after accidental or routine oil discharges into aquatic environments and the microbial community of this particular eco-system, dealt with in a vast literature, are not considered in the present paper. The reader is referred to recent reviews of this subject; e.g., Bartha and Atlas (1977) treated pertinent problems, presented lists of dominating microor-ganisms, and discussed changes in the size of microbial populations and in the species diversity of the microbial community after exposure to oil pollution.

6. Phytoneuston Layers: Some Characteristics

Already the size of phytoplankton exclude them from the *few-micron layer* samples. Several investigators, however, show their presence in screen and drum microlayer samples (60–250 μm thick) by microscopical work and/or deduce their presence from analysis of chlorophyll *a* and phaeophytin in marine environ-ments (Harvey, 1966; Brockmann *et al.*, 1976; Sieburth *et al.*, 1976), flagellates, dinoflagellates, ciliates, and diatoms being represented.

Parker and Hatcher (1974) suggest a nonlinear change in algal community densities within a depth corresponding to the microlayers, thus suggesting some sort of microstrata. They found the seasonal distribution of algal genera to be random, with no genus consistently inhabiting either the surface microlayers or the subsurface water to a depth of 10 cm. However, for a given sampling, each genus was mainly concentrated either in the surface or in the subsurface water. Differences in the organic and inorganic chemistry of surface microlayers relative to subsurface water may contribute to differences in growth rates and resulting populations. Estimations of *in situ* activity, however, expressed as, e.g., C assimi-lated/mg photopigment and hour, belong, as many other works, to the future. Daumas *et al.* (1976) also suggest a stratification of the surface algal layers. Whether this partition of the biological material is due to the sampling procedure or varies with oxygenation, irradiation at the air–sea interface, or other environ-mental conditions was not decided.

As previously mentioned, Brockmann *et al.* (1976) followed the develop-ment of a large sea slick off the island of Sylt and studied the associated biolog-ical and physicochemical changes. They identified and counted phytoplankton and bacteria and evaluated fatty acids as an indication of the biogenic surface active material. They found an active and dense accumulation of phytoplankton, especially of the dinoflagellate *Prorocentrum micans*, but also of some diatoms, mainly *Ceratium* spp. At the maximum, built up during a few hours at daytime, *Prorocentrum* amounted to more than 90% of the calculated biomass and occurred in concentrations of 2.7×10^6 cells/ml. In the water just beneath the 250-μm surface layers, an increasing concentration of *Prorocentrum* cells was also observed, whereas a decrease occurred at greater depths. Simultaneously, the percentage of C 16:1 increased in the surface film. This fatty acid is the main component of *Prorocentrum* lipids. Its transport by bubbles and vesicles to the surface film seems very probable. The authors also recorded an increase by a factor of two in the number of bacteria.

In the previous sections, evidence has shown that the microbial community of the surface microlayers differs from that of the underlaying water. We have shown that the initial adhesion of bacteria to the air-water interface material involves a selection of bacteria from the subsurface water. Furthermore, some microorganisms, mainly spore formers, bacteria, and fungi, reach the surface from the air. We have suggested a series of physiological and biochemical properties which may be of value for microbial survival in the surface microlayers and which lead to a selection of the best-fitted individuals. To obtain energy for growth, the heterotrophic surface bacteria must necessarily possess uptake mechanisms and enzymic makeup for degrading the substances which remain at the interface. From laboratory experiments, one might conclude that they differ from subsurface bacteria as regards uptake mechanisms. The highly hydrophobic substances which remain at the surface microlayers may be so refractory that only very slow growth is possible. A great deal of the autochthonous flora may even be dormant. When offered fresh new material in case of a heavy accumulation, as in the described sea slick formation of *Prorocentrum*, the autochthonous flora may also flourish, but it has to compete with an entering allochthonous flora. The microbial structure of the microsurface layers has generally been dealt with phenomenologically. We hope that the increasing interest in interface studies will lead to a greater understanding of the microbial contribution to the dynamic processes within this very specific ecosystem.

References

Adam, N. K., 1937, A rapid method for determining the lowering of exposed water surfaces, with some observations on the surface tension of the sea and inland waters, *Proc. R. Soc. London Ser., B* **122**:134–139.

Baier, R. E., 1972, Organic films on natural waters: Their retrieval, identification, and modes of elimination, *J. Geophys. Res.* **77**:5062–5075.

Baier, R. E., 1975, Applied chemistry at protein interfaces, *Adv. Chem. Ser.* **145**:1–25.

Baier, R. E., Goupil, D. W., Perlmutter S., and King, R., 1974, Dominant chemical composition of sea-surface films, natural slicks, and foams, *J. Rech. Atmos.* **8**:571–600.

Bangham, A. D., 1975, Models of cell membrane, in: *Cell Membranes: Biochemistry, Cell Biology and Pathology* (G. Weisman and R. Claiborne, eds.), pp. 24–34, HP Publishing, New York.

Barcelo, J. A., and Calkins, J., 1979, Positioning of aquatic microorganisms in response to visible light and simulated solar UV-B irradiation, *Photochem. Photobiol.* **29**:75–83.

Bartha, R., and Atlas, R. M., 1977, Microbiology of aquatic oil spills, *Adv. Appl. Microbiol.* **22**:225–266.

Bezdek, H. F., and Carlucci, A. F., 1972, Surface concentration of marine bacteria, *Limnol. Oceanogr.* **17**:566–569.

Björkman, O., Pearcy, R. W., Harrison, A. T., and Mooney, H., 1972, Photosynthetic adaptation to high temperatures: A field study in Death Valley, California, *Science* 175: 786-789.

Blanchard, D. C., and Parker, B. C., 1977, The freshwater to air transfer of microorganisms and organic matter, in: *Aquatic Microbial Communities* (J. J. Cairns, ed.), pp. 627-658, Garland, New York.

Blanchard, D. C., and Syzdek, L. D., 1970, Mechanism for the water-to-air transfer and concentration of bacteria, *Science* 170:626-628.

Blanchard, D. C., and Syzdek, L. D., 1978, Seven problems in bubble and jet drop researches, *Limnol. Oceanogr.* 23:389-400.

Blumer, M., 1970, Dissolved organic compounds in sea water: Saturated and olefinic hydrocarbons and singly branched fatty acids, in: *Symposium on Organic Matter in Natural Water*, Sept. 2-4, 1968 (D. W. Hood, ed.), pp. 153-167, University of Alaska, Fairbanks.

Blumer, M., Guillard, R. R. L., and Chase, T., 1971, Hydrocarbons of marine phytoplankton, *Mar. Biol.* 8:183-189.

Bocard, C., Gatellier, C., Petroff, N., Renault, Ph., and Roussel, J. C., 1977, Biogenic hydrocarbons and petroleum fractions, *Rapp. P. V. Réun. Cons. Int. Explor. Mer* 171:91-93.

Bowden, W. B., 1977, Comparison of two direct count techniques for enumerating aquatic bacteria, *Appl. Environ. Microbiol.* 33:1229-1232.

Brockmann, U. H., Kattner, G., Hentschel, G., Wandschneider, K., Junge, H. D., and Hühnerfuss, H., 1976, Natürliche Oberflächenfilme im Seegebiet vor Sylt, *Mar. Biol.* 36:135-146.

Bührer, H., 1977, Verbesserte acridineorange Methode zur Direktzählung von Bakterien aus Seesediment, *Schweiz. Z. Hydrol.* 39:99-103.

Carlucci, A. F., and Williams, P. M., 1965, Concentration of bacteria from sea water by bubble scavenging, *J. Cons. Perm. Int. Explor. Mer.* 30:28-33.

Carty, C., and Colwell, R. R., 1975, A microbiological study of air and surface water microlayers in the open ocean, *J. Wash. Acad. Sci.* 65:148-153.

Christian, J. H. B., and Hall, J. M., 1972, Water relations of *Salmonella oranienburg:* Accumulation of potassium and amino acids during respiration, *J. Gen. Microbiol.* 70: 497-506.

Costerton, J. W., Ingram, J. M., and Cheng, K.-J., 1974, Structure and function of cell envelope of gram-negative bacteria, *Bacteriol. Rev.* 38:87-110.

Cronan, J. E., and Gelmann, E. P., 1975, Physical properties of membrane lipids: Biological relevance and regulation, *Bacteriol. Rev.* 39:232-256.

Crow, S. A., Ahearn, D. G., Cook, W. L., and Bourquin, A. W., 1975, Densities of bacteria and fungi in coastal surface films as determined by membrane-adsorption procedure, *Limnol. Oceanogr.* 10:602-605.

Daley, R. J., and Hobbie, J. E., 1975, Direct counts of aquatic bacteria by a modified epifluorescence technique, *Limnol. Oceanogr.* 20:875-882.

Danielsson, A., Norkrans, B., and Björnsson, A., 1977, On bacterial adhesion—The effect of certain enzymes on adhered cells of a marine *Pseudomonas* sp., *Bot. Mar.* 20:13-17.

Daumas, R. A., LaBorde, P. L., Marty, J. C., and Saliot, A., 1976, Influence of sampling method on the chemical composition of water surface film, *Limnol. Oceanogr.* 21: 319-326.

David, P. M., 1965, The surface fauna of the ocean, *Endeavour* 24:95-100.

Decad, G. M., and Nikaido, H., 1976, Outer membrane of gram-negative bacteria XII. Molecular-sieving function of cell wall, *J. Bacteriol.* 128:325-336.

Dexter, S. C., Sullivan, J. D., Jr., Williams, J., III, and Watson, S. W., 1975, Influence of substrate wettability on the attachment of marine bacteria to various surfaces, *Appl. Microbiol.* 30:298-308.

Dietz, A. S., Albright, L. J., and Tuominen, T., 1976, Heterotrophic activities of bacterioneuston and bacterioplankton, *Can. J. Microbiol.* **22**:1699-1709.

Dietz, A. S., Albright, L. J., and Tuominen, T., 1977, Alternative model and approach for determining microbial heterotrophic activities in aquatic systems, *Appl. Environ. Microbiol.* **33**:817-823.

Dommergues, Y. R., Belser, L. W., and Schmidt, E. L., 1978, Limiting factors for microbial growth and activity in soil, in: *Advances in Microbial Ecology*, Vol. 2 (M. Alexander, ed.), pp. 49-104, Plenum Press, New York.

Duce, R. A., Quinn, J. G., Olney, C. E., Piotrowicz, S. R., Ray, B. J., and Wade, T. L., 1972, Enrichment of heavy metals and organic compounds in the surface microlayer of Narragansett Bay, Rhode Island, *Science* **176**:161-163.

Duce, R. A., Hoffman, G. L., Ray, B. J., Fletcher, I. S., Wallace, G. T., Fasching, J. L., Piotrowicz, S. R., Walsh, P. R., Hoffman, E. J., Miller, J. M., and Heffter, J. L., 1976, Trace metals in tne marine atmosphere: Sources and fluxes, in: *Marine Pollutant Transfer* (H. L. Windom and R. A. Duce, eds.), pp. 77-119, Lexington Books, Lexington.

Dutka, B. J., and Kwan, K. K., 1978, Health-indicator bacteria in water-surface microlayers, *Can. J. Microbiol.* **24**:187-188.

Eisenreich, S. J., Elzerman, A. W., and Armstrong, D. E., 1978, Enrichment of micronutrients, heavy metals, and chlorinated hydrocarbons in wind-generated lake foam, *Environ. Sci. Technol.* **12**:413-417.

Elzerman, A. W., and Armstrong, D. E., 1979, Enrichment of Zn, Cd, Pb, and Cu in the surface microlayer of Lakes Michigan, Ontario, and Mendota, *Limnol. Oceanogr.* **24**:133-144.

Fehon, W. C., and Oliver, J. D., 1977, Degradation of crude oil by mixed populations of bacteria from the surface microlayer in an estuarine system, *J. Elisha Mitchell Soc.* **93**:72-73.

Fenchel, T. M., and Jørgensen, B. B., 1977, Detritus food chains of aquatic ecosystems: The role of bacteria, in: *Advances in Microbial Ecology*, Vol. 1 (M. Alexander, ed.), pp. 1-58, Plenum Press, New York.

Fletcher, M., and Floodgate, G. D., 1973, An electron-microscopic demonstration of an acidic polysaccharide involved in the adhesion of a marine bacterium to solid surfaces, *J. Gen. Microbiol.* **74**:325-334.

Fletcher, M., and Loeb, G. I., 1979, Influence of substratum characteristics on the attachment of a marine pseudomonad to solid surfaces, *Appl. Environ. Microbiol.* **37**:67-72.

Fogg, G. E., 1977, Excretion of organic matter by phytoplankton, *Limnol. Oceanogr.* **22**:576-577.

Francisco, D. E., Mah, R. A., and Rabin, A. C., 1973, Acridine orange-epifluorescence technique for counting bacteria in natural waters, *Trans. Am. Microsc. Soc.* **92**:416-421.

Gaines, G. L., 1966, *Insoluble Monolayers at Liquid-Gas Interfaces*, John Wiley & Sons, New York.

Garrett, W. D., 1965, Collection of slick forming material from the sea surface, *Limnol. Oceanogr.* **10**:602-605.

Garrett, W. D., 1967, The organic chemical composition of the ocean surface, *Deep-Sea Res.* **14**:221-227.

Garrett, W. D., 1972, Impact of natural and man-made surface films on the properties of the air-sea interface, in: *The Changing Chemistry of the Oceans*, Nobel Symposium 20, (D. Dyrssen and D. Jagner, eds.), pp. 75-91, Almqvist & Wiksell, Stockholm.

Gates, D. M., 1973, Plants. I. Plant temperatures and energy budget, in: *Temperature and Life* (H. Precht, J. Christophersen, H. Hensel, and W. Larcher, eds.), pp. 87-101, Springer-Verlag, Berlin.

Gibbons, R. J., and van Houte, J., 1975, Bacterial adherence in oral microbial ecology, *Annu. Rev. Microbiol.* **29**:19-44.

Gocke, K., and Hoppe, H.-G., 1977, Description and application of dehydrogenase activity (DHA) as a measure for potential respiration, in: *Microbial Ecology of a Brackish Water Environment* (G. Rheinheimer, ed.), pp. 67–70, Springer–Verlag, Berlin.

Grimes, D. J., and Morrison, S. M., 1975, Bacterial bioconcentration of chlorinated insecticides from aqueous systems, *Microb. Ecol.* 2:43–59.

Gunkel, W., 1970, Light. Bacteria, fungi, and blue-green algae, in: *Marine Ecology* (O. Kinne, ed). Vol. 1, pp. 103–124, John Wiley & Sons, New York.

Hagström, Å., 1978, Bacterial growth *in situ*, a study in aquatic microbiology, Ph.D. Thesis, University of Umeå, Sweden.

Hamilton, E. I., and Clifton, R. J., 1979, Techniques for sampling the air–sea interface for estuarine and coastal waters, *Limnol. Oceanogr.* 24:188–193.

Harden, V. P., and Harris, J. O., 1953, The isoelectric point of bacterial cells, *J. Bacteriol.* 65:198–202.

Harvey, G. W., 1966, Microlayer collection from the sea surface: A new method and initial results, *Limnol. Oceanogr.* 11:608–613.

Harvey, G. W., and Burzell, L. A., 1972, A simple microlayer method for small samples, *Limnol. Oceanogr.* 17:156–157.

Hatcher, R. F., and Parker, B. C., 1974a, Microbiological and chemical enrichment of fresh water-surface microlayers relative to the bulk-subsurface water, *Can. J. Microbiol.* 20:1051–1057.

Hatcher, R. F., and Parker, B. C., 1974b, Laboratory comparisons of four surface microlayer samplers, *Limnol. Oceanogr.* 19:162–165.

Hermansson, M., Kjelleberg, S., and Norkrans, B., 1979, Interaction of pigmented wildtype and pigmentless mutant of *Serratia marcescens* with lipid surface film, *FEMS Microbiol. Lett.* 6:129–132.

Hirsch, P., 1972, Neue Methoden zur Beobachtung und Isolierung ungewöhnlicher oder wenig bekannter Wasserbakterien, *Z. Allg. Mikrobiol.* 12:203–218.

Hirsch, P., and Pankratz, St.H., 1970, Study of bacterial populations in natural environments by use of submerged electron microscope grids, *Z. Allg. Mikrobiol.* 10:589–605.

Hoober, J. K., 1977, Photodynamic induction of a bacterial cell surface polypeptide, *J. Bacteriol.* 131:650–656.

Hoope, H.-G., 1974, Untersuchungen zur Analyse mariner Bakterienpopulationen mit einer autoradiographischen Methode, *Kiel. Meeresforsch.* 30:107–116.

Hoppe, H.-G., 1977, Analysis of actively metabolizing bacterial populations with the autoradiographic method, in: *Microbial Ecology of a Brackish Water Enivornment* (G. Rheinheimer, ed.), pp. 179–197, Springer–Verlag, Berlin.

Jannasch, H. W., and Jones, G. E., 1959, Bacterial populations in seawater as determined by different methods of enumeration, *Limnol. Oceanogr.* 4:128–139.

Jannasch, H. W., and Pritchard, P. H., 1972, The role of inert particulate matter in the activity of aquatic microorganisms, *Mem. 1st Ital. Idrobiol. Suppl.* 29:289–308.

Jarvis, N. L., 1967, Adsorption of surface-active material at the sea-air interface, *Limnol. Oceanogr.* 12:213–221.

Jones, G. E., and Jannasch, H. W., 1959, Aggregates of bacteria in sea water as determined by treatment with surface active agents, *Limnol. Oceanogr.* 4:269–276.

Kates, M., Adams, G. A., and Martin, S. M., 1964, Lipids of *Serratia marcescens*, *Can. J. Biochem.* 42:461–479.

Kattner, G. G., and Brockmann, U. H., 1978, Fatty-acid composition of dissolved and particulate matter in surface films, *Mar. Chem.* 6:233–241.

King, J. D., and White, D. C., 1977, Muramic acid as a measure of microbial biomass in estuarine and marine samples, *Appl. Environ. Microbiol.* 33:777–783.

Kjelleberg, S., and Håkansson, N., 1977, Distribution of lipolytic, proteolytic, and amylolytic marine bacteria between the lipid film and the subsurface water, *Mar. Biol.* 39:103–109.

Kjelleberg, S., and Stenström, T. A., 1980, Lipid surface films: Interaction of bacterial strains with free fatty acids and phospholipids at the air/water interface, *J. Gen. Microbiol.* 116:417–423.

Kjelleberg, S., Norkrans, B., Löfgren, H., and Larsson, K., 1976, Surface balance study of the interaction between microorganisms and lipid monolayer at the air/water interface, *Appl. Environ. Microbiol.* 31:609–611.

Kjelleberg, S., Stenström, T. A., and Odham, G., 1979, A comparative study of different hydrophobic devices for sampling lipid surface films and adherent microorganisms, *Mar. Biol.* 53:21–26.

Kjelleberg, S., Lagercrantz, C., and Larsson, T., 1980, Quantitative analysis of bacterial hydrophobicity studied by the binding of dodecanoic acid, *FEMS Microbiol. Lett.* 7:41–44.

Krinsky, N. I., 1976, Cellular damage initiated by visible light, *Symp. Soc. Gen. Microbiol.* 26:209–239.

Kushner, D. J., 1978, Life in high salt and solute concentrations: Halophilic bacteria, in: *Microbial Life in Extreme Environments* (D. J. Kushner, ed.), pp. 317–368, Academic Press, New York.

Lange, P., and Hühnerfuss, H., 1978 Drift response of monomolecular slicks to wave and wind action, *J. Phys. Oceanogr.* 8:142–150.

Larcher, W., 1973. "Ökologie der Pflanzen," Verlag Eugen Ulmer, Stuttgart.

Larsson, K., 1973, Lipid multilayers, *Surf. Colloid Sci.* 6:261–285.

Larsson, K., Odham, G., and Södergren, A., 1974, On lipid films on the sea. I. A simple method for sampling and studies of composition, *Mar. Chem.* 2:49–57.

Larsson, K., Weibull, C., and Cronberg, G., 1978, Comparison of light and electron microscopic determinations of the number of bacteria and algae in lake water, *Appl. Environ. Microbiol.* 35:397–404.

Ledet, E. J., and Laseter, J. L., 1974, Alkanes at the air–sea interface from offshore Louisiana and Florida, *Science* 186:261–263.

Liss, P. S., 1975, Chemistry of the sea surface microlayer, in: *Chemical Oceanography*, Vol. 2, 2nd ed. (J. P. Riley and G. Skirrow, eds.), pp. 193–243, Academic Press, New York.

Liss, P. S., 1977, Effect of surface films on gas exchange across the air–sea interface, *Rapp. P. V. Réun. Cons. Int. Explor. Mer* 171:120–124.

Longton, R. W., Cole, J. S., and Quinn, P. F., 1975, Isoelectric focusing of bacteria: Species location within an isoelectric focusing column by surface charge, *Arch. Oral Biol.* 20:103–106.

MacIntyre, F., 1968, Bubbles: A boundary-layer "microtome" for micron-thick samples of liquid surface, *J. Phys. Chem.* 72:589–592.

MacIntyre, F., 1974a, The top millimeter of the ocean, *Sci. Am.* 230:62–77.

MacIntyre, F., 1974b, Chemical fractionation and sea-surface microlayer processes, in: *The Sea*, Vol. 5 (E. D. Goldberg, ed.), pp. 245–299, Wiley–Interscience, New York.

MacLeod, R. A., 1965, The question of the existence of specific marine bacteria, *Bacteriol. Rev.* 29:9–23.

MacLeod, R. A., 1971, Salinity. Bacteria, fungi, and blue-green algae, in: *Marine Ecology*, Vol. 1 (O. Kinne, ed.), pp. 689–703, John Wiley & Sons, New York.

Marshall, K. C., 1975, Clay mineralogy in relation to survival of soil bacteria, *Annu. Rev. Phytopathol.* 13:357–373.

Marshall, K. C., 1976, *Interfaces in Microbial Ecology*, pp. 1–156, Harvard University Press, Cambridge, Massachusetts.

Marshall, K. C., Stout, R., and Mitchell, R., 1971, Mechanism of the initial events in the sorption of marine bacteria to surfaces, *J. Gen. Microbiol.* 68:337–348.

Marumo, R., Tuga, N., and Nakai, T., 1971, Neustonic bacteria and phytoplankton in surface microlayers of the equatorial waters, *Bull. Plankton Soc. Jpn* 18:36–41.

Mathews, M. M., and Sistrom, W. R., 1959, Function of carotenoid pigments of *Sarcina lutea, Arch. Mikrobiol.* 35:139–146.

Measures, J. C., 1975, Role of amino acid in osmoregulation of non-halophilic bacteria, *Nature (London)* 257:398–400.

Measures, J. C., 1976, Reactions of microbial cells to osmotic stress, Colloq. Int. C.N.R.S. No. 246, *L'eau et les Systèmes Biologique*, pp. 303–309.

Meyer-Reil, L. A., 1978, Autoradiography and epifluorescence microscopy combined for the determination of number and spectrum of actively metabolizing bacteria in natural waters, *Appl. Environ. Microbiol.* 36:506–512.

Miget, R., Kator, H., Oppenheimer, C., Laseter, J. L., and Ledet, E. J., 1974, New sampling device for the recovery of petroleum hydrocarbons and fatty acids from aqueous surface films, *Anal. Chem.* 46:1154–1157.

Nasim, A., and James, A. P., 1978, Life under conditions of high irradiation, in: *Microbial Life in Extreme Environments* (D. J. Kushner, ed.), pp. 409–439, Academic Press, New York.

Naumann, E., 1917, Beiträge zur Kenntnis des Teichnannoplanktons, II. Über das Neuston des Süsswassers, *Biol. Zentralbl.* 37:98–106.

Norkrans, B., 1979, Role of surface microlayers, in: *Microbial Degradation of Pollution in Marine Environments*, (A. W. Bourquin and P. H. Pritchard, eds.), U.S. Environmental Protection Agency, EPA-600/9-79-012. Gulf-Breeze, Florida, pp. 201–213.

Norkrans, B., and Sörensson, F., 1977, On the marine lipid surface microlayer–Bacterial accumulation in model systems, *Bot. Mar.* 20:473–478.

Norkrans, B., and Stehn, B. O., 1978, Sediment bacteria in the deep Norwegian Sea, *Mar. Biol.* 47:201–209.

Odham, G., Norén, B., Norkrans, B., Södergren, A., and Löfgren, H., 1978, Biological and chemical aspects of the aquatic lipid surface microlayer, *Prog. Chem. Fats Other Lipids* 16:31–44.

Olsson, J., Glantz, P. O., and Krasse, B., 1976, Surface potential and adherence of oral streptococci to solid surfaces, *Scan. J. Dent. Res.* 84:240–242.

Oppenheimer, C. H., 1970, Temperature. Bacteria, fungi, and blue-green algae, in: *Marine Ecology*, Vol. 1 (O. Kinne, ed.), pp. 347–361, John Wiley & Sons, New York.

Ottow, J. C. G., 1975, Ecology, physiology, and genetics of fimbriae and pili, *Annu. Rev. Microbiol.* 29:79–108.

Paris, D. F., and Lewis, D. L., 1976, Accumulation of methoxychlor by microorganisms isolated from aqueous systems, *Bull. Environ. Contam. Toxicol.* 15:24–32.

Parker, B., and Barsom, G., 1970, Biological and chemical significance of surface microlayers in aquatic ecosystems, *BioScience* 20:87–93.

Parker, B. C., and Hatcher, R. F., 1974, Enrichment of surface freshwater microlayers with algae, *J. Phycol.* 10:185–189.

Pellenbarg, R. E., and Church, T. M., 1979, The estuarine surface microlayer and trace metal cycling in a salt marsh, *Science* 203:1010–1012.

Quickenden, T. I., and Barnes, G. T., 1978, Evaporation through monolayers–Theoretical treatment of the effect of chain length, *J. Colloid Interface Sci.* 67:415–422.

Quinn, J. A., and Otto, N. C., 1971, Carbon dioxide exchange at the air-sea interface: Flux augmentation by chemical reaction, *J. Geophys. Res.* 76:1539–1549.

Setlow, R. B., 1974, The wavelengths in sunlight effective in producing skin cancer: A theoretical analysis, *Proc. Natl. Acad. Sci. U.S.A.* 71:3363–3366.

Sharp, J. H., 1977, Excretion of organic matter by marine phytoplankton: Do healthy cells do it?, *Limnol. Oceanogr.* 22:381–399.

Sieburth, J. McN., 1965, Bacteriological samplers for air-water and water-sediment interfaces, Ocean Science and Ocean Engineering, *Transactions of the Joint Conference, MTS and ASLO, Washington, D.C.*, pp. 1064–1068.

Sieburth, J. McN., 1971, Distribution and activity of oceanic bacteria, *Deep-Sea Res.* 18: 1111–1121.

Sieburth, J. McN., Willis, P.-J., Johnson, K. M., Burney, C. M., Lavoie, D. M., Hinga, K. R., Caron, D. A., French, F. W., III, Johnson, P. W., and Davis, P. G., 1976, Dissolved organic matter and heterotrophic microneuston in the surface microlayers of the North Atlantic, *Science* 194:1415–1418.

Smith, R. S., and Baker, K. S., 1979, Penetration of UV-B and biologically effective dose-rates in natural waters, *Photochem. Photobiol.* 29:311–323.

Szekielda, K.-H., Kupperman, S. L., Klemas, V., and Polis, D. F., 1972, Element enrichment in organic films and foam associated with aquatic frontal systems, *J. Geophys. Res.* 77: 5278–5282.

Tanford, C., 1979, Interfacial free energy and the hydrophobic effect. *Proc. Natl. Acad. Sci. U.S.A.* 76:4175–4176.

Tempest, D. W., Meers, J. L., and Brown, C. M., 1970, Influence of environment on the content and composition of microbial free amino acid pools, *J. Gen. Microbiol.* 64:171–185.

Tsiban, A. V., 1975, Bacterioneuston and problem of degradation in surface films of organic substances released into the sea, *Prog. Water Technol.* 7:793–799.

Tweedy, J. M., Park, R. W. A., and Hodgkiss, W., 1968, Evidence for the presence of fimbriae (pili) on *Vibrio* species, *J. Gen. Microbiol.* 51:235–244.

Wangersky, P. J., 1976, The surface film as a physical environment, *Annu. Rev. Ecol. Syst.* 7:161–176.

Wangersky, P. J., 1978, Production of dissolved organic matter, in: *Marine Ecology*, Vol. 4 (O. Kinne, ed.), pp. 115–220, John Wiley & Sons, New York.

Watson, S. W., Novitsky, T. J., Quinby, H. L., and Valois, F. W., 1977, Determination of bacterial number and biomass in the marine environments, *Appl. Environ. Microbiol.* 33:940–946.

Weeks, O. B., and Andrewes, A. G., 1972, Naturally occurring nonapreno and decapreno carotenoids, in: *The Chemistry of Plant Pigments* (C. O. Chichester, ed.), *Advances in Food Research Supplement 3*, pp. 23–32, Academic Press, New York.

Wheeler, J. R., 1975, Formation and collapse of surface films, *Limnol. Oceanogr.* 20:338–342.

Young, L. Y., 1978, Bacterioneuston examined with critical point drying and transmission electron microscopy, *Microb. Ecol.* 4:267–277.

Zimmermann, R., 1977, Estimation of bacterial number and biomass by epifluorescence microscopy and scanning electron microscopy, in: *Microbial Ecology of a Brackish Water Environment* (G. Rheinheimer, ed.), pp. 103–120, Springer-Verlag, Berlin.

Zimmermann, R., and Meyer-Reil, L. A., 1974, A new method for fluorescence staining of bacterial populations on membrane filters, *Kiel. Meeresforsch.* 30:24–27.

Zimmermann, R., Iturriaga, R., and Becker-Birck, J., 1978, Simultaneous determination of the total number of aquatic bacteria and the number thereof involved in respiration, *Appl. Environ. Microbiol.* 36:926–935.

ZoBell, C. E., 1943, The effect of solid surfaces on bacterial activity, *J. Bacteriol.* 46:38–59.

ZoBell, C. E., 1946, *Marine Microbiology*, Chronica Botanica, Waltham, Massachusetts.

3

Factors Affecting Survival of Rhizobium in Soil

HENRY S. LOWENDORF

1. Introduction

Legumes are already widely used as an extremely important source of protein for human and animal consumption and of nitrogen to maintain soil fertility. Cowpeas, lentils, gram (usually chick-peas or mung beans), and pigeon peas are major crops used for edible seeds (pulses); beans, peas, peanuts (groundnuts), and soybeans are important vegetable or industrial crops. Clovers, alfalfa (lucerne), and lupines stand out as common pasture plants, whereas *Centrosema*, *Desmodium*, and *Stylosanthes* are becoming increasingly important in the tropics. Enormous land areas are devoted to production of pulses and oils. In protein-poor localities, legumes are especially important because they are excellent sources of protein, and the crude protein content often ranges as high as 34% (as in winged bean; see National Academy of Sciences, 1975) compared to 5% in grasses.

Because of their symbiotic relationship with *Rhizobium*, the root-nodule bacterium which fixes atmospheric nitrogen, legumes are often able to fulfill their own nitrogen requirements and therefore require the addition of little or no nitrogen fertilizers. Where these fertilizers are too costly for the farmer, especially the small farmer, legumes are of substantial benefit both in producing food and also in adding usable nitrogen to the soil, thus benefiting the next crop.

Despite the actual and potential advantages of legumes to tropical agriculture, there are many serious problems in growing them. Often excellent nitrogen fixation and crop yields are found in areas adjacent to farms getting poor yields. Large acreages produce small yields, and often farmers apply nitrogen fertilizers to legumes rather than depend on symbiotic nitrogen fixation.

HENRY S. LOWENDORF • Department of Agronomy, Cornell University, Ithaca, New York 14853.

Many of these problems can be resolved by employing proper farm management or by using improved legume varieties. Other problems are specifically related to the bacterial symbiont. Among legume-production problems related to *Rhizobium* are the following: (1) there are few or no rhizobia in the soil, those in the soil do not nodulate the crop that is planted, or the indigenous bacterial strains are not effective nitrogen fixers with the particular legume under cultivation; (2) the commercial *Rhizobium* inoculum provided does not contain the proper symbiont or, if proper, then there are insufficient live bacteria to bring about satisfactory rates of infection; (3) any of a number of difficulties with the handling of the inoculum by the farmer; (4) the inoculated rhizobia do not survive on the seeds or in the soil long enough to infect the first crop; or (5) the rhizobia infect the first crop but do not survive to infect subsequent plantings.

Some of these problems can be solved by establishing regional programs to train indigenous technicians. Once trained, these individuals could determine which problems are of significance in the region. They would know how to handle and test commercial inocula, and they could verify that the commercial inocula are prepared properly (Schall *et al.*, 1970).

A more difficult task will be to deal with those field problems arising because the required rhizobia do not persist in the soil. In such instances, the many soil types, climates, and legume (and therefore rhizobial) species are each variables and potential causes of poor survival. Poor survival may be a problem even where the proper inoculum is available since the farmer may not continue to use inoculum because of variable results, presumed lack of time, or even cost considerations. Thus, assuring survival of rhizobia in the soil, along with other required characteristics of the symbiosis, is critical for producing high yields of legumes, guaranteeing a satisfactory supply of edible protein, and raising the fertility of many problem soils.

Although books and reviews on the subject of legumes, the legume *Rhizobium* symbiosis, and nitrogen fixation are numerous, there is no comprehensive review of work specifically relating to rhizobial survival in soil. Nutman (1975) and Alexander (1977a) have dealt briefly with selected topics regarding rhizobial ecology, and there are related reviews touching on or covering certain aspects of the subject (Vincent, 1965, 1974; Chowdhury, 1977; Parker *et al.*, 1977). The objectives of this report are to describe the research done on rhizobial survival, cite the major results obtained, determine if generalizations can be made, and make recommendations ultimately for appropriate farm practice.

2. Terminology and Methods Used for Enumerating and Distinguishing Strains of *Rhizobium* in the Soil

2.1. Terminology

In the initial studies of *Rhizobium*, the only certain identification of members of that genus was to assess their capacity to cause nodulation of legume roots. Species identification was then made according to which genus of legume the bacteria infected. The two symbionts (the plant and bacterium) were considered a *homologous pair*. Scientists constructed *cross-inoculation groups* for intergeneric homology, but the validity of these groups was severely questioned because both bacteria and plants were sometimes found to be promiscuous in their relationships (Wilson, 1944).

Effectiveness refers to a successful symbiosis in which the host plant provides nutrients that the microsymbiont requires, the bacteria in return provide fixed nitrogen i.e., assimilate or "fix" N_2, converting it to a form the plant uses. Effectiveness is measured by assessing the weight of inoculated plants, total nitrogen uptake by the plants (compared to uninoculated legumes or to plants grown on nitrate or ammonium), or by determinations of nitrogen-fixation activity using ^{15}N-N_2 or the acetylene-reduction technique. Effectiveness is a relative term because plant vigor is quite variable, because there is a continuous spectrum of symbiotic efficiencies, and because a given variety of legume (or strain of *Rhizobium*) forms an effective symbiosis with some but not necessarily all of its homologous rhizobia (or legume hosts) (Vincent, 1954; Bjälfve, 1963).

2.2. Methods

Detailed methods for studying the ecology of *Rhizobium* are provided in Vincent's (1970) manual. Rhizobia are generally grown on yeast extract-mannitol medium (YEM), and their numbers, either in culture or from sterile soil, may be counted on YEM agar plates. Dilution–nodulation is a method devised by Wilson (1926) to assess the number of rhizobia in nonsterile soil. Serial dilutions of the soil are made and inoculated into sterile soil. After 10–14 days to allow for colonization of the soil, surface sterilized host–legume seeds are planted. At the appropriate time, the legume roots are observed for nodulation. Presence of nodulation is presumed to indicate at least one viable *Rhizobium* in the original sample, and from the greatest dilution leading to nodulation, the approximate number of rhizobia in the soil can be obtained. This method has been simplified by growing the test plants on agar or in sand or gravel culture, and most-probable-number (MPN) tables can be used to improve the precision of numerical estimates (Tuzimura and Watanabe, 1961; Date and Vincent, 1962; Brockwell, 1963; Weaver and Frederick, 1974). These same analyses have shown that one or two rhizobia present in the tube receiving a soil dilution are sufficient to initiate nodulation. Thompson and Vincent (1967) have, however, pointed out that this method may fail when large populations of other organisms interfere with nodulation. The MPN technique also suffers from being time consuming and laborious.

A second method, which is much quicker and easier to carry out than the above, has been devised to count rhizobia that have been introduced into nonsterile soils. It is based upon the selection of antibiotic-resistant mutants (Obaton, 1971; Danso *et al.*, 1973; Schwinghamer and Dudman, 1973), adding them to the natural soil, and then following changes in population density by plating serial dilutions of the soil onto YEM agar containing the antibiotic and counting the colonies that appear. This method cannot be used to do field surveys unless the resistant rhizobia were earlier introduced. It can potentially be used to count each in a mixture of strains of bacteria if various markers (e.g., mutations to resistance to different inhibitors) are used to distinguish between strains. Because fast-growing rhizobia (in general, *R. trifolii, R. meliloti, R. leguminosarum*, and *R. phaseoli*) develop to visible colonies on agar plates within about 2 days, the growth of fungi usually presents no problem in enumerating the bacteria, provided the rhizobia are present in large numbers. However, with slow-growing species (*R. lupini, R. japonicum*, and cowpea-type rhizobia), or with soils containing few rhizobia, fungi may overgrow the plates before bacterial colonies appear; under these conditions, an antifungal compound such as cycloheximide is used. It may prove important to include a more potent fungicide (or several antifungal compounds) because cycloheximide is not always completely effective in suppressing the growth of fungi. Also it is important to test antibiotic-resistant mutants to assure that infectivity and effectiveness are not impaired (Schwinghamer and Dudman, 1973).

Serological methods may be used for identifying strains of *Rhizobium*. One of these methods, the fluorescent-antibody technique (Schmidt *et al.*, 1968; Jones and Russell, 1972), can serve for counting bacteria and visualizing them in soil. The problem of nonspecific adsorption and autofluorescence associated with this method in soil (Nutman, 1975) seems to have been overcome with proper counterstaining (Bohlool and Schmidt, 1968). However, the instrumentation and supporting facilities are often not available in the poorer countries (Alexander, 1977a).

A second serological method is immunogel diffusion (Vincent, 1970; Dudman, 1971). Both of the serological techniques are used primarily to identify isolates, particularly from root nodules. Because of antigen-antibody specificity, serological techniques are well suited to distinguish between many different strains. Thus, the serological techniques and the antibiotic-resistant techniques complement each other. Brockwell *et al.* (1977) have compared the antibiotic resistance and gel-diffusion procedures in field studies over 3 years and demonstrated that both were equally and highly reliable for identifying strains recovered from soil.

3. Factors Affecting Rhizobial Survival

3.1. Observed Persistence in the Field

Although legumes have been recognized for thousands of years as crops which improve the soil, only since the late 19th century has *Rhizobium* been known as the agent of the improvement and of nitrogen fixation (Fred *et al.*,

1932). It is obvious that these bacteria have survived in the soil for untold ages, yet direct evidence of field persistence has come, of course, only in about the last half century. For example, in several studies, *R. leguminosarum*, *R. lupini*, *R. japonicum*, *R. meliloti*, and *R. trifolii* have been found to survive at least 10–125 years after the cultivation of their respective homologous host (Norman, 1942; Jensen, 1969; de Escuder, 1972; Nutman, 1975). In particular, Nutman (1969) tested samples of soil taken in 1968 that had grown wheat continuously since 1843. Following serial dilutions of the soil, he checked nodulation on representative host plants. He found *R. leguminosarum* to be the most abundant, averaging 28,000/g dry soil, and all were effective fixers. *R. trifolii* was the next most abundant, averaging 320/g, but 10% were poor fixers. *R. lupini* occurred sparsely and sporadically, averaging 310/g in 13 of 20 plots in which it was found, and all but one strain were effective. *R. meliloti* was the most sparse, being absent in nearly half the plots, and the average was 13/g in the rest; more than half of the strains were fully effective.

Nutman and Ross (1970) found roughly the same survival in fields under continuous root crops since 1843. They also determined the numbers of rhizobia in fields maintained fallow for 8 years; here, the numbers of *R. trifolii* fluctuated between 1000 and 100,000/g, *R. leguminosarum* fluctuated between 10 and 1000, and *R. meliloti* fluctuated considerably but seemed to be declining to zero. These results were obtained at two separate sites in Great Britain.

Norman (1942) observed soybean yield and nodulation in fields that had not been planted to soybeans for over 24 years. His results with a less fertile and partly eroded sandy loam and a more fertile loam showed no significant difference between numbers of nodules on the plants or yields, whether or not the seeds had been inoculated, but he did not look at numbers of *Rhizobium*. Thus, *R. japonicum* had persisted in these soils for over 24 years in numbers high enough to be agronomically significant. Weaver *et al.* (1972) determined MPN counts of *R. japonicum* in 52 Iowa fields and reported that numbers were largely correlated with whether or not soybeans had been grown at the site (within the previous 13 years); 55% of the fields which had at any time been cropped to soybeans had over 100,000 *R. japonicum*/g soil, whereas only 5% of the fields which had not been cropped to soybeans had that many root-nodule bacteria. However, no significant correlation was found between rhizobial numbers and the length of time since the last crop had been grown (even if present at the time of sampling) or between rhizobial population and the number of times soybeans were grown. Elkins *et al.* (1976) reported comparable results. Tuzimura and Watanabe (1959a), on the other hand, found that *R. japonicum* numbers decreased in the absence of the host.

It thus seems that *R. japonicum* can survive quite well in the absence of its host legume. The contrary has been reported with certain strains of *R. trifolii* (see below), which in many regions of Australia were incapable of persisting in high enough numbers to reinfect a clover crop seeded in the second year. Thus, it is possible that some rhizobia persist in cropped land for periods long enough to get food infection of a crop without reinoculation, but it is not yet possible to predict survival for specific fields and specific strains.

3.2. Stimulation of *Rhizobium* in the Rhizosphere

It has long been known that a given species of *Rhizobium* may be found in significantly higher numbers under plants than in unplanted soil (Wilson and Lyon, 1926; Wilson, 1934a). The question of the influence of the growth of the legume host, nonhost legume, and nonlegume on the numbers and persistence of *Rhizobium* has been studied in an extensive series of experiments by Tuzimura and Watanabe (1962a,b) and Tuzimura *et al.* (1966).

In one set of experiments, Tuzimura *et al.* (1966) placed nonsterile volcanic ash soil in porcelain pots, planted either ladino clover, alfalfa, soybeans, tomatoes, or Sudan grass, and inoculated each with soil containing both alfalfa and clover rhizobia. They then tested samples of inner-rhizosphere (root zone) soil and soil without plants using ladino clover or alfalfa to provide MPN values. Their results revealed that clover rhizobia were stimulated by clover, alfalfa, and tomato plants, whereas alfalfa rhizobia were stimulated only by alfalfa plants. They also counted numbers of rhizobia in the soils 2 months after the plant tops had been removed and noted that the abundance of clover bacteria was higher in pots which had grown clover, alfalfa, or soybeans than in soil without plants. The numbers of alfalfa rhizobia were higher only in soil which had grown alfalfa. Thus, the authors concluded that certain hosts, nonhost legumes, and nonlegumes can be stimulatory to the growth of some but not all species of *Rhizobium* and that the soil under inquiry was more favorable to the clover than the alfalfa rhizobia.

In a second set of experiments, Tuzimura *et al.* (1966) observed the effects of the presence or absence of plants on rhizobial numbers in two different soils, the volcanic ash soil used in the previously described experiment (pH 5.5), which initially had counts of less than 10 clover rhizobia per gram and no detectable alfalfa rhizobia, and an alluvial clay loam from a well-drained paddy field (pH 6.2), which initially had 180 clover rhizobia per gram and also no detectable alfalfa rhizobia. The soils were inoculated with strains of *R. trifolii* and *R. meliloti* and left fallow or planted separately with ladino clover, alfalfa, wheat, red radish (*Rhapus sativus*), or common vetch (*Vicia sativa*). In both fallow soils, counts of clover rhizobia were nearly constant for the 3.5-month experiment, fluctuating between 2 and 80/g; although counts of alfalfa rhizobia varied in the fallow clay loam between 7 and 300/g, they seemed to increase in the volcanic soil from 340 to 7000/g. The authors found that alfalfa and clover stimulated both *Rhizobium* species in the two soils, but the populations were stimulated more dramatically under the homologous host, often to greater than 10^5 or 10^6/g. Vetch stimulated both species in the two soils. Wheat, however, promoted the development of alfalfa rhizobia in both soils (4 to 7×10^4/g, 62 days after seeding) but stimulated clover rhizobia only in the clay loam (1.4×10^4

compared to 8/g in the volcanic ash). Thus, the stimulation by wheat depended both on *Rhizobium* species and soil type. Radish favored the two species of *Rhizobium* but only in the clay loam. It is interesting that in the first experiment, clover did not enhance the proliferation of alfalfa rhizobia in the ash soil, but in the second experiment it did; however, the background counts in unplanted soil were both high and variable, and the *Rhizobium* strains may well have been different.

Robinson (1967) reported field results similar to those of Tuzimura *et al.* (1966). Thus *R. trifolii* was found in high numbers (up to 100,000/g) in soil near the host plant (*Trifolium subterraneum*), nonhost legume (*Medicago sativa*), and nonlegume (*Poa australis*). *R. meliloti*, on the other hand, was found in high numbers only near roots of its host (*M. sativa*) and the nonlegume (*P. australis*) and was not detected under *T. subterraneum*. Jensen (1969) found *R. trifolii* densities of 100,000/g of soil in many clover-free fields, but similar densities of *R. meliloti* were confined to existing alfalfa fields. According to the results of Hely and Brockwell (1962), the apparent numbers of *R. meliloti* at different sites in Australia depended on the species of *Medicago* used as test host. This indicated that the population density varied with particular strain and soil. The bacteria were often found in appreciable numbers in the absence of the host, and in some cases the presence of the host plant did not lead to an increased population. Grasses also usually stimulated the proliferation of *R. meliloti*.

It would seem from the above discussion that *R. trifolii* is more generally stimulated by nonhost plants than *R. meliloti*, although strain and soil differences seem to play a major role in their responses to plants. The reasons for these differences are unknown.

Wilson (1934b) found no relationship between the numbers of *R. trifolii*, *R. leguminosarum*, or *R. japonicum* and the crop on the soil. Elkan (1962) reported that there are differences in rhizosphere stimulation between a nodulating and nonnodulating line of soybean, and he noted that detecting a stimulation or inhibition of *R. japonicum* depended on when after germination the samples were collected.

Peters and Alexander (1966) found no specificity regarding the stimulation of various species of *Rhizobium* when testing legumes growing in solution culture. Vincent (1967) made similar observations. Hence, while rhizobia are favored by the presence of plant roots, the degree of stimulation depends on plant variety, soil type, *Rhizobium* species, and even strain. Furthermore, there is still controversy concerning the extent of specificity of the stimulation.

The effect of plant species on the density of rhizobia in the soil is obviously important to the farmer. It also has some important consequences for the investigator studying survival in the field. For example, de Escuder (1972) noted that in fields that had grown barley for 9 years and also in pastures of perennial rye grass, the numbers of *R. trifolii* were 10^6/g or more. However, she also noted that white clover was a widespread weed, approaching 15% of the flora in the pasture. Thus, a conclusion from her results that *R. trifolii* persisted in high numbers under rye grass and barley must be considerably tempered because many host plants were also present. It is worth pointing out that Nutman (1969) and Nutman and Ross (1970), in the studies previously mentioned, looked for the presence of leguminous weeds in the fields that had grown wheat or root crops for 120 years. They reported that there was no record of leguminous weeds in the fields with root crops, and the distribution of rhizobia in the wheat fields did not correspond with that of the leguminous weeds. Thus, their results are not complicated by the presence of the host legume. However, in a survey of a meadowland cut for hay for over 100 years, Nutman and Ross (1970) found a "striking correspondence" between the occurrence of *Rhizobium* species in the soil of each plot and the presence of its host legume.

3.3. Soil Factors

In the preceding section, consideration was given to ways in which various plant species influence numbers and persistence of rhizobia. In this section are discussed the effects on *Rhizobium* of general and specific soil factors.

3.3.1. Soil Management

Walker and Brown (1935) investigated the influence of soil-management practices on rhizobia. Iowa fields (original soil pH 5.7) which were cropped in 3-year rotations (corn, oats, and red clover and alfalfa) were compared to fields cropped to corn and oats in 2-year rotations. The fields either were not fertilized or were treated with various combinations of manure, limestone, rock phosphate, and crop residues. Rhizobial density was determined using Wilson's dilution–nodulation technique. The results showed that the numbers of *R. trifolii* and *R. meliloti* were approximately 10-fold greater in the 3-year rotation (samples were taken during growth of a nonlegume crop) than in the 2-year rotation. The authors also noted that soil amended with manure, lime, and rock phosphate (pH 6.9) contained 100- to 1000-fold more rhizobia than fields without these amendments regardless of the kind of rotation. They therefore concluded that, although the kind of rotation affected the root-nodule bacteria, soil management was more important in promoting high rhizobial numbers than the frequency of the host crop.

Numerous other reports, on the other hand, have indicated no significant differences in rhizobial populations as a result of various types of soil-management practices (Wilson, 1931, 1934b; Weaver *et al.*, 1972). However, the effect of soil management on rhizobial survival has not been systematically studied.

3.3.2. Physical and Chemical Stresses

3.3.2a. Acidity. Because many agricultural soils are naturally somewhat acidic and because large tracts of tropical soils are often highly leached and acidic, the effect of low pH on the survival of *Rhizobium* is of considerable importance.

Bryan (1923) has done the most comprehensive experiments in this regard. He treated five acid soils (one sand, one fine sandy loam, and three silt loams) with different amounts of lime, placed portions in test tubes, and inoculated these soils with suspensions of alfalfa, clover, or soybean rhizobia. The tubes were incubated for 75 days at room temperature, at which time suspensions of the soil were poured over sterile sand planted to alfalfa, red clover, or soybeans. After 5 to 7 weeks, the presence of nodules was assessed. The study disclosed that no nodules appeared on alfalfa inoculated with soils below pH 4.6–5.0, on red clover from soils below pH 4.5–4.9, and on soybeans from soils below pH 3.5–4.2. Above these pH values, nodulation occurred. Bryan (1923) tested alfalfa rhizobia further on fine sand, peat, and silt loam and recorded the same results. He then concluded that the above pH ranges were critical for survival of the respective rhizobia (Table I). Since Bryan (1923) made no bacterial counts, the initial rhizobial densities, the rate of decline or increase in population, and whether nodulation reflected the presence of one or many rhizobia are unknown.

The critical pH values are similar to those found for pure cultures by Fred and Davenport (1918), however. These investigators inoculated a defined, mannitol–salts medium with 35,000 cells/ml, and they observed that different strains did not grow below a critical pH value. For alfalfa and sweet clover rhizobia, the critical pH was 4.9; for garden pea, field pea, and vetch, pH 4.7; for red clover and common bean, pH 4.1; for soybean and velvet bean, pH 3.3; for lupines, pH 3.2. Thus Fred and Davenport's (1918) results for alfalfa rhizobia in pure culture correspond to those of Bryan (1923) in soil, and their results for red clover and soybean give critical pH values slightly below the respective values found by Bryan (1923). On the other hand, Graham and Parker (1964) showed that sensitivity of a given species of *Rhizobium* to pH in culture varied up to 1 pH unit according to individual strain.

Additional values for the lower pH limit to rhizobial growth or survival are given in Table I. All but Bryan (1923) and Vincent and Waters (1954) derived the limit for growth by using culture media. The latter authors inoculated sterile soils and observed growth of *R. trifolii* between pH 5.5 and 8.0 and loss of viability below pH 5.5.

Jensen (1969) carried out laboratory studies on survival in soils of varying (and somewhat variable) pH for up to 159 days. At pH values of about 6.1 and above, the numbers of both *R. trifolii* and *R. meliloti* appeared stable (3000–100,000/g) for this period. At about pH 5.5 and below, the abundance of both species decreased with time, but *R. meliloti* appeared to decline more quickly than *R. trifolii*; the lower the pH, the faster the decline. Had Jensen used Bryan's nodulation-or-not technique and stopped the experiment at 76 days, his results would have been similar to Bryan's (1923): the "critical" pH for *R. meliloti* would be 4.5–4.9 and, for *R. trifolii*, just below pH 4.5. Jensen (1969) actually

Table I. pH Below Which *Rhizobium* Does Not Grow or Survive

				Lower pH for growth or survival			
Bacterium	Bryan (1923)	Fred and Daven- port (1918)	Holding and Lowe (1971)	Jensen (1942)	Mendez- Castro and Alexander (1976)	Vincent and Waters (1954)	Graham and Parker (1964)
R. meliloti	4.6–5.0	4.9		5.3–5.5			4.5–5.0
R. trifolii	4.5–4.9	4.2	4.5	5.1–5.3	4.6	5.5	4.5–5.0
R. legumin- osarum		4.7					4.0–5.0
R. phaseoli		4.2					4.0–5.0
R. lupini		3.15					4.0–5.0
R. japonicum	3.5–4.2	3.3			4.2–4.6		3.5–4.5
Rhizobium sp. (velvet bean)[a]		3.3					
Rhizobium spp.							4.0–5.0

[a]*Rhizobium* from the cowpea-type cross-inoculation group is designated for clarity as *Rhizobium* sp. In parentheses is given the plant species from which the strain was isolated.

concluded that the "critical" pH for *R. meliloti* was 1 pH unit above that for *R. trifolii*.

Jensen (1969) also took samples from 214 field sites in Denmark. He reported that *R. trifolii* was almost ubiquitous in cultivated soils. In 27% of the samples, the density of *R. trifolii* was 100,000/g or more, and only a minor fraction of these samples came from clover fields. Similarly high densities of *R. meliloti*, however, were confined to existing alfalfa fields of pH above 6.0. Below pH 6.0, the occurrence of *R. meliloti* was sporadic, and its population did not seem to exceed 1000/g.

The above differences in *R. meliloti* and *R. trifolii* population densities in the field have been underscored by Rice *et al.* (1977). They found that bacterial numbers under host plants were between 100,000 and 1,000,000/g of soil at approximately neutral pH and fell off sharply below pH 6.0 for *R. meliloti* and 4.9 for *R. trifolii*.

Peterson and Goodding (1941) sampled 316 Nebraska soils to establish the effect of pH on *R. meliloti*. Samples of the soil were added to sterile sand planted to alfalfa, which was then observed for nodulation. In these studies, 40% of all samples contained rhizobia. When the percentage of samples containing rhizobia was plotted against the pH of the soil, a straight line resulted, running from pH 5.6 (10% contained rhizobia) to pH 8.6 (100% contained rhizobia). Thus, the frequency of *R. meliloti* occurring in these soils was directly related to the pH. It is worth noting that all the pH values in this study are above the critical pH of Bryan (1923) for *R. meliloti* (>pH 5.6 vs. pH 4.6–5.0). Peterson and

Goodding (1941) did not determine the numbers of rhizobia in the soil nor is it possible to establish how long the rhizobia inhabited the soil. From their results, one can conclude that the higher the soil pH, the greater is the probability of finding surviving alfalfa root-nodule bacteria.

Wilson (1934a) similarly sampled soils of different pH values, but he counted *R. japonicum* by the dilution-nodulation procedure. He found no correlation between pH and the number of soils with *R. japonicum*. As was the case with the soils used by Peterson and Goodding (1941), all of Wilson's samples were almost 2 pH units above Bryan's critical pH value (>5.0 vs. 3.5-4.2). Yet the two sets of results are quite dissimilar, possibly reflecting differences between *R. japonicum* and *R. meliloti*.

Liming of acid soils often enhances rhizobial survival. For example, Vincent (1958) failed to find *R. meliloti* in soil of pH 5.3-5.7, whereas counts of 10^4 - 10^5/g were found if the soil pH had been raised to 6.8. He also found that *R. trifolii* failed to spread beyond rows of lime- and superphosphate-pelleted clover seeds, presumably because the environment between the rows was too acid. From such data, it it not possible to determine if raising the pH leads to growth or just favors persistence. In other cases, adding lime or superphosphate increased nodulation (Robson and Loneragan, 1970), but again no information was obtained about survival of *Rhizobium*.

On the other hand, Mulder *et al.* (1966) found much higher numbers of *R. trifolii* under red clover in a soil at pH 5.1 than in a soil which had been limed (7800/g vs. 0) but was unplanted. Thus, the rhizosphere may protect rhizobia from the harmful effects of acidity.

Norris (1965) advanced the hypothesis that slow-growing or "tropical" rhizobia are more resistant to acidity in soils than fast-growing or temperate rhizobia. However, no supporting evidence exists apart from the results of Bryan (1923). For example, the work of Loustalot and Telford (1948) on tropical kudzu was cited by Norris (1959) as indicating that tropical rhizobia survive at low pH values; however, although kudzu grew better and contained a higher percentage of nitrogen at pH 4 than at higher pH values, Loustalot and Telford (1948) provided no evidence that rhizobia survived outside the nodules at that pH or even that nodulation can take place at pH 4 since the plants were transplanted to sand at the appropriate pH when they were 3 months old.

The results described above lead to the conclusion that strains of *R. meliloti* are less tolerant of acid soils than those of other rhizobial species, few of which have been studied in any detail. The definition of "critical" pH is not clear. One can imagine that the rhizobia either grow or at least persist in reasonably constant numbers above some acid pH value. Below this value, the bacteria probably decrease in numbers, very likely with greater rapidity as the pH decreases. It is important to establish this critical pH and to know the extent of survival in soils of lower pH values. If the soil pH is below the critical level but is high enough to allow sufficient rhizobia to persist into the next growing season, the bacteria may still bring about adequate nodulation the following years.

3.3.2b. Mn and Al. Frequent reference is made to the possibility of Mn and Al toxicity to *Rhizobium*, especially in acidic soils in which Mn and Al are usually more available (Holding and King, 1963; Vincent 1965, 1974; Franco and Döbereiner, 1971; Graham and Hubbell, 1975; Alexander, 1977), and several studies have indeed dealt with the inhibition of nodulation by Mn (Van Schreven, 1958; Franco and Döbereiner, 1971). The effect of subtoxic levels of Mn on the growth of *Rhizobium* in culture media has been tested, but the results are contradictory (Holding and King, 1963; Sherwood, 1966; Holding and Lowe, 1971).

The only study of Mn effects on rhizobial numbers in soil, albeit in the presence of a host plant, is that of Lowe and Holding (1970). They determined the numbers of several strains of *Rhizobium* in soil after harvesting 17-week-old white clover plants. Raising the concentration of Mn from 5.7 to 20 ppm resulted in a 12- to 29-fold decrease in numbers, the extent of the decline varying with the *Rhizobium* strain. It is not clear whether these decreases in density result directly from an influence of Mn on rhizobia in the soil or indirectly from changes in the plant rhizosphere or the release of bacteria from the nodules.

Thus, the direct effects of Al and Mn on survival of *Rhizobium* in soils have yet to be investigated.

3.3.2c. Salinity and Alkalinity. Saline and alkaline soils are found in arid regions which are poorly drained or in places where salts wash down through the profile because of heavy rains and are returned with water to the surface during the dry period of the year (Pillai and Sen, 1966). Such saline and alkaline soils often do not support good growth of legumes.

Bhardwaj (1972) studied dhaincha (*Sesbania cannabinea*), a salt-tolerant legume grown in saline-alkaline soils in India, and rhizobia isolated from dhaincha plants growing in normal soil or in saline-alkaline soil. The soil was sterilized, left unamended or amended with 1% gypsum or 1% manure, and inoculated with the bacterial isolates. After 7 and 14 days, the abundance of *Rhizobium* was determined. Both strains derived from saline-alkaline soil but only one of the two strains from normal soil survived in the unamended test soil. All four strains survived in the amended test soils, although the strains isolated from saline-alkaline soils persisted in approximately 100-fold higher numbers. Bhardwaj (1972) concluded that inocula for dhaincha growing in saline-alkaline soils should be derived from rhizobia isolated from like soils.

Bhardwaj (1975) subsequently showed that berseem clover, cluster bean, alfalfa, and Indian clover growing in sterile, neutral soil nodulated when inoculated with some saline–alkaline soil. Soybeans, peas, black gram, bengal gram, and cowpeas did not nodulate under these circumstances. He then tested rhizobia isolated from plants (*Trifolium alexandrinum*, *Melilotus sativum*, *M. parviflora*, *Glycine max*, *Phaseolus mungo*, and *Trifolium foenum-grecum*) growing on neutral soil to determine whether they would survive for 21 days in sterilized alkaline soils. A majority of the isolates from each species survived, although no

numbers were given. He also found no obvious differences in survival of 30
R. leguminosarum strains, 10 of which had been isolated from acid, 10 from
neutral, and 10 from saline-alkaline soils (pH 6.3, 7.5, and 9.2 respectively).
These results are not consistent with those obtained with dhaincha *Rhizobium*
(Bhardwaj, 1972).

Rhizobia may in fact survive in saline-alkaline soils better than their legume
host. Thus, Subba Rao *et al.* (1972) noted that most test strains of *R. meliloti*
grew in YEM medium containing 1% NaCl, whereas the host *M. sativa* was considerably inhibited when grown in agar with 0.5% NaCl or less. El Essawi and
Abdel-Ghaffer (1967), comparing *R. trifolii* isolated in Egypt with those from
Europe and the United States, found that several Egyptian but only one foreign
strain grew in culture at 5% NaCl. Mendez-Castro and Alexander (1976) were
able to grow *R. japonicum* and *R. trifolii* on NaCl concentrations as high as 3-4%.
On the other hand, Wilson and Norris (1970) found *R. japonicum* would not
grow in 1% NaCl.

Thus, it is possible that rhizobial survival does not limit the rhizobial-legume
symbiosis in saline-alkaline soils, although the data are far from convincing because none of the studies was done in nonsterile soil and none dealt with growth
or survival for time periods long enough to determine the rate of decline of
bacteria in these soils.

3.3.2d. Drying. In many areas of the world, a substantial part of the year
passes with little or no rainfall, and in the absence of irrigation, agricultural soils
dry out. This desiccation has been recognized for a long time as a potential stress
on rhizobial survival, especially in the absence of the host plant. An additional
and often accompanying stress, particularly in the tropics, is high temperatures.
Because drying of soils and high temperatures often occur concomitantly, both
have frequently been studied together. However, drying and high-temperature
stress will be considered separately in order to determine what role each plays.

For purposes of this discussion, high temperatures are taken to be those
above 40°C and normal temperatures to be between 20 and 30°C. This distinction is made because *Rhizobium* grows in culture at temperatures up to about
40°C and because most experiments have been done either at 20-30°C or at
40°C and above.

It is clear that some rhizobia can survive in dry soils for periods of time long
enough to be agronomically useful. For example, in the 1890s, Kansas farmers
could not get soybeans to nodulate, so the agricultural experiment station sent
to Massachusetts for some soil on which well-nodulated soybeans had grown.
It arrived "in a dry, pulverized condition, not unlike the dust in our roads during
a dry season" (Cottrel *et al.*, 1900). This soil apparently served satisfactorily to
nodulate soybeans from 1896 to at least 1898. Richmond (1926) stored some
Illinois soils for 3.5 years in an air-dry state and was rewarded with satisfactory
nodulation of soybeans and cowpeas provided the soil pH was above 5.8 and 4.8,
respectively. Sen and Sen (1956) planted surface-sterilized seeds in Indian soils

which had originally shown good soybean nodulation and then had been stored in an air-dry state for 19 years. Only two of the nine soils gave nodulation. These investigators measured hygroscopic moisture, pH, P content, and percentage of sand, silt, clay, organic carbon, and organic nitrogen. The only parameter which possibly correlated with nodulation was hygroscopic moisture, the least-dry soils apparently giving the longest survival. The numbers of rhizobia were not determined in these experiments, and therefore the rates of bacterial decline are unknown.

In other studies, however, counts have been made. Thus, Jensen (1961) reported on R. meliloti stored in air-dried soils for 30–45 years and R. trifolii and R. leguminosarum for 10–14 years, but he found no correlation between length of storage and bacterial density. Pant and Iswaran (1970, 1972) observed that survival of R. japonicum and a Rhizobium strain for peanut was shortest in the more alkaline soils of pH 8.1 and 8.5. Danso and Alexander (1974) used a streptomycin-insensitive strain of R. meliloti to measure resistance to drying in a silt loam (pH 5.6) and loamy fine sand (pH 5.0), both nonsterile, and in sterile quartz sand (pH 6.5). The sampled soils were inoculated and incubated at 28°C in desiccators containing $CaCl_2$. The abundance of R. meliloti dropped initially from about 10^8 to 10^5–10^6/g in the dry soils, and the population level was then maintained for the 8-week measurement period. The results were similar in moist soils. In dry sand, however, the die-off was quite marked, and no viable cells remained at 7 weeks. In moist sand, on the other hand, no loss of viability took place in 8 weeks. Thus, rhizobia seem to resist drying in nonsterile soils but not in sterile sand.

In the examples above, the water content of the soils was not determined, so it is not clear how the numbers of Rhizobium vary with that parameter. This question was dealt with by Foulds (1971), who used several different Australian soils known to contain R. trifolii, R. meliloti, and Rhizobium sp. nodulating Lotus. The soils were allowed to air dry at 22°C, and soil moisture and Rhizobium numbers (using the MPN method) were determined. In the first soil, the water content had fallen from 50 to 5.1%, and R. trifolii had declined from 10^4 to 10^3/g dry soil, R. meliloti from 10^3 to 10^1, the lotus Rhizobium from 10^5 to 10 by 115 days. In these tests, there appeared to be no correlation between rhizobial numbers and water content. In a second soil, the moisture was initially 32%; this dropped to about 3% at 7 days and remained there until the final sample at day 42. R. trifolii counts, however, remained at their initial value (23×10^4/g) throughout the decline in soil moisture, for at least 14 days. By day 21, the numbers had declined 25-fold to 92×10^2; the cell density was slightly higher about day 42. Parallel declines, also unrelated to water content, were shown by R. meliloti and the lotus Rhizobium.

Hedlin and Newton (1948) followed numbers of R. leguminosarum and soil water content and reported that, in a nonsterile Canadian soil, rhizobial counts rose from 30×10^6 at the start to 287×10^6 at 38 days, but fell to 0 by 148 days. The water contents at the same times were 29, 22, and 6.6%, respectively. Comparable results were found with sterile soil. Vandecaveye (1927) also observed no clear correlation between soil water content and the number of surviving *R. meliloti*, *R. leguminosarum*, and *R. trifolii* introduced into sterilized silt loam (pH 5.5-6.5) from Washington. Results not entirely consistent with the above are those of Sen and Sen (1956) described previously and those of Peña-Cabriales (1978). In the latter was revealed a complex relationship between soil water content and survival of *Rhizobium*. Peña-Cabriales (1978) inoculated streptomycin-resistant strains of several species into 10 g of a silt loam and allowed it to air dry at 30°C. Upon determining the numbers of bacteria remaining at regular intervals, he found a fast stage of die-off followed by a slower stage, both apparently exponential. The time for half the population to die ranged from 0.4-0.8 days during the fast stage and 2-10 days during the slow stage. Although no differences in half-time for the fast stage were found among any of the species tested, a population of *R. japonicum* declined at two-thirds the rate constant of *R. leguminosarum* during the slow stage. What significance the break point between stages has or how soil parameters might influence the break point was not determined.

By holding the soil water content relatively constant at various levels of dryness, Peña-Cabriales (1978) was able to show that the exponential decline was slowest at high and low soil moisture and greatest at intermediate levels. It would be of interest if these results can be generalized to many strains and soils. Vincent *et al.* (1962) also found an exponential decline in numbers of *R. trifolii* when dried on glass beads.

Thus, in natural soils, evidence for a correlation between numbers of rhizobia and soil moisture is contradictory. However, *Rhizobium* is quite persistent when certain soils are subjected to drying. The following question is thus relevant: what soil types, amendments, or bacterial features provide resistance to drying?

Some evidence has already been cited that rhizobial resistance to desiccation differs between soils containing much clay and silt and those containing mostly sand (Danso and Alexander, 1974). Giltner and Longworthy (1916) considered that hygroscopic water retained by colloidal substances in heavier textured soil was responsible for the resistance to drying. They added a suspension of rhizobia to sterilized sand or clay loam. After 10 days, the soil was air dry. The counts at 0, 17, and 27 days were 1.6×10^6, 25, and 0, respectively, in sand, and 1.6×10^6, 42,000, and 33,000, respectively, in the clay loam.

K. C. Marshall has done the most extensive research on the protective influence of clay in connection with his studies of "second year mortality" of annual clovers in Western Australia. In second-year mortality, nodulation and growth of inoculated clovers were satisfactory in the first year, but severe mortality was found in the second season, with greater than 80% of the plants being free of nodules (Marshall *et al.*, 1963). This high mortality was primarily found in grey sands, was slightly less common in yellow sandy soils, and was absent from red sandy soils containing 6-8% silt and clay. MPN counts of clover rhizobia in the problem areas were less than 10/g soil, whereas in adjacent nonproblem areas the counts were 10^4-10^5/g (Date and Vincent, 1962).

Marshall and Roberts (1963) amended some plots of these problem soils with a montmorillonite clay or fly ash from a powerhouse, the latter consisting mainly of burnt clay. They observed that the percentage nodulation in the second season was 28, 54, and 73% in unamended, montmorillonite-treated, and fly-ash-amended soils, respectively. They reasoned that "second year mortality" did not result from the pH of the soil, which was 5.8, but might result from the prolonged drought and high temperatures during summer. They found that most of the nodules on clovers in the second season were 5-8 cm below the soil surface and that MPN counts in midsummer gave no rhizobia at 2.5 cm. They concluded that montmorillonite and fly ash provided protection only at depths greater than 5 cm, where maximum temperatures are lower than at the surface. For example, in shallow sand in this region, temperatures in 1957-1961 at 3 p.m. averaged 50.5°C (maximum of 65.5°C) at 2.5 cm and 37.8°C (maximum of 47.2°C) at 10 cm. They stated that it was possible that other microorganisms, antibiotics, soil texture, or nutritional status might be altered by the clay amendments. In fact, Chatel *et al.* (1968), in a subsequent review of the problem of second-year clover mortality, concluded that it resulted from toxins produced by other microbial inhabitants of the soil.

Mechanical, chemical, and mineralogical analyses of problem (yellow sand and grey sand) and nonproblem (red sand, brown loamy sand, and brown sandy loam) soils indicated that the major differences between them were the content in the former of much less clay, potassium, calcium, and iron. In addition, the problem soils contained no illite clay, whereas the nonproblem soils contained trace to moderate amounts. All soils contained kaolinite (Marshall, 1964).

Marshall autoclaved and dried samples of the soils, inoculated them and then dried them overnight at 30°C with forced air. He noted that the strain involved in the second-year-mortality problem, *R. trifolii* TA1, survived drying slightly better in the nonproblem soils than in the problem soils. Problem sand amended with montmorillonite or fly ash also gave about 10-fold better survival. Amendment with haematite or goethite, two iron oxides, gave no protection to drying at 30°C. In addition, montmorillonite provided no distinct protection to *R. meliloti*, *R. japonicum*, or *R. lupini.*

Bushby and Marshall (1977a) reported on more detailed studies using a sandy soil from Tasmania (pH 6.7). The methods were similar to the previous study (Marshall, 1964), except that drying was carried out at 28°C instead of 30°C. For example, *R. leguminosarum* TA101 numbers dropped from an initial 10^6 to 676/g in soil without amendment, to 1410 in soil amended with wet montmorillonite, and to 14,100 in soil amended with dry montmorillonite. The reasons for the difference in protection given by wet and dry clay are not clear.

Certain clays protect a number of species of *Rhizobium* from the effects of drying at temperatures between 20 and 30°C. Other substances also give protection or alter susceptibility to drying. In the experiments of Bushby and Marshall (1977a), polyethylene glycol 6000, glucose, sucrose, and maltose (20% solutions) mixed with soil in a ratio of 1:1 (v/v) resulted in counts after drying of 6,920, 30,900, 45,700, and 41,700, respectively, compared to 676/g for unamended soil. Polyvinyl pyrrolidone 40,000 (20%) and glycerol (5%) afforded no protection, and polyethylene glycol 400 or 1500 (each 20%) were actually detrimental to survival. Vincent *et al.* (1962) also found that maltose protected rhizobia against the effects of drying on glass beads.

Other factors important in resistance to drying have also been reported. Pant and Iswaran (1972) found a correlation between soil phosphorus and ability to survive drying. It also appeared that high soil pH made a *Rhizobium* strain for peanut and *R. japonicum* more susceptible to drying (Pant and Iswaran, 1970; 1972). An earlier report indirectly implicated low pH in the susceptiblity of *R. japonicum* and cowpea *Rhizobium* to desiccation (Richmond, 1926).

Some evidence exists that sterile soils present no safer haven for survival of rhizobia to drying than nonsterile soils (Hedlin and Newton, 1948), and bacterial capsules appear to provide no protection against the consequences of drying. The strain of *R. japonicum* recovered by Sen and Sen (1956) after 19 years from dry soil had no capsule. Bushby and Marshall (1977a) state that capsulated strains of *R. trifolii* were not consistently better protected against drying than their noncapsulated variants, although the data presented in that report do not uphold their statement.

Species differences to drying have also been established. Foulds (1971) found *R. trifolii* to be much more resistant to drying in soil than *R. meliloti* or the bacteria nodulating *Lotus*. The results of Peña-Cabriales (1978) described above point to no species differences during drying, but to potential differences in drought tolerance at constant low soil moisture. Bushby and Marshall (1977a) showed that many strains of slow-growing rhizobia (*R. japonicum, R. lupini*, and the cowpea rhizobia) survived short-term drying better than the fast growers (*R. meliloti, R. leguminosarum*, and *R. trifolii*), although within each group there was much variation. To the contrary, however, Delin (1971) observed that a strain of *R. meliloti* took eight-fold longer to be reduced to 1% of its initial number on glass beads than strains of *R. japonicum* or *R. lupini.*

Certain treatments have been claimed to increase resistance to drying. Delin (1971) holds a patent on a method to produce rhizobial strains resistant to desiccation. He selected strains that required seven times as long as the parent to die off to 1% of the initial number on glass beads. Chen and Alexander (1973) reported that bacterial growth in solutions of low water activity could confer drought resistance. Bushby and Marshall (1977a), on the other hand, found no relationship between the ability to grow in solutions of low water activity and resistance to desiccation, although the methods used for drying soil were quite different.

All told, the mechanism relating soil water loss to the numbers of rhizobia surviving desiccation is just beginning to come to light. The bacteria appear more susceptible to the effects of drying in sand than in heavier textured soils, and certain clays such as montmorillonite and a number of sugars can often provide protection. Susceptibility does not decline in sterile soil, so other soil micro-organisms are not a factor. Evidence, not entirely consistent, suggests that the group of slow-growing rhizobia is more resistant to drying or survives better in dry soil than the fast-growing rhizobia. Bacterial capsules are reported to afford no protection to drying.

3.3.2e. High Temperature. Soil temperatures in some parts of the tropics and even in some temperate areas have been found to be quite high compared to the overlying air; e.g., 65.5°C at 2.5 cm below the surface (Marshall and Roberts, 1963). Survival of rhizobia in soil at high temperatures is strongly conditioned by whether or not the soil is wet or dry, and, whereas survival at moderate temperature is greater when soil is moist (Foulds, 1971; Danso and Alexander, 1974), survival at temperatures above 40°C is greatest when soils are dry. Poor nodulation of various legume seedlings in southeastern Queensland following mid- to late-summer rains was observed by Bowen and Kennedy (1959), although watering strains of the same bacteria onto rows several weeks later gave good nodulation. Since absence of moisture and nutrition were not problems in these trials, they decided to test survival at high temperatures, which they defined upon examining the maximum temperatures that allowed growth in culture. These maxima as well as those reported by Ishizawa (1953) and Mendez-Castro and Alexander (1976) are given in Table II.

Table II. Maximum Temperature for Growth of *Rhizobium* in Three Studies

Cross-inoculation group	Bowen and Kennedy (1959)			Ishizawa (1953)		Mendez-Castro and Alexander (1976)
	Number of strains	Maximum range (°C)	Average	Number of strains	Maximum range (°C)	Maximum range (°C)
Alfalfa	8	36.5–42.5	41.0	28	39–42.5	
Clover	9	31.0–38.4	33.2	5	35–37	38–39
Pea	2	32.0–32.7	32.3	8	35–37	
Cowpea group, etc.	68	30.0–42.0	35.4	49	32–42.5	
Lupine				9	32–33	
Soybean				10	32–35	42–43
Bean				8	35–37	

There are clear differences in maximum temperatures for growth, and to date alfalfa rhizobia show the highest average. Accordingly, Bowen and Kennedy (1959) chose 40°C at which to test survival, and they found that in moist sand only alfalfa rhizobia survived 24 hr, whereas all other strains tested had lost viability or were dying off in that time. They also found that, of two *Rhizobium* strains which nodulated *Centrosema pubescens*, the one with the higher maximum temperature for growth (38.5°C) apparently survived at 40°C better than one with a lower maximum temperature for growth (35.2°C); in this experiment, however, they measured nodulation and not numbers of bacteria.

In order to test adaptation to different climatic conditions, Wilkins (1967) took soils, all containing rhizobia, from warmer and cooler regions of New South Wales and subjected them to high temperatures. Although she measured nodulation and not counts, she concluded that rhizobia from warmer regions survived higher temperatures better than rhizobia from cooler regions; the conclusion is questionable, however, since initial rhizobial numbers in these different soils may not have been the same. On the other hand, her comparisons of survival at high temperatures under moist and dry conditions are noteworthy. In moist soil, alfalfa rhizobia were found to survive 50° but not 55°C for 5 hr. In dry soil, by contrast, these bacteria survived over 100°C for 5 hr. *Psoralea* rhizobia survived 60°C in moist soil for 5 hr, but some cells were still viable at 100°C in dry soil for the same period of time. Moisture, which protects root-nodule bacteria at moderate temperature, is thus a deleterious factor at elevated temperatures. The same conclusion is reached by a consideration of routine sterilization procedures used in the laboratory. Means to amend moist soils to protect rhizobia from high temperatures have not been sought, but Bowen and Kennedy (1959) lowered temperatures significantly by covering soil with a plant canopy or plant residues.

Marshall has extensively studied rhizobial survival at high temperatures in dry soil and has not only found amendments that protect rhizobia, but he also tried to explain how they work. Marshall (1964) first demonstrated that *R. trifolii* TA1 was much less susceptible to high temperatures in dry, sterile, non-problem soils than in problem ("second year mortality") soils of Western Australia, noting that after 5 hr at 70°C, rhizobial counts were 10^2-10^3 greater in the nonproblem soils. Clays such as montmorillonite and illite protected (gave 10^4-fold greater counts), but kaolinite did not. Similarly, of the iron oxides tested, haematite protected rhizobia but goethite did not. Montmorillonite also protected against four or more exposures to 50°C. In addition, Marshall showed that fast-growing *R. meliloti* SU47, like *R. trifolii* TA1, was protected by montmorillonite or fly ash, but that slow-growing strains *R. japonicum* QA372 and *R. lupini* CP1 survived in high numbers in the absence of these clays and were not protected further by their addition.

In a series of experiments, Marshall (1967, 1968a,b, 1970) and Bushby and Marshall (1977a,b) attempted to elucidate the mechanism by which montmorillonite protects *Rhizobium* against heat in dry soils. Initially, Marshall (1967, 1968b) sought to determine whether montmorillonite absorbed preferentially to the fast-growing rhizobia. This was accomplished by measuring, through a microscope, the time it took suspended bacteria to drift a given distance across an electrophoretic cell with a known voltage, with and without added montmorillonite. Furthermore, by varying the pH of the bacterial suspension and noting any changes in mobility, Marshall determined at what pH values chemical groups on the bacterial surface became associated with H^+ or dissociated, thus providing information on the surface charge. He concluded that (1) two types of surface groups characterized the many rhizobia, a simple carboxyl surface found on all the fast growers tested except one (*R. trifolii* TA1) and a complex carboxyl-amino surface found on all the slow growers and *R. trifolii* TA1; and (2) at the highest montmorillonite densities, the strains with the simple carboxyl surface held about twice as much clay per unit area as the others. He then postulated that a closely packed clay envelope conserved cell water and protected the bacteria from drying out.

A number of criticisms can be raised against these conclusions. First is the anomaly of *R. trifolii* TA1. This strain—which sometimes did not survive in soil (thereby resulting in second-year clover mortality), was sensitive to drying and high temperatures, and was protected from the effects of high temperatures by montmorillonite—had a surface charge and responded to montmorillonite, just as did the slow-growing strains which resisted drying, high temperatures, and the protection of clay. Second, these experiments were conducted at 25°C, a temperature at which montmorillonite is only slightly effective. Third, no comparable experiments have been reported using kaolinite, which does not protect *R. trifolii* from high temperatures. An additional criticism relates to the previously discussed findings that rhizobia are more resistant to heat when in dry than in moist soils. One might expect that protection from the effects of heat would be greater if the protectant helped remove moisture from the cells rather than if it preserved the cell water, as Marshall proposed.

The last criticism was dealt with by Bushby and Marshall (1977a,b). The authors measured water sorption isotherms (that is, the amount of water held by bacteria, or Ca montmorillonite, at numerous different relative water vapor pressures) at 40°C. They found that at moderate and low relative vapor pressures, fast-growing bacteria absorbed more water than slow growers, although less than Ca montmorillonite. They also calculated that the water sorbed per gram of cells, and the water-absorbing surface, was twice as great for the fast growers as for the slow growers. Thus, the temperature-sensitive fast growers retained water better.

Since montmorillonite had a greater affinity for water than the root-nodule bacteria, they proposed that the clay envelope lowered the equilibrium water held by rhizobia, rather than slowing its removal as suggested by Marshall (1968b). The basis for this proposal is subject to a number of criticisms. (1)

Because the effect of montmorillonite on water absorption by bacteria was not determined, the authors have no evidence that the clay does in fact lower the amount of water retained by the organisms. (2) The authors also have no data on the effect of montmorillonite on rates of water loss by the rhizobia. (3) Furthermore, the anomalous strain, R. trifolii TA1, was not included in their measurements of water adsorption. It would be of interest to know whether R. trifolii TA1—which is susceptible to drying and high temperatures and is protected by montmorillonite (like other fast growers) but has a surface charge and affinity for montmorillonite like the slow growers—has an adsorption isotherm like the other fast growers or like the slow growers. It should be pointed out that Bushby and Marshall considered that montmorillonite primarily protects rhizobia from desiccation, but the most convincing data relate to protection against heat in dry soil.

Bushby and Marshall (1977b) also demonstrated that the intracellular osmotic pressure of the fast growers was consistently lower than in the slow growers; therefore, the greater water-holding capacity of the fast growers that they found did not result from a greater intracellular solute concentration.

Despite unresolved questions on the mechanism of protection by some materials, it is clear that certain clays and iron oxides do protect some rhizobia from high temperatures in dry soils, that fast growers are more susceptible to heat in dry soils, and that heat in moist soil is much more detrimental to the root-nodule microorganisms than in dry soil.

3.3.2f. Low Temperature. Low temperatures are of no concern in the tropics but are important to farmers in temperate regions. Rhizobia found in fields year after year in northern climates obviously survive freezing. Few studies of this question have been carried out, but Fred (1920) has shown that these bacteria do live through the winter. Furthermore, Wilson (1930) did nodulation-dilution counts of R. trifolii and R. leguminosarum during a winter in Ithaca, N. Y. fields, and he found no obvious decline. Ek-Jander and Fahraeus (1971) compared strains of R. trifolii isolated from northern and southern Scandinavia and observed that cultures of R. trifolii isolated in the north grew better at 10°C in culture and gave earlier nodulation at that temperature. These results, along with those of Wilkins (1967), Bowen and Kennedy (1959), and others, are clear indications that resistance to temperature stress is an adaptable characteristic of Rhizobium and should be considered in choosing inoculant strains.

3.3.2g. Flooding. Flooding of soil as a stress to rhizobia has hardly been tested, and early reports seemed to indicate that it is not a problem (Vandecaveye, 1927; Tuzimura and Watanabe, 1959b). More recent studies lead to the same conclusion: Osa-Afiana (1979) characterized two strains of Rhizobium, one each of R. trifolii and R. japonicum, for susceptibility to flooding in a silt loam, pH 6.6. Both strains survived well in flooded soils at 30°C. Although both strains declined over a 6-week period when inoculated at rather high densities, 10^9 cells/g of dry soil, their numbers leveled off at about 10^5/g for R. trifolii and

$10^7/g$ for *R. japonicum*. These final densities were independent of initial inoculum size (down to at least $10^5/g$). At 4°C, flooding was no greater stress to survival than other moisture regimes tested (10, 22, 35, and 45% water held/g of dry soil). Maximum water-holding capacity for this soil was 45%. At 40°C, die-off of both strains was more rapid for moist than for air-dry soil (see Section 3.3.2e), although *R. japonicum* seemed less sensitive to these combined stresses than did *R. trifolii*.

3.3.3. Biotic Stresses

As reported with physical and chemical stresses such as pH, drought, and high temperature, rhizobia have certain limits beyond which they survive poorly or not at all. This is true both in culture and soil, although the boundaries of survival may and often do differ between the two. In the case of stresses on *Rhizobium* brought about directly or indirectly by other organisms, however, the limits are less clear, and conflicting results have usually been obtained in culture medium and in soil (Vandecaveye *et al.*, 1940; Robison, 1945; Fogle and Allen, 1948; Anderson, 1957; Holland and Parker, 1966; Chhonkar and Subba-Rao, 1966). This is partly a consequence of the fact that the two habitats are acting both on *Rhizobium* and its living antagonists. In addition, the effects of antagonistic organisms on rhizobial survival in culture have generally been compared in the soil environment to nodulation or plant growth. Thus, attempts to establish whether antibiotics, predators, or parasites affect rhizobial survival in soil are complicated by effects on the host plant and on the rhizobia–legume symbiosis. Some investigations have pinpointed effects on rhizobia, and for the purpose of this report these studies will be emphasized.

3.3.3a. Plant-Associated Toxins. The seeds of subterranean clover, and to lesser extents white clover and alfalfa, contain a material toxic to *R. trifolii* TA1 on agar plates (Thompson, 1960). This toxin is extractable from the seed coat but not the embryo. Field tests showed that soaking seeds prior to inoculation, applying the inoculum directly to soil, or separating the rhizobia from seeds by coating the latter with an inert material eliminated the influence of the toxin.

Plants important in the pioneer stages of old-field successions inhibit growth of various species of *Rhizobium* on agar and in some cases reduce nodulation and fresh weight of plants (Rice, 1964, 1968). The inhibition of rhizobia by roots and root extracts of nonnodulating legumes has been reported by Ranga Rao *et al.* (1973); thus, *Cassia fistula* roots inhibited the rhizobia of *Cajanus cajan* on agar plates, and root extracts of *C. fistula*, *C. occidentalis*, and *Leucaena leucocephala* suppressed rhizobia in broth, although pea roots stimulated pea rhizobia.

Microorganisms in the rhizosphere of a nonnodulating line of soybeans were compared with those under a normally nodulating near-isogenic line (Elkan, 1962). During the first 40 days after seedling emergence, the rhizosphere of nonnodulating soybeans supported more rhizobia, actinomycetes, and total microorganisms, but the stimulation disappeared until, when the soybeans reached maturity, the rhizosphere of these plants contained considerably fewer total microorganisms. Differences in nutrition of the bacteria were also observed between the two rhizosphere communities. In none of the above cases, however, have specific toxins yet been isolated and tested in soils, and thus the state of knowledge on stresses from plant toxins on *Rhizobium* is quite limited.

3.3.3b. Microbial Toxins. The "second year mortality" problems of subterranean clover in Western Australia resulted primarily from the numbers of *R. trifolii* falling off in the second and subsequent years. No such mortality was found for lupines and serradellas or for *R. lupini* (Chatel *et al.*, 1968). The major cause of *R. trifolii* TA1 die-off was traced to a water-soluble soil factor which strongly inhibited *R. trifolii* TA1 and *R. meliloti* SU47 but had no such effect on *R. lupini* WU7. The extracted factor was removed from soil by centrifuging the soil sample over a micropore filter. The toxin was found in the growing season in soils drying out after a light rain, and the inhibitory substances were quite labile, lasting usually 1 day. Of 59 microorganisms isolated from affected soils, 9 (including an actinomycete and many fungi) inhibited both *R. trifolii* and *R. lupini* on agar plates and 19 inhibited *R. trifolii* alone. No isolate inhibited *R. lupini* without also inhibiting *R. trifolii*. Based on these data, it was postulated that the "second year clover mortality" problem resulted from a microbial toxin inhibitory to *R. trifolii* TA1. The problem was solved by isolating strains of *R. trifolii* which survived in these fields and using them for the preparation of inoculants. The toxins or extracts are not known to have been tested against the presumably resistant strains or to determine their effect on the survival of bacteria in soil.

Beggs (1961, 1964) described another problem of clover establishment in certain soils in New Zealand. The problem was related to oversowing of seeds on "sun-bleached" silt loam soils, especially in the presence of danthonia plants, although clover could be established by cultivating and liming the soil or sterilizing it with formalin. Parle (1964) added nitrogen fertilizer and got good clover growth until the available nitrogen level fell. He found that the problem was related to danthonia plants present in the area; with live danthonia present, clover failed to nodulate, and no clover rhizobia could be found in the soil. When the live danthonia plants were removed, Parle (1964) observed that 23% of the test clover plants nodulated and 1670 rhizobia were present per gram of soil; with only dead plants present, 45% of the test clovers were nodulated and there were 14,300 rhizobia per gram of soil. Thus, there was apparently a potent toxin against *Rhizobium* in the danthonia rhizosphere.

Parle (1964) compared a *Rhizobium* strain from a small patch of clover established in the problem fields with the commercial strain. In pot experiments, he found 2–3 times as many nodules formed by the new isolate in the presence of danthonia. Extracts of danthonia roots and adhering soil gave a zone of inhibition on agar plates 50% greater with the commercial strain than with the isolate. It is not known how this toxin acts against the rhizobia or whether it is formed by the plant roots or from microorganisms associated with the roots. Because attempts to extract toxins from ground danthonia roots were unsuccessful, whereas colonies of bacteria associated with the roots were antagonistic to rhizobia, Parle (1964) suggested that the toxin was of microbiological origin.

These two examples of problems presumably caused by microbial toxins appear to have been solved by finding strains of rhizobia which survive where the originally inoculated strain did not. A third case of difficulty in establishing clover has been described by Holland (1962) and Holland and Parker (1966). On certain newly cleared fields in Western Australia, subterranean clover failed to nodulate when inoculated with *R. trifolii* strains Na30 and TA1. However, when the soil was fallowed or cropped to cereals for 3–5 years, normal nodulation took place. The soil was a grey sand (pH 6.8) over a bright-yellow clayey sand. Aqueous extracts from newly cleared fields applied to agar plates seeded with *R. trifolii* showed the presence of a toxin, but this finding was obtained only twice with certainty; however, the toxin could also be extracted from leaves of affected clover plants, which had turned purplish red.

The authors compared nodulation, growth of inoculated clover, and rhizobial counts in nonsterilized soil and soil previously treated with a fumigant. In these pot experiments, 42 of 42 plants nodulated in the treated soil, but only 4 of 43 were nodulated in the untreated control. The numbers of rhizobia in the untreated and fumigated soil were 0 and 3000/g, respectively. Of a large number of microorganisms isolated from the problem and nonproblem fields, 27% from newly cleared fields and only 10% from established clover fields produced a toxin against *R. trifolii*. When these microorganisms were inoculated along with *R. trifolii* into sterile sand cultures of clover, nodulation and growth were normal, and no toxicity was observed. However, if sucrose or straw was added to the sand together with individual microorganisms, many fungi were seen to inhibit growth and nodulation and to cause purplish-red leaves to develop on the clover. None of the microorganisms from established clover fields inhibited the clover, only those from the newly cleared fields. Chatel and Parker (1972) pointed out that it remains to be shown that the soil toxins are produced and inhibit rhizobia in unamended soils (or that the problem fields are comparable to the amended experimental soils) and that the substances are microbial metabolites.

Hely *et al*. (1957) were unable to identify a presumed microbial antagonist to clover rhizobia at another location in Australia, but they obtained good results by sterilizing the soil or amending it with charcoal (100 lb/acre). It seems as if problems resulting from microbial toxins have been especially prominent with, or perhaps even limited to, newly cleared fields and fields newly cropped to legumes, at least in the case of clovers. The established microbial populations may be antagonistic to newly inoculated rhizobia.

Fungi. There are numerous studies of fungal toxins against rhizobia, but the information gained is generally not directly related to field problems. Thus, Thornton *et al.* (1949) isolated strains of *Penicillium* which inhibited *R. meliloti* on agar plates. Chhonkar and Subba-Rao (1966) found that some species of fungi isolated from nodules of common legumes produced toxins against several species of *Rhizobium* on agar, and they demonstrated that one fungus inhibited growth and nitrogen fixation of inoculated *Trifolium alexandrinum*.

Actinomycetes. Van Schreven (1964) tested 24 actinomycetes on agar against many strains and species of *Rhizobium* and reported that some of the bacteria were susceptible to antibiotics produced by the actinomycetes. Abdel-Ghaffer and Allen (1950) tested 300 strains of *Streptomyces* for antagonism to several species of *Rhizobium* in culture. However, they were unable to demonstrate the production of antibiotics in soil containing many streptomycetes. Landerkin and Lochhead (1948), Fogle and Allen (1948), Damirgi and Johnson (1966), and others also found actinomycete antagonism in culture and in some cases inhibition of nodulation in soil. Van Schreven (1964) has summarized the early literature.

Bacteria. Soil bacteria have been isolated which produce toxins or antibiotics to rhizobia in culture (Hattingh and Louw, 1969; Smith and Miller, 1974). Rhizobia themselves may synthesize bacteriocins and antibiotics (Roslycky, 1967; Schwinghamer and Belkengren, 1968; Schwinghamer, 1971). No convincing evidence exists that one species of *Rhizobium* is more susceptible to microbial antagonism than another (Landerkin and Lochhead, 1948; Fogle and Allen, 1948; Abdel-Ghaffer and Allen, 1950; Van Schreven, 1964; Chhonkar and Subba-Rau, 1966). Moreover, no evidence was given in any of these papers that microbial antagonists in question have any effect on rhizobial survival in soil.

3.3.4. Parasites and Predators

3.3.4a. Bacteriophages. Bacteriophages can rapidly decimate a host bacterial population in laboratory media, but no data exist to show convincingly that these parasites do appreciable harm to rhizobia in soil. Fatigue of alfalfa soils has been ascribed to a bacteriophage which attacks alfalfa rhizobia, and Demolon and Dunez (1935, 1939) isolated a bacteriophage-resistant *Rhizobium* which in some years as much as doubled the alfalfa yields on problem soils in France. However, no evidence was offered that the beneficial effect of the resistant strain was attributable to its bacteriophage resistance. An attempt by Vandecaveye *et al.* (1940) to repeat these experiments in Washington failed; thus, inoculation of alfalfa with *R. meliloti* resistant to a bacteriophage isolated from fatigued fields did not improve yields, although nitrogen fertilizer resulted in nearly a three-fold increase in yield. Bacteriophages against *R. meliloti* have been isolated in every alfalfa field examined in New York, and these viruses appear to be common in such fields but rare in land not cropped to alfalfa (Katznelson and Wilson, 1941). Bacteriophages active against clover and pea rhizobia were isolated from soil in England maintained for 100 years in clover and yet no "clover sickness" appeared, and these viruses were like bacteriophages found in pots with "sick" clover both in numbers and host range (Kleczkowska, 1957). Bacteriophages active against the *Rhizobium* nodulating *Lotus* and other rhizobia appear equally without effect in soil (Bruch and Allen, 1955; Laird, 1932). It thus seems that bacteriophages parasitic on *Rhizobium* threaten no immediate harm to growth of legume crops in the field.

3.3.4b. Bdellovibrio. Another parasite, *Bdellovibrio*, has been found to reduce rhizobial density in culture. Despite this, however, rhizobia persist in relatively large numbers in the presence of this parasite in both culture and soil (Keya and Alexander, 1975). Furthermore, this parasite has so far not been implicated in problems of growing legumes (Parker and Grove, 1970).

3.3.4c. Protozoa. Predation is one of the many ways rhizobial populations are decreased in soil, and predation by protozoa in liquid culture and in sterile and nonsterile soils has been studied extensively at Cornell University (Danso *et al.*, 1975; Danso and Alexander, 1975; Habte and Alexander 1977, 1978).

In culture and in soil, neither of two test protozoa, *Tetrahymena pyriformis* or *Colpoda* sp., was able to reduce the initially high numbers of *R. japonicum* (10^9-10^{10} cells per milliliter or per gram) to fewer than 10^5-10^7 in 1-4 weeks (Habte and Alexander, 1978). Although these numerical reductions are significant, the remaining rhizobia are probably sufficiently abundant to nodulate soybean seedlings (Weaver and Frederick, 1974). Habte and Alexander (1978) determined that the persistence of rhizobia did not result from a decline in the rate of protozoan feeding caused by rhizobial excretions or by the lack of energy for the predators. Rhizobial persistence also did not come about because of the availability of alternate prey or because of the inedibility of the surviving bacteria. Rather, survival of the rhizobia appeared to reflect a steady state resulting from reduction of the rhizobial population by predation and replacement of the cells thus consumed through reproduction by the remaining population. This was demonstrated in sterile soil by preventing rhizobial reproduction upon addition of chloramphenicol: in the absence of predators, *R. japonicum* numbers remained high, but in their presence, the bacterial density fell to 10^2/g or less. In natural soil without inhibitor, moreover, *R. japonicum* dropped slightly from 2×10^8 to 6×10^6/g in 1 month; in the presence of chloramphenicol, the bacterial abundance dropped to 6×10^2/g because the protozoa continued to feed whereas the bacteria could no longer reproduce. Hence, rhizobia appear capable of reproducing in natural soil, and as long as this is possible, the root-nodule bacteria can remain in substantial numbers in the presence of protozoan predators.

The evidence to date points to microbial and perhaps plant toxins as being major biotic factors governing rhizobial survival, at least in certain soils. Methods of overcoming these stresses include removal of plant parts (physically or by burning), adding charcoal to newly cleared fields (Hely *et al.*, 1957; Turner, 1955; Beggs, 1961, 1964; Holland and Parker, 1966), and especially the selection of toxin-resistant *Rhizobium* strains (Parle, 1964; Chatel *et al.*, 1968; Schwinghamer and Belkengren, 1968). The importance of selecting strains of *Rhizobium* which survive well in soils newly cropped to the host legume will be discussed in the next section.

3.4. *Rhizobium* Strain Differences

The soil and the legume host, singly and together, effect a great selective pressure on the *Rhizobium* symbiont. It is possible to select *Rhizobium* strains in the laboratory that are resistant to drying (Delin, 1971), antibiotics (Obaton, 1971; Danso *et al.*, 1973), and high salt concentrations, low pH values, and moderately high temperatures (Mendez-Castro and Alexander, 1976). Natural conditions, however, are much more complex than culture media, and in any field newly planted to a legume, it is not clear what characteristics must be selected for. The traditional criterion for choosing an inoculant strain, effectiveness, has in the past been quite narrowly defined and, as Brockwell *et al.* (1968) and Chatel *et al.* (1968) have shown, must be supplemented by several more criteria: ability to invade the roots in competition with ineffective, indigenous strains; ability to grow outside the rhizosphere; and capacity to withstand environmental extremes and persist in the soil in adequate numbers. It should be clear from the previous sections that strains meeting these criteria frequently are unavailable in commercial inoculants. An obvious question then is how to find these strains.

In consideration of the importance of the above-mentioned criteria, it is worth stressing that the inoculation of legumes with a presumed effective strain does not always lead to good nitrogen fixation. In part, this may result from a mismatch between host legume and *Rhizobium* strain. For example, Vincent (1954) found that rhizobia isolated from red and white clovers in a field survey were effective on red and white but ineffective on subterranean and crimson clovers. Effective isolates from subterranean and crimson clovers, on the other hand, showed variable effectiveness on red and white clovers. Brockwell *et al.* (1968) tested rhizobia isolated from subterranean clover grown in the Mediterranean region on several varieties of that clover used in Australia, and they observed that effectiveness varied considerably depending on variety of clover, rhizobial isolate, and temperature. It has also been shown that indigenous white clover in Ireland and Scotland gave higher yields when inoculated with rhizobia isolated from the wild ecotypes than when inoculated with commerical clover strains (Lowe and Holding, 1970; Sherwood and Masterson, 1974). Thus, crucial to high yields is proper matching of the two symbionts.

An effective *Rhizobium* strain may be unable to compete with indigenous, ineffective strains for nodulation. In experiments with subterranean clover, Harris (1954) inoculated seeds with a combination of two effective and two ineffective strains. One effective strain in all combinations gave good nodulation and growth in the presence or absence of the two ineffective strains. In contrast, the second strain, which was effective when inoculated alone, gave poor nodulation and growth when the two ineffective strains were added at the same time. Robinson (1969a,b), however, found that clover plants tended to select for effective *Rhizobium* strains in agar, even when the effective strains were outnumbered by a factor of 10,000. Bell and Nutman (1971) were able to suppress effective native strains of *R. meliloti* for only 1 year by massive inoculation (50,000 rhizobia/seed) of alfalfa with ineffective rhizobia. In the second year, the indigenous, effective strains had overgrown the plots. Brockwell *et al.* (1972) reported similar results. In all probability, when more than one infective strain is present, factors in addition to effectiveness play a role in determining which strains predominate in nodulation, but effectiveness very likely plays an important role.

It is also possible that an inoculant strain gives excellent nitrogen fixation for a short time, but then it declines in abundance. Such was the case with *R. trifolii* TA1 inoculated onto subterranean clover in Australia, its inability to persist in certain soils leading to "second year clover mortality" (Chatel *et al.*, 1968). Although Brockwell *et al.* (1968) recommended early nodulation as one criterion for strain selection because early nodulation is correlated with high initial yields, it may be more important to determine yields for a long period, especially for those annuals which are planted seasonally but the growth of which is interrupted by winter, summer drought, or other conditions unfavorable to the plant (Lie, 1971). Some evidence has been gathered that yields of clover (Mulder *et al.*, 1966; Sherwood and Masterson, 1974) and soybean (Scudder, 1975) improve with time, yet no clear connection between yields and rhizobial effectiveness or survival has been made.

Certain "volunteer" strains of *R. trifolii* have proven to survive better in the field and give better yields than commercial inocula (Chatel *et al.*, 1968; Gibson *et al.*, 1976; Roughley *et al.*, 1976). Gibson *et al.* (1976) inoculated clover with five serologically distinct strains known to form effective symbioses with *Trifolium subterraneum*, and they sampled nodules present on the plants for four seasons. No naturally occurring strains were evident the first year. By the third season 10% and by the fourth season 9% of the isolates were not identifiable as any of the inoculant strains. Of the newly appearing rhizobia, over three-fourths were as effective or more effective than the bacteria used for inoculation. The authors suggested three sources for these new strains: (1) they were present at the beginning of the experiment in undetectable numbers and responded to a favorable environment in the rhizosphere; (2) genetic changes in the indigenous ineffective population; or (3) contamination from external sources. In addition, the inoculated strains may have changed their effectiveness and serological patterns, although the results of Brockwell *et al*. (1977) tend to rule this out. Chatel *et al*. (1968) suggested that the ultimate source of the indigenous *R. trifolii* in Australia (legumes with fast-growing rhizobia not being native) was the first settlers from Europe, and that therefore these strains had 130 years to adapt and be selected in the soils. Regardless of their ultimate source, the volunteer strains show characteristics that fit the more complex criteria for successful inoculant strains (Brockwell *et al*., 1968; Chatel *et al*., 1968), including the ability to colonize the rhizosphere and infect legumes, participate in an effective symbiosis, and persist. The factors that allow strains to persist in these soils are at present almost entirely unknown.

Because of the considerable variation in tropical soils, weather patterns, and legume cultivars, an important concern is how to select and isolate such successful volunteer strains simply and rapidly. Can persistent strains be selected in laboratory experiments? In field experiments, should one sow uninoculated seeds and wait for a few vigorous plants to appear, or should the seeds be inoculated with an effective nonpersistent strain? Of 18 new isolates of *R. trifolii* found by Gibson *et al*. (1976) in the fourth season on uninoculated plots, all were as effective as or more effective than the commercial strains they used. Also, of 10 new isolates found in plots originally inoculated with a fairly persistent and effective strain, all were effective. Of 11 isolates recovered from plots originally inoculated with a poorly persisting strain, 5 were effective and 6 ineffective. Thus, the preceding evidence does not provide a clear strategy to use in trying to find high-performance *Rhizobium* strains. Nevertheless, such a procedure or set of procedures should be developed. Screening procedures have been made a high priority in some regions (Brockwell *et al*., 1968; Bergersen, 1970; Lie, 1971; Graham and Hubbell, 1975), and possibly it will not be necessary to test every soil and wait until the third season before effective isolates appear.

4. Conclusions and Recommendations

After over three-quarters of a century of study, scientists are still only beginning to probe *Rhizobium* ecology. Many reports make no distinction between legume nodulation and rhizobial survival or confuse effects on the host plant or on the symbiosis with effects on the microsymbiont. Numerous stresses have been uncovered which potentially limit rhizobial survival, but there is little convincing evidence that these stresses are actually responsible for the microorganisms' failure to persist. Only a very few studies have looked at causal relationships. In most cases in which poor establishment of a legume results from poor rhizobial survival, scientific inquiry ended once a persistent strain was discovered. The great potential for determining the mechanism of persistence by comparing strains which survive with closely related strains which do not has not been pursued. For regional agricultural reasons, most research on *Rhizobium* ecology has been carried out with *R. trifolii*, *R. meliloti*, and *R. japonicum*, whereas virtually nothing is known about the survival of the other species, expecially the tropical and cowpea-type rhizobia.

Yet, despite the lack of progress in answering the scientific questions, it is important to make recommendations for field practice, and thus certain tentative conclusions must be drawn from the limited information now available. The most serious natural environmental threats to rhizobial persistence appear to be: (1) pH, especially in the acid range; (2) drying, primarily in sandy soils; (3) high temperatures, particularly in moist or sandy soils; and (4) microbial toxins, especially in newly cleared land.

Obviously it is important to establish first whether a problem in obtaining high legume yields can be traced to poor rhizobial survival or to some other possibility. But to avoid expected problems, certain steps and precautions can be followed.

When in doubt about whether effective strains of *Rhizobium* are present, farmers should be encouraged to inoculate. In choosing an inoculum, priority should be given to a strain proven to be effective with the plant variety and known to persist in the soil type under consideration. If the soil pH is low (e.g., under pH 5.5-6.0 for *R. trifolii*, *R. meliloti*, *R. leguminosarum*, and *R. phaseoli* or below pH 5.0-5.5 for *R. lupini*, *R. japonicum*, and members of the cowpea group of rhizobia) and difficulty in establishing the legume is a good possibility, then the recommendation is to lime the soil or pellet the seed with lime. If soils are very alkaline, gypsum might be recommended. Inoculant strains for legumes in saline, alkaline, or acid soils should be isolated from effective nodules or plants growing on similar soils.

Dry sandy soils are a potential threat to survival of some species of *Rhizobium*, and the suggested remedies, aside from irrigation, are using a resistant strain of *Rhizobium* or, possibly and if practical, adding montmorillonite, illite, or ash to the soil. High temperatures pose a substantial threat, especially if soil is likely to be moist, or in light-textured soils that are dry. Means of alleviating high-temperature stresses on rhizobia are to adjust planting date, to maintain a plant cover, or to cover the exposed soil with plant remains. However, caution is in order if the field is newly cropped to a legume because leaving plant residues may be counterproductive. Changing the soil microflora is not as easy as removing one crop and replacing it with another, and in several instances, enough microorganisms which colonized a field produced toxins antagonistic to newly inoculated *R. trifolii*. The old plant residues possibly stimulated the growth of these antagonistic organisms. Thus, in this case, removal of plant residues may be desirable. Of primary importance is defining more precisely the reasons for poor survival of many rhizobia or of rhizobia in certain circumstances. At the same time, but particularly with this information, strains should be sought that are capable of enduring natural stresses on the root-nodule bacteria so that farmers and consumers can get greater benefit from the nitrogen-fixing activity of *Rhizobium* living in association with leguminous plants.

ACKNOWLEDGMENT. Thanks are due to M. Alexander for much valuable criticism. The author also acknowledges many helpful comments by M. Habte, L. B. Lennox, M. T. Lieberman, L.O. Osa-Afiana, J. J. Peña-Cabriales, C. Ramirez-Martinez, and J. T. Wilson. Support for the author to write this paper was provided by a grant (AID/CSD-2834) to Cornell University by the U.S. Agency for International Development.

References

Abdel-Ghaffer, A. S., and Allen, O. N., 1950, The effects of certain microorganisms on the growth and function of rhizobia, *Trans. 4th Int. Cong. Soil Sci.* 3:93–96.

Alexander, M., 1977a, Ecology of nitrogen-fixing organisms, in: *Biological Nitrogen Fixation in Farming Systems of the Tropics* (A. Ayanaba and P. J. Dart, eds.), pp. 99–114, Wiley-Interscience, New York.

Alexander, M., 1977b, *Introduction to Soil Microbiology*, John Wiley & Sons, New York.

Anderson, K. J., 1957, The effect of soil microorganisms on the plant–rhizobia association, *Phyton* 8:59–73.

Beggs, J. P., 1961, Soil sterilant aid to clover establishment indicates antinodulation factor in soils, *N. Z. J. Agric.* 103:325.

Beggs, J. P., 1964, Growth inhibitor in soil, *N. Z. J. Agric.* 108:529–535.

Bell, F., and Nutman, P. S., 1971, Experiments on nitrogen fixation by nodulated lucerne, *Plant Soil, Spec. Vol.* 231–264.

Bergersen, F. J., 1970, Some Australian studies relating to the long-term effects of the inoculation of legume seeds, *Plant Soil* 32:727–736.

Bhardwaj, K. K. R., 1972, Note on the growth of *Rhizobium* strains of dhaincha (*Sesbania cannabina* (Retz) Pers.) in a saline–alkaline soil, *Indian J. Agric. Sci.* 42:432–433.

Bhardwaj, K. K. R., 1975, Survival and symbiotic characteristics of *Rhizobium* in saline-alkali soils, *Plant Soil* 43:377–385.

Bjälfve, G., 1963, The effectiveness of nodule bacteria, *Plant Soil* 18:70–76.

Bohlool, B. B., and Schmidt, E. L., 1968, Nonspecific staining: Its control in immunofluorescence examination of soil. *Science* 162:1012–1014.

Bowen, G. D., and Kennedy, M. M., 1959, Effect of high soil temperatures on *Rhizobium* spp., *Qd. J. Agric. Sci.* 16:177–197.

Brockwell, J., 1963, Accuracy of a plant-infection technique for counting populations of *Rhizobium trifolii*, *Appl. Microbiol.* 11:377–383.

Brockwell, J., Dudman, W. F., Gibson, A. H., Hely, F. W., and Robinson, A. C., 1968, An integrated programme for the improvement of legume inoculant strains, *Trans. 9th Int. Cong. Soil Sci. Adelaide* 2:103–114.

Brockwell, J., Bryant, W. G., and Gault, R. R., 1972, Ecological studies of root-nodule bacteria introduced into field environments. 3. Persistence of *Rhizobium trifolii* in association with white clover at high elevations, *Aust. J. Exp. Agric. Anim. Husb.* 12:407–413.

Brockwell, J., Schwinghamer, E. A., and Gault, R. R., 1977, Ecological studies of root-nodule bacteria introduced into field environments. V. A critical examination of the stability of antigenic and streptomycin-resistance markers for identification of *Rhizobium trifolii*, *Soil Biol. Biochem.* 9:19–24.

Bruch, C. W., and Allen, O. N., 1955, Description of two bacteriophages active against *Lotus* rhizobia, *Soil Sci. Soc. Am. Proc.* 19:175–179.

Bryan, O. C., 1923, Effect of acid soils on nodule-forming bacteria, *Soil Sci.* 15:37–40.

Bushby, H. V. A., and Marshall, K. C., 1977a, Some factors affecting the survival of root-nodule bacteria on desiccation, *Soil Biol. Biochem.* 9:143–147.

Bushby, H. V. A., and Marshall, K. C., 1977b, Water status of rhizobia in relation to their susceptibility to desiccation and to their protection by montmorillonite, *J. Gen. Microbiol.* **99**:19-27.

Chatel, D. L., and Parker, C. A., 1972, Inhibition of rhizobia by toxic soil-water extracts, *Soil Biol. Biochem.* **4**:289-294.

Chatel, D. L., Greenwood, R. M., and Parker, C. A., 1968, Saprophytic competence as an important character in the selection of *Rhizobium* for inoculation, *Trans. 9th Int. Cong. Soil Sci. Adelaide* **2**:65-73.

Chen, M., and Alexander, M., 1973, Survival of soil bacteria during prolonged desiccation, *Soil Biol. Biochem.* **5**:213-221.

Chhonkar, P. K., and Subba-Rao, N. S., 1966, Fungi associated with legume root nodules and their effect on rhizobia, *Can. J. Microbiol.* **12**:1253-1261.

Chowdhury, M. S., 1977, Effects of soil antagonists on symbiosis, in: *Exploiting the Legume-Rhizobium Symbiosis in Tropical Agriculture* (J. M. Vincent, A. S. Whitney, and J. Bose, eds.), pp. 385-411, University of Hawaii, Honolulu.

Cottrel, H. M., Otis, D. H., and Haney, J. G., 1900, Soil inoculation for soybeans, *Bull. Kans. Agric. Exp. Sta.* **96**:97-116.

Damirgi, S. M., and Johnson, H. W., 1966, Effect of soil actinomycetes on strains of *Rhizobium japonicum, Agron. J.* **58**:223-224.

Danso, S. K. A., and Alexander, M., 1974, Survival of two strains of *Rhizobium* in soil, *Soil Sci. Soc. Am. Proc.* **38**:86-89.

Danso, S. K. A., and Alexander, M., 1975, Regulation of predation by prey density: The protozoan-*Rhizobium* relationship, *Appl. Microbiol.* **29**:515-521.

Danso, S. K. A., Habte, M., and Alexander, M., 1973, Estimating the density of individual bacterial populations introduced into natural ecosystems, *Can. J. Microbiol.* **19**:1450-1451.

Danso, S. K. A., Keya, S. O., and Alexander, M., 1975, Protozoa and the decline of *Rhizobium* populations added to soil, *Can. J. Microbiol.* **21**:884-895.

Date, R. A., and Vincent, J. M., 1962, Determination of the number of root-nodule bacteria in the presence of other organisms, *Aust. J. Exp. Agric. Anim. Husb.* **2**:5-7.

de Escuder, A. M. Q., 1972, A survey of rhizobia in farm soils at Wye College, Kent., *J. Appl. Bacteriol.* **35**:109-118.

Delin, P. S., 1971, Production of *Rhizobium* strains resistant to drying, U.S. Patent 3,616, 236, Oct. 26, 1971.

Demolon, A., and Dunez, A., 1935, Recherche sur le rôle du bactériophage dans la fatigue des luzernières, *Ann. Agron.* **5**:89-111.

Demolon, A., and Dunez, A., 1939, Sur la lyso-résistance du *B. radiciola* et son importance pratique, *Trans. 3rd Comm. Int. Soc. Soil Sci. New Brunswick* **A**:39-42.

Dudman, W. F., 1971, Antigenic analysis of *Rhizobium japonicum* by immunodiffusion, *Appl. Microbiol.* **21**:973-985.

Ek-Jander, J., and Fahraeus, G., 1971, Adaptation of *Rhizobium* to sub-arctic environment in Scandinavia, *Plant Soil, Spec. Vol.* 129-137.

El Essawi, T. M., and Abdel-Ghaffer, A. S., 1967, Cultural and symbiotic properties of rhizobia from Egyptian clover (*Trifolium alexandrinum*), *J. Appl. Bacteriol.* **30**:354-361.

Elkan, G. H., 1962, Comparison of rhizosphere microorganisms of genetically related nodulating and non-nodulating soybean lines, *Can. J. Microbiol.* **8**:79-87.

Elkins, D. M., Hamilton, G., Chan, C. K. Y., Briskovich, M. A., and Vandeventer, J. W., 1976, Effect of cropping history on soybean growth and nodulation and soil rhizobia, *Agron. J.* **68**:513-517.

Fogle, C. E., and Allen, O. N., 1948, Associative growth of actinomycetes and rhizobia, *Proc. Meet. Soc. Am. Bacteriol.* **1**:53.

Foulds, W., 1971, Effect of drought on three species of *Rhizobium*, *Plant Soil* **35**:665-667.

Franco, A. A., and Döbereiner, J., 1971, Toxidez de manganês de un solo ácido na simbiose soja-*Rhizobium*, *Pesq. Agropec. Bras. Sér. Agron.* **6**:57-66.

Fred, E. B., 1920, Are legume bacteria killed by freezing?, *Wisc. Agric. Exp. Sta. Bull.* **323**: 36.

Fred, E. B., and Davenport, A., 1918, Influence of reaction on nitrogen-assimilating bacteria, *J. Agric. Res.* **14**:317-336.

Fred, E. B., Baldwin, I. L., and McCoy, E., 1932, *Root Nodule Bacteria and Leguminous Plants*, University of Wisconsin Press, Madison.

Gibson, A. H., Date, R. A., Ireland, J. A., and Brockwell, J., 1976, A comparison of competitiveness and persistence amongst five strains of *Rhizobium trifolii*, *Soil Biol. Biochem.* **8**:395-401.

Giltner, W., and Longworthy, V. H., 1916, Some factors influencing the longevity of soil microorganisms subjected to desiccation with special reference to soil solution, *J. Agric. Res.* **5**:927-942.

Graham, P. H., and Hubbell, D. H., 1975, Soil-plant-*Rhizobium* interaction in tropical agriculture, in: *Soil Management in Tropical America* (E. Bornemisza and A. Alvarado, eds.), pp. 211-227, North Carolina State University, Raleigh.

Graham, P. H., and Parker, C. A., 1964, Diagnostic features in the characterization of the root-nodule bacteria of legumes, *Plant Soil* **20**:383-396.

Habte, M., and Alexander, M., 1977, Further evidence for the regulation of bacterial populations in soil by protozoa, *Arch. Microbiol.* **113**:181-183.

Habte, M., and Alexander, M., 1978, Mechanisms of persistence of low numbers of bacteria preyed upon by protozoa, *Soil Biol. Biochem.* **10**:1-6.

Harris, J. R., 1954, Rhizosphere relationships of subterranean clover. I. Interactions between strains of *Rhizobium trifolii*, *Aust. J. Agric. Res.* **5**:247-270.

Hattingh, M. J., and Louw, H. A., 1969, Clover rhizoplane bacteria antagonistic to *Rhizobium trifolii*, *Can. J. Microbiol.* **15**:361-364.

Hedlin, R. A., and Newton, S. D., 1948, Some factors influencing the growth and survival of *Rhizobium* in humus and soil cultures, *Can J. Res. Sect. C* **26**:174-187.

Hely, F. W., and Brockwell, J., 1962, An exploratory survey of the ecology of *Rhizobium meliloti* in inland New South Wales and Queensland, *Aust. J. Agric. Res.* **13**:864-879.

Hely, F. W., Bergersen, F. J., and Brockwell, J., 1957, Microbial antagonism in the rhizosphere as a factor in the failure of inoculation of subterranean clover, *Aust. J. Agric. Res.* **8**:24-44.

Holding, A. J., and King, J., 1963, The effectiveness of indigenous populations of *Rhizobium trifolii* in relation to soil factors, *Plant Soil* **18**:191-198.

Holding, A. J., and Lowe, J. F., 1971, Some effects of acidity and heavy metals on the *Rhizobium*-leguminous plant association, *Plant Soil, Spec. Vol.* 153-166.

Holland, A. A., 1962, The effects of indigenous saprophytic fungi upon nodulation and establishment of subterranean clover, *Antibiotics in Agriculture* (M. Woodbine, ed.), pp. 147-164, Butterworths, London.

Holland, A. A., and Parker, C. A., 1966, Studies on microbial antagonism in the establishment of clover pasture. II. The effect of saprophytic soil fungi upon *Rhizobium trifolii* and the growth of subterranean clover, *Plant Soil* **25**:329-340.

Ishizawa, S., 1953, Studies on the root nodule bacteria of leguminous plants. I. Characters in artificial media. Part 5. Effect of temperature on the growth of rhizobia, *J. Sci. Soil Manure (Tokyo)* **24**:227-230.

Jensen, H. L., 1942, Nitrogen fixation in leguminous plants. I. General characters of root-nodule bacteria isolated from species of *Medicago* and *Trifolium* in Australia, *Proc. Linn. Soc. N. S. W.* **67**:98-108.

Jensen, H. L., 1961, Survival of *Rhizobium meliloti* in soil culture, *Nature* **192**:682-683.

Jensen, H. L., 1969, The distribution of lucerne and clover rhizobia in agricultural soils in Denmark, *Tidsskr. Planteavl* **73**:61–72 (in Danish).

Jones, D. G., and Russell, P. E., 1972, The application of immunofluorescence techniques to host plant/nodule bacteria selectivity experiments using *Trifolium repens*, *Soil Biol. Biochem.* **4**:277–282.

Katznelson, H., and Wilson, J. K., 1941, Occurrence of *Rhizobium meliloti* bacteriophage in soils, *Soil Sci.* **51**:59–63.

Keya, S. O., and Alexander, M., 1975, Factors affecting growth of *Bdellovibrio* on *Rhizobium*, *Arch. Microbiol.* **103**:37–43.

Kleczkowska, J., 1957, A study of the distribution and the effects of bacteriophage of root-nodule bacteria in the soil, *Can. J. Microbiol.* **3**:171–180.

Laird, D. G., 1932, Bacteriophage and root nodule bacteria, *Arch. Mikrobiol.* **3**:159–193.

Landerkin, G. B., and Lochhead, A. G., 1948, A comparative study of the activity of fifty antibiotic-producing actinomycetes against a variety of soil bacteria, *Can. J. Res. Sect. C.* **26**:501–506.

Lie, T. A., 1971, Symbiotic nitrogen fixation under stress conditions, *Plant Soil, Spec. Vol.* 117–127.

Loustalot, A. J., and Telford, E. A., 1948, Physiological experiments with tropical kudzu, *J. Am. Soc. Agron.* **40**:503–511.

Lowe, J. F., and Holding, A. J., 1970, Influence of clover source and of nutrient manganese concentrations on the *Rhizobium*/white clover association, in: *Proceedings of the Symposium on White Clover Research* (J. Lowe, ed.), pp. 79–89, British Grassland Society, Hurley, Maidenhead, Berkshire.

Marshall, K. C., 1964, Survival of root nodule bacteria in dry soils exposed to high temperatures, *Aust. J. Agric. Res.* **15**:273–281.

Marshall, K. C., 1967, Electrophoretic properties of fast and slow-growing species of *Rhizobium*, *Aust. J. Biol. Sci.* **20**:429–438.

Marshall, K. C., 1968a, Nature of bacterium–clay interactions and its significance in survival of *Rhizobium* under arid conditions, *Trans. 9th Int. Cong. Soil Sci. Adelaide* **3**:275–280.

Marshall, K. C., 1968b, Interaction between colloidal montmorillonite and cells of *Rhizobium* species with different ionogenic surfaces, *Biochim. Biophys. Acta* **156**:179–186.

Marshall, K. C., 1970, Methods of study and ecological significance of *Rhizobium*–clay interactions, in: *Methods of Study in Soil Ecology* (J. Phillipson, ed.), pp. 107–110, UNESCO, Paris.

Marshall, K. C., and Roberts, F. J., 1963, Influence of fine particle materials on survival of *Rhizobium trifolii* in sandy soil, *Nature* **198**:410–411.

Marshall, K. C., Mulcahy, M. J., and Chowdhury, M. S., 1963, Second-year clover mortality in Western Australia—a microbiological problem, *J. Aust. Inst. Agric. Sci.* **29**:160–164.

Mendez-Castro, F. A., and Alexander, M., 1976, Acclimation of *Rhizobium* to salts, increasing temperature and acidity, *Rev. Latinoam. Microbiol.* **18**:155–158.

Mulder, E. G., Lie, T. A., Dilz, K., and Houwers, A., 1966, Effect of pH on symbiotic nitrogen fixation of some leguminous plants, *Proc. 9th Int. Cong. Microbiol.*, Moscow, pp. 133–151.

National Academy of Sciences, 1975, *Underexploited Tropical Plants with Promising Economic Value*, National Academy of Science, Washington, D.C.

Norman, A. G., 1942, Persistence of *Rhizobium japonicum* in soil, *J. Am. Soc. Agron.* **34**:499.

Norris, D. O., 1959, Legume bacteriology in the tropics, *J. Aust. Inst. Agric. Sci.* **25**:202–207.

Norris, D. O., 1965, Acid production by *Rhizobium*. A unifying concept, *Plant Soil* **22**:143–166.

Nutman, P. S., 1969, Symbiotic Nitrogen Fixation: Legume Nodule Bacteria, Rothamsted Report for 1968, Part 2, pp. 179-181.

Nutman, P. S., 1975, *Rhizobium* in the soil, in: *Soil Microbiology, A Critical Review*. (N. Walker, ed.), pp. 111-131, Butterworths, London.

Nutman, P. S., and Ross, G. J. S., 1970, *Rhizobium* in the Soils of the Rothamsted and Woburn Farms, Rothamsted Report for 1969, Part 2, pp. 148-167.

Obaton, M., 1971, Utilization de mutants spontanés résistants aux antibiotiques pour l'etude écologique des *Rhizobium*, *C. R. Hebd. Seances Acad. Sci. Ser. D. Sci. Nat. (Paris)* **272**:2630-2633.

Osa-Afiana, L. O., 1979, Moisture relation in the persistence of *Rhizobium* in soil, Masters Thesis, Cornell University.

Pant, S. D., and Iswaran, V., 1970, Survival of groundnut *Rhizobium* in Indian soils, *Mysore J. Agric. Sci.* **4**:19-25.

Pant, S. D., and Iswaran, V., 1972, Survival of *Rhizobium japonicum* in Indian soils, *Philipp. J. Sci.* **101**:81-91.

Parker, C. A., and Grove, P. L., 1970, *Bdellovibrio bacteriovorus* parasitizing *Rhizobium* in Western Australia, *J. Appl. Bacteriol.* **33**:253-255.

Parker, C. A., Trinick, M. J., and Chatel, D. L., 1977, Rhizobia as soil and rhizosphere inhabitants, in: *A Treatise on Dinitrogen Fixation* (R. W. F. Hardy and A. H. Gibson, eds.), Sect. IV, pp. 311-352, Wiley-Interscience, New York.

Parle, J. N., 1964, Some aspects of the nodulation problem occuring in the Wither Hills, *Proc. N. Z. Grassland Assoc.* **26**:123-126.

Peña-Cabriales, J. J., 1978, Survival of *Rhizobium* in soil during drying, Masters Thesis, Cornell University.

Peters, R. J., and Alexander, M., 1966, Effect of legume exudates on the root nodule bacteria, *Soil Sci.* **102**:380-387.

Peterson, H. B., and Goodding, T. H., 1941, The geographic distribution of *Azotobacter* and *Rhizobium meliloti* in Nebraska soils in relation to certain environmental factors, *Univ. Nebr. Agric. Exp. Sta. Bull. 121.*

Pillai, R. N., and Sen. A., 1966, Salt tolerance of *Rhizobium trifolii, Indian J. Agric. Sci.* **36**:80-84.

Ranga Rao, V., Subba Rao, N. S., and Mukerji, K. G., 1973, Inhibition of *Rhizobium in vitro* by non-nodulating legume roots and root extracts, *Plant Soil* **39**:449-452.

Rice, E. L., 1964, Inhibition of nitrogen fixing and nitrifying bacteria by seed plants, *Ecology* **45**:824-837.

Rice, E. L., 1968, Inhibition of nodulation of inoculated legumes by pioneer plant species from abandoned fields, *Bull. Torrey Bot. Club* **95**:346-358.

Rice, W. A., Penney, D. C., and Nyborg, M., 1977, Effects of acidity on rhizobia numbers, nodulation and nitrogen fixation of alfalfa and red clover, *Can. J. Soil. Sci.* **57**:197-203.

Richmond, T. E., 1926, Longevity of the legume nodule organisms, *J. Am. Soc. Agron.* **18**:414-416.

Robinson, A. C., 1967, The influence of host on soil and rhizosphere populations of clover and lucerne root-nodule bacteria in the field, *J. Aust. Inst. Agric. Sci.* **33**:207-209.

Robinson, A. C., 1969a, Competition between effective and ineffective strains of *Rhizobium trifolii* in the nodulation of *Trifolium subterraneum, Aust. J. Agric. Res.* **20**:827-841.

Robinson, A. C., 1969b, Host selection for effective *Rhizobium trifolii* by red clover and subterranean clover in the field, *Aust. J. Agric. Res.* **20**:1053-1060.

Robison, R. S., 1945, The antagonistic action of the by-products of several microorganisms on the activities of the legume bacteria, *Soil Sci. Soc. Am. Proc.* **10**:206-210.

Robson, A. D., and Loneragan, J. F., 1970, Nodulation and growth of *Medicago truncatula* on acid soils. I. Effect of calcium carbonate and inoculation level on the nodulation of *Medicago truncatula* on a moderately acid soil, *Aust. J. Agric. Res.* **21**:427-434.

Roslycky, E. B., 1967, Bacteriocin production in the rhizobia bacteria, *Can. J. Microbiol.* **13**:431-432.

Roughley, R. J., Blowes, W. M., and Herridge, D. G., 1976, Nodulation of *Trifolium subterraneum* by introduced rhizobia in competition with naturalized strains, *Soil Biol. Biochem.* **8**:403-407.

Schall, E. D., Schenberger, L. C., and Swope, A., Jr., 1970, Inspection of Legume Inoculants and Pre-inoculated Seeds, Purdue Univ. Agric. Exp. Sta. Inspection Report 85.

Schmidt, E. L., Bankole, R. O., and Bohlool, B. B., 1968, Fluorescent-antibody approach to study of rhizobia in soil, *J. Bacteriol.* **95**:1987-1992.

Schwinghamer, E. A., 1971, Antagonism between strains of *Rhizobium trifolii* in culture, *Soil Biol. Biochem.* **3**:355-363.

Schwinghamer, E. A., and Belkengren, R. P., 1968, Inhibition of rhizobia by a strain of *Rhizobium trifolii:* Some properties of the antibiotic and of the strain, *Arch. Mikrobiol.* **64**:130-145.

Schwinghamer, E. A., and Dudman, W. E., 1973, Evaluation of spectinomycin resistance as a marker for ecological studies with *Rhizobium* subspecies, *J. Appl. Bacteriol.* **36**:263-272.

Scudder, W. T., 1975, *Rhizobium* inoculation of soybeans for sub-tropical and tropical soils. I. Initial field trials, *Soil Crop Sci. Soc. Fla. Proc.* **34**:79-82.

Sen, A., and Sen, A. N., 1956, Survival of *Rhizobium japonicum* in stored air dry soils, *J. Indian Soc. Soil Sci.* **4**:215-220.

Sherwood, M. T., 1966, Effects of Metals on Effectiveness of *Rhizobium trifolii*, Foras Taluntais, Dublin, Soils Division Research Report, pp. 64-65.

Sherwood, M. T., and Masterson, C. L., 1974, Importance of using the correct test host in assessing the effectiveness of indigenous populations of *Rhizobium trifolii*, *Irish J. Agric. Res.* **13**:101-108.

Smith, R. S., and Miller, R. H., 1974, Interactions between *Rhizobium japonicum* and soybean rhizosphere bacteria, *Agron. J.* **66**:564-567.

Subba Rao, N. S., Lakshmi-Kumari, M., Singh, C. S., and Magu, S. P., 1972, Nodulation of lucerne (*Medicago sativa L.*) under the influence of sodium chloride, *Indian J. Agric. Sci.* **42**:384-386.

Thompson, J. A., 1960, Inhibition of nodule bacteria by an antibiotic from legume seed coats, *Nature* **187**:619.

Thompson, J. A., and Vincent, J. M., 1967, Methods of detection and estimation of rhizobia in soil, *Plant Soil* **26**:72-84.

Thornton, G. D., Alencar, J., and Smith, F. B., 1949, Some effects of *Streptomyces albus* and *Penicillium* spp. on *Rhizobium meliloti*, *Soil Sci. Soc. Am. Proc.* **14**:188-191.

Turner, E. R., 1955, The effect of certain adsorbents on the nodulation of clover plants, *Ann. Bot.* **19**:149-160.

Tuzimura, K., and Watanabe, I., 1959a, Saprophytic life of *Rhizobium* in soils free of the host plant, *J. Sci. Soil Manure* (*Tokyo*) **30**:506-516 (in Japanese).

Tuzimura, K., and Watanabe, I., 1959b, Estimation of number of root-nodule bacteria by nodulation-dilution frequency method and some applications. I. Ecological studies of *Rhizobium* in soils, *J. Sci. Soil Manure* (*Tokyo*) **30**:292-296 (in Japanese).

Tuzimura, K., and Watanabe, I., 1961, Estimation of root-nodule bacteria by a nodulation-dilution frequency method, *Soil Sci. Plant Nutr. Tokyo* **7**:61-65.

Tuzimura, K., and Watanabe, I., 1962a, The growth of *Rhizobium* in the rhizosphere of the host plant. Ecological studies of root-nodule bacteria (Part 2), *Soil Sci. Plant Nutr. Tokyo* **8**:19-24.

Tuzimura, K., and Watanabe, I., 1962b, The effect of rhizosphere of various plants on the growth of *Rhizobium*. Ecological studies of root-nodule bacteria (Part 3), *Soil Sci. Plant Nutr. Tokyo* 8:153–157.

Tuzimura, K., Watanabe, I., and Shi, J. F., 1966, Different growth and survival of *Rhizobium* species in the rhizosphere of various plants in different sorts of soil. Ecological studies on root-nodule bacteria in soil (Part 4), *Soil Sci. Plant Nutr. Tokyo* 12:99–106.

Vandecaveye, S. C., 1927, Effect of moisture, temperature, and other climatic conditions on *R. leguminosarum* in the soil, *Soil Sci.* 23:355–362.

Vandecaveye, S. C., Fuller, W. H., and Katznelson, H., 1940, Bacteriophage of rhizobia in relation to symbiotic nitrogen fixation by alfalfa, *Soil Sci.* 50:15–27.

Van Schreven, D. A., 1958, Some factors affecting the uptake of nitrogen by legumes, in *Nutrition of the Legumes.* (E. G. Hallsworth, ed.), pp. 137–163, Butterworths, London.

Van Schreven, D. A., 1964, The effect of some actinomycetes on rhizobia and *Agrobacterium radiobacter*, *Plant Soil* 21:283–302.

Vincent, J. M., 1954, The root-nodule bacteria as factors in clover establishment in the red, basaltic soils of the Lismore District, New South Wales. I. A survey of "native" strains, *Aust. J. Agric. Res.* 5:55–60.

Vincent, J. M., 1958, Survival of the root-nodule bacteria, in: *Nutrition of the Legumes* (E. G. Hallsworth, ed.), pp. 108–123, Butterworths, London.

Vincent, J. M., 1965, Environmental factors in the fixation of nitrogen by the legume, in: *Soil Nitrogen* (W. V. Bartholomew and F. E. Clark, eds.), pp. 384–439, American Society of Agronomy, Madison, Wisconsin.

Vincent, J. M., 1967, Symbiotic specificity, *Aust. J. Sci.* 29:192–197.

Vincent, J. M., 1970, *A Manual for the Study of the Root-Nodule Bacteria*, Blackwell Scientific Publishers, Oxford.

Vincent, J. M., 1974, Root-nodule symbioses with *Rhizobium*, in: *The Biology of Nitrogen Fixation* (A. Quispel, ed.), pp. 265–341, North-Holland, Amsterdam.

Vincent, J. M., and Waters, L. M., 1954, The root-nodule bacteria as factors in clover establishment in the red, basaltic soils of the Lismore District, New South Wales. II. Survival and success of inocula in laboratory trials, *Aust. J. Agric. Res.* 5:61–76.

Vincent, J. M., Thompson, J. A., and Donovan, K. O., 1962, Death of root-nodule bacteria on drying, *Aust. J. Agric. Res.* 13:258–270.

Walker, R. H., and Brown, P. E., 1935, The numbers of *Rhizobium meliloti* and *Rhizobium trifolii* in soils as influenced by soil management practices, *J. Am. Soc. Agron.* 27:289–296.

Weaver, R. W., and Frederick, L. R., 1974, Effect of inoculum rate on competitive nodulation of *Glycine max*. L. Merrill. I. Greenhouse studies, *Agron. J.* 66:229–232.

Weaver, R. W., Frederick, L. R., and Dumenil, L. C., 1972, Effect of soybean cropping and soil properties on numbers of *Rhizobium japonicum* in Iowa soils, *Soil Sci.* 114:137–141.

Wilkins, J., 1967, The effects of high temperatures on certain root-nodule bacteria, *Aust. J. Agric. Res.* 18:299–304.

Wilson, J. K., 1926, Legume bacteria population of the soil, *J. Am. Soc. Agron.* 18:911–919.

Wilson, J. K., 1930, Seasonal variation in the numbers of two species of *Rhizobium* in soil, *Soil Sci.* 30:289–296.

Wilson, J. K., 1931, Relative numbers of two species of *Rhizobium* in soils, *J. Agric. Res.* 43:261–266.

Wilson, J. K., 1934a, Longevity of *Rhizobium japonicum* in Relation to Its Symbiont in the Soil, Cornell Univ. Agric. Exp. Sta. Memoir 162.

Wilson, J. K., 1934b, Relative numbers of three species of *Rhizobium* in Dunkirk silty clay soils, *J. Am. Soc. Agron.* 26:745–748.

Wilson, J. K., 1944, Over five hundred reasons for abandoning the cross-inoculation groups of the legumes, *Soil Sci.* 58:61–69.

Wilson, J. K., and Lyon, T. L., 1926, The Growth of Certain Microorganisms in Planted and Unplanted Soil, Cornell Univ. Agric. Exp. Sta. Mem. 103.

Wilson, J. R., and Norris, D. O., 1970, Some effects of salinity on *Glycine javanica* and its *Rhizobium* symbiosis, *Proc. 11th Int. Grassland Cong., Surfers Paradise, Queensland,* pp. 455–458.

4

Microbial Ecology of Flooded Rice Soils

I. WATANABE AND C. FURUSAKA

1. Introduction

More than 90% of the rice fields in Asia are cultivated with the soil submerged during most of the growth period. Wetland rice fields are of three types: irrigated, shallow-water rain fed, and deep-water rain fed. Management practices in an irrigated rice field include: (1) leveling the land and constructing levees to impound water; (2) puddling of wet soil; (3) maintenance of 5–10 cm of standing water during rice growth; (4) draining and drying at harvest of rice; and (5) reflooding after an interval of from a few weeks to as long as 8 months. In shallow-water rain-fed fields, the soil is submerged as in the irrigated fields, but the rice crop is frequently subjected to drought or to flooding deeper than the height of rice. In deep-water rice fields, the rice is sown on dry land before the onset of the monsoon, and the depth of the floodwater gradually increases as the rice grows, sometimes reaching 2–5 m. Near the harvest, the floodwater is gradually drained. Rain-fed rice fields in monsoon Asia may be subjected to severe desiccation during the dry season.

Despite the fluctuation of the water regime, the wetland rice field is characterized by submergence during most of the growing season. After flooding of the dry soil, microbial activities in soil are encouraged, leading to vigorous O_2 consumption. Because O_2 moves 10,000 times slower through a water phase than through a gas phase, the capacity of the soil to exchange gases with the atmosphere decreases as the soil becomes water saturated (Greenwood and Goodman, 1964) and deprived of O_2, leading to an anaerobic condition.

I. WATANABE • The International Rice Research Institute, Los Baños, Laguna, Philippines. C. FURUSAKA • Institute of Agricultural Research, Tohoku University, Sendai, Japan.

After flooding, a bloom of microbial activity occurs. Following that, microbial activities are retarded gradually due to exhaustion of initially available organic substances, and a steady state is reached between O_2 supply through floodwater and its consumption beneath the water. Thus, the soil is differentiated into an oxidative layer at the surface, which is several millimeters thick, and an underlying reductive layer (Pearsall and Mortimer, 1939; Shioiri, 1941).

In the oxidized layer, NO_3^-, SO_4^{2-}, MnO_2, and Fe^{3+} are stable chemical forms, whereas in the reduced layer, NH_4^+, S^{2-}, Mn^{2+}, and Fe^{2+} are stable (Pearsall and Mortimer, 1939). Among soil scientists, the oxidized layer is referred to as the layer where ferric ion is the predominant form of iron.

Beneath the plow pan layer, a zone of iron and maganese illuviation is developed. This layer is generally oxidative. Sandwiched between the surface and subsurface oxidative layers, there is the reduced plow layer with patches of oxidized sites adjacent to the rice roots. The ability of wetland rice to transport air from the atmosphere and oxidize the rhizosphere was first recognized by van Raalte (1941).

In short, a flooded rice field is composed of five sites: (1) floodwater, (2) surface oxidized layer, (3) reduced plow layer, (4) oxidized subsoil, and (5) rhizosphere of rice, as shown diagramatically in Fig. 1.

This review presents the principal characteristics of each site and the ecology of its microorganisms. The chemistry of the submerged soil was reviewed by Ponnamperuma (1972). The biochemistry of anaerobic soil was reviewed by Yoshida (1975). The description of chemical and biochemical aspects is included only when necessary for the understanding of changes brought about by the microorganisms described. For the same reason, this review includes some of the knowledge of the microorganisms in other anaerobic sediments. An attempt is

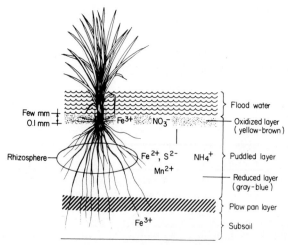

Figure 1. Profiles of paddy soil.

made here to present current knowledge on the microbiology of flooded rice soils and to delineate areas in which more detailed quantitative research may be undertaken. The terms flooded rice soil, paddy, paddy soil, and submerged rice soil are used synonymously. Rice grown in flooded soils is called wetland rice, and the rice grown in dryland soils in called dryland rice.

2. Floodwater

The floodwater in the irrigated rice field is maintained at a 2- to 10-cm depth. The chemical nature of floodwater fluctuates greatly because of variation in the composition of irrigation water and rain and in the dispersion of the surface soil in the water. The surface soil and floodwater must be considered as continuous sites because the biota on the soil surface migrates to the floodwater due to phototaxis or mechanical dispersion of soil particles, and the biota in the floodwater moves into the soil. For convenience, floodwater and the surface oxidized soil are described separately.

2.1. Organisms Inhabiting the Floodwater

2.1.1. Algal Flora

Many reports include the floristic description and succession of algae in flooded rice soils (Kurasawa, 1956, 1957; Bunt, 1961; Singh, 1961; Pandey, 1965; Gupta, 1966; Jutono, 1973; Pantastico and Suayan, 1973; Kikuchi *et al.*, 1975; Venkataraman, 1975; and Al-Kaisi, 1976), but most of these reports are qualitative descriptions of the flora. Kurasawa (1956) determined the cell number of diatoms, Cyanophyceae, and Conjugateae. After transplanting, Cyanophyceae and diatoms appeared first, followed by the bloom of Conjugateae, amounting to 0.54 g/m^2 dry weight of phytoplankton. The biomass of phytoplankton decreased as the water surface was shaded by the rice crop, and diatoms became dominant.

Comprehensive studies on the succession and biomass of algae in Senegal rice soils were conducted by Roger and Reynaud (1976, 1977) and Reynaud and Roger (1978a), who used improved plating techniques for algal counting. The plating technique was improved by (1) selective media that permit enumeration of algae classified as N_2 fixing, procaryotic, or eucaryotic (Reynaud and Roger, 1977); and (2) determining the mean volume of each count unit (cell, filament, or colony, according to species) by direct examination of the first dilution and multiplying the results by the corresponding *volume unit*. This method of expressing algal biomass is accurate, but it is tedious and time-consuming. The succession in the predominant algal flora in Senegal rice soils is summarized as follows: (1) from planting to maximum tillering—diatom and unicellular green algae, (2) from tillering to panicle initiation—filamentous green algae and homo-

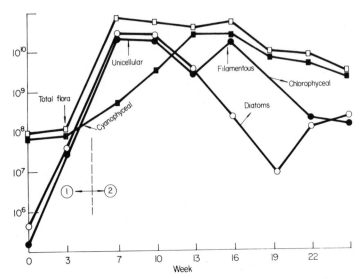

Figure 2. Variation in biomass of different components in the total algal flora during the vegetative cycle. From Roger and Reynaud (1976).

cystous blue-green algae (cyanobacteria), and (3) after panicle initiation—heterocystous and homocystous Cyanophyceae under dense plant canopy or filamentous green algae and homocystous Cyanophyceae under thin plant canopy (Fig. 2). The similar successional pattern starting from diatoms and unicellular green algae to filamentous green algae, to blue-green algae or green algae was reported by Pantastico and Suayan (1973) in the Philippines and Kikuchi *et al.* (1975) in Japan.

Factors controlling the succession are complicated. Roger and Reynaud (1978) present a review of them. They state that high temperature, high pH, low ammonium concentration, and low light intensity are more favorable for the growth of blue-green algae than for green algae.

2.1.2. Heterotrophic Flora and Fauna

In the floodwater and at the surface soil, a community of consumers is maintained.

Suzuki (1967) reported 10^5–10^6 viable cells of aerobic bacteria/ml in the floodwater. The aerobic bacterial density in the floodwater reported by Rangaswami and Narayanaswami (1965) fluctuated from 10^5 to 10^6/ml and was highly correlated with the concentration of solid matter in the water, suggesting that bacteria in the floodwater come partly from the suspended surface soil particles. Little is known about the sources and quantity of energy to support the heterotrophic microflora.

There must be a close relationship between primary producers and consumers in floodwater. Kurasawa (1956, 1957) reported that zooplankton in floodwater increased 1 week after the onset of phytoplankton growth after flooding. The maximum biomass of zooplankton appeared 2 weeks later than the peak of phytoplankton growth. The maximum dry weight of zooplankton was 5.7 mg/m^2, about one-hundredth of the maximum biomass of phytoplankton. When the maximum biomass of zooplankton was obtained, respiration rates per unit weight of plankton were the highest, suggesting the consumer's maximum activity at that time.

The decrease of phytoplankton biomass was parallel to the decrease of zooplankton after the maximum of zooplankton was attained. This finding suggests a possible producer–consumer relation in paddy fields. Kurasawa (1957) suggested a possible producer–consumer relation between green algae and rotifers. However, little is known about the energy balance between producers and consumers in the aquatic community of flooded rice soil.

2.2. Photosynthesis and Respiration

Because of photosynthetic activity by phytoplankton and submerged weeds, the O_2 concentration in floodwater exceeds saturation during the daytime, and the pH of the water sometimes increases to 9.5 because of CO_2 depletion. By contrast the O_2 concentration becomes lower than the saturation point and the pH decreases at night due to the respiration of the aquatic community.

Photosynthetic and respiratory activity in floodwater has been determined by the dark- and light-bottle method (Kurasawa, 1956, 1957; Ichimura, 1954) and by diurnal curves of O_2 concentration (Saito and Watanabe, 1978). At the early stage of rice growth, when the surface of the water was not heavily shaded, positive net production was observed, but at the later stage of rice growth, net production in floodwater (besides rice) was negative (Kurasawa, 1957; Ichimura, 1954). Both researchers put water samples in dark–light bottles to obtain productivity data that excluded the contribution from the benthic biota. On the other hand, Saito and Watanabe (1978) used a diurnal curve technique to determine productivity of floating and sediment biomass.

Saito and Watanabe (1978) recorded negative net productivity (respiration > photosynthesis) in most cases during rice growth. Because oxidizable substances may diffuse into the floodwater, apparent net productivity in the floodwater does not mean negative production of the aquatic community *per se*. After the disturbance of soil by plowing or weeding, respiration greatly surpasses photosynthesis, presumably due to the release of reduced substances from the soil. Daily gross production was as high as 0.6–3.3 g O_2/m^2, which included production from submerged weeds. Because the growth of submerged weeds compensates for the decrease in algal biomass, the decrease of productivity is not clearcut as the rice plant grows.

2.3. N_2 Fixation by Photoautotrophs

De (1936) attributed the maintenance of nitrogen fertility of a submerged rice soil to the N_2-fixing activity of phototrophic N_2 fixers in the floodwater.

Since then, many studies on autotrophic N_2 fixation and on N_2-fixing agents and their inoculation into paddy soil have been reported. Reviews were written by Singh (1961), Mague (1977), Roger and Reynaud (1979), Watanabe and Cholitkul (1979), and Watanabe (1978). There are limited quantitative studies on the N_2-fixing rates by photoautotrophs in the field. Use of the *in situ* acety-lene-reduction technique enabled an analysis of factors that affect phototrophic N_2 fixation in the paddy (Belandreau *et al.*, 1974; Alimagno and Yoshida, 1977; Watanabe *et al.*, 1978a,b). Reynaud and Roger (1978a) presented an estimate of N_2-fixing rates in relation to the N_2-fixing algal biomass in Senegal rice soils. The N_2-fixing algal biomass ranged from 50–5000 kg fresh weight/ha, and the corresponding N_2-fixing rate was 0.4–40 kg N/ha/one crop of rice.

Watanabe *et al.* (1978a) presented results of N_2-fixing (acetylene-reduction) activity assays; the maximum observed daily value was 0.5 kg N/ha in an algal community composed mainly *Gloeotrichia*.

The acetylene-reduction assay can, however, give only semiquantitative data.

3. Surface Oxidized Layer

Bouldin (1968) and Howeler and Bouldin (1971) studied the differentiation process and the depth of the oxidized layer of a flooded soil in relation to O_2 consumption and O_2 content in the atmosphere. The O_2 consumption was best described by models that included consumption by microbial respiration in the aerobic zone and oxidation of mobile and nonmobile reductants such as Fe^{2+} in the anaerobic zone.

In the oxidized layer, oxidizable substances accumulate, not only from the underlying layer but also from the overlying floodwater as planktonic debris and weed residue.

Kobo and Uehara (1943) found that mineralizable nitrogen (as an index of the amount of decomposable organic matter) accumulated on the surface of the paddy soil during the growth period of rice. Hirano (1958) also demonstrated that the total nitrogen and total carbon content in the surface layer (0–1 cm) in-creased during the rice growing season. Both results indicate that the surface soil in a paddy is the site of organic enrichment.

To study the vertical distribution of microbes near the soil surface, thin sec-tioning of an undisturbed soil core is necessary (Reddy and Patrick, 1976). Few data on the microbial distribution near the soil surface are available as a result of limited thin-section studies (see Section 3.1). Reynaud and Roger (1978b) studied the vertical distribution of algal biomass in the algal mat developed on the surface of a submerged Senegal sandy soil. They made 2-mm sections by solidifying the algal mat and soil with silica gel. The highest algal density was 5-10 mm beneath the surface. In the soil layer beneath the algal mat, a second maximum density of algal biomass was found, but the density was not as high as that of the overlying algal mat. Reynaud and Roger (1978b) attributed a spindle-shaped distribution of the algal biomass to a protective mechanism against high light intensity.

3.1. Nitrification at the Surface Soil Layer

Shioiri (1941) reported that ammonia was converted to nitrate at the oxidized soil layer, and the nitrate was reduced to N_2 in the reduced layer, resulting in a loss of nitrogen. The increase in N_2 gas evolution was demonstrated from a flooded soil incubated at a 1-cm depth with an ammonium salt (Shioiri, 1941). This loss process was later reconfirmed by using ^{15}N (Patrick and Tusneem, 1972). Despite the well-documented mechanism of nitrogen loss from the paddy, the population involved and the kinetics of nitrification at the soil surface have had limited study.

Suzuki (1967) reported the number of ammonium oxidizers in the surface soil (0-0.5 cm) layer and in the deep layers. During a waterlogged period, the nitrifiers increased more in the surface layer than in the lower layer, and the numbers decreased after drainage, indicating that nitrification by chemautotrophs took place in a flooded soil.

Using a laboratory-incubated soil, Takai and Uehara (1973) and Uehara et al. (1978) reported an enrichment of ammonium oxidizers and nitrite oxidizers at the soil surface (0-1 cm, Takai and Uehara, 1973; 0-0.3 cm, Uehara et al., 1978). The number of ammonium oxidizers at a depth of 0-0.3 cm increased from $10^2/g$ at the start to 10^6 at 28 days. The loss of nitrogen from the surface layer and from layers beneath it occurred only after the enrichment of nitrifiers. The daily nitrification rate after incubation for 28-60 days was calculated to be about 5 μg N/g soil at the 0- to- 0.3-cm layer. The nitrification rate per unit number of ammonium oxidizers fitted quite well with the known kinetics constant of the nitrifiers (Watanabe, 1974; Focht and Verstraete, 1977). The potential denitrifying capacity of each soil layer determined by the ^{15}N method showed the highest capacity at the 0- to 0.3-cm layer. The numbers of denitrifiers were also highest at the surface 0- to 0.3-cm layer. This capacity for denitrification was much higher than the observed rate of ammonium loss, indicating that the nitrification process was rate limiting in the nitrification–denitrification sequence. Because the 3-mm depth of a surface soil may include the uppermost layer of the reduced zone, it is not clear whether the oxidized layer has the highest denitrifying capacity. Sørensen (1978) found the highest denitrifying capacity in the oxidized layer of lake sediment. It is likely that denitrification occurs partly in the oxidized layer. Partial denitrification in the oxidized layer is also predicted from the E_h required for denitrification. Because the differentiation between oxidized and reduced layers is based on the oxidation–reduction of iron and the redox potential for denitrification is higher than for iron reduction (see 4.1), denitrification may take place in the oxidized layer. Chen and Chou (1961) presented the view that nitrifiers and denitrifiers are in close contact in the soil because denitrifying bacteria were always found during enrichment culture of nitrifying bacteria. This aspect needs to be verified.

Reddy et al. (1976) and Patrick and Reddy (1977) presented a kinetic analysis of the nitrification–denitrification loss process in soil columns incubated in the laboratory. They developed a mathematical model to explain ammonium diffusion, nitrification, and the subsequent diffusion and denitrification of the nitrate in the soil column. Their model was the first kinetic approach to denitrification in submerged soil.

The assumption that nitrification proceeds uniformly in the oxidative layer and denitrification takes place only in the reductive layer is unrealistic as discussed above. The O_2 uptake necessary for a nitrification rate of 3–7 μg N/cm^3/day was too high to account for a 1.5-cm depth of the aerobic layer maintained by O_2 diffusion. Certainly, the nitrification site must have been located at the surface of the oxidized layer in Reddy and Patrick's experiment.

Focht (1978) further developed a kinetic consideration of the nitrification in submerged soil. Oxygen diffusion and ammonium concentration in soil solution determine the nitrification rate at the surface. Focht predicted a maximum nitrification rate in the submerged soil based on known kinetic constants. This value is about 1 kg N/ha/day (assuming a liquid volume ratio of 0.5). Little is known about the roles of heterotrophic nitrifiers in paddy soil.

3.2. Methane Oxidation

Harrison and Aiyer (1915) drew attention to methane oxidation at the surface of the paddy soil as a possible source of CO_2 for algal growth, and they isolated methane-oxidizing bacteria from the algal film on the surface.

As acetylene is a competitive inhibitor of methane oxidation (De Bont and Mulder, 1976), the difference in methane evolution between acetylene-treated and untreated plots in the field can be ascribed to the oxidation of methane in the soil. By applying this principle, De Bont et al. (1978) obtained data showing that methane oxidation is taking place in situ. In an unplanted plot without acetylene, there was no increase in the methane content of the atmosphere above the pots with flooded soil; presumably, all of the methane diffusing into the aerobic zone was oxidized at the soil surface. With acetylene, methane evolution from the unplanted plot was distinctly observed. At the aerobic top layer, the most-probable-number (MPN) of methane oxidizers was measured as 2–18 × 10^6/g dry soil.

3.3. Photosynthetic Bacteria

High populations of nonsulfur purple bacteria (as high as 10^6/g soil) and sulfur bacteria (mostly 10^2–10^3) were found in Japanese paddy soils (Okuda et al.,

1975) and in tropical soils (Kobayashi *et al.*, 1967). Kobayashi and Haque (1971) reported that the populations of both groups in the rhizosphere of wetland rice approached their peaks near the crop-heading stage.

Photosynthetic bacteria require anaerobic conditions and organic acids or sulfide, which are formed by anaerobic metabolism, for their light-dependent growth. The surface soil or floodwater may not meet these requirements. However, the association of photosynthetic bacteria and aerobic heterotrophic bacteria, as experimentally shown by Okuda *et al.* (1960) and Kobayashi *et al.* (1965a,b), does not require anaerobic conditions. If we assume the presence of this association in a natural condition, the role of photosynthetic bacteria could not be denied. But this association was studied only in the laboratory, and nothing is known about the role of these bacteria in N_2 fixation in the paddy field. Reynaud and Roger (1978b) estimated the biomass of photosynthetic bacteria (Athiorhodaceae) as making up only 2% of the N_2-fixing biomass in the surface soil (see Section 3).

4. Reduced Layer

4.1. Sequential Reductions and Microorganisms Responsible for Each Reduction Process

After a submerged soil is deprived of O_2, reduction takes place roughly in the sequence predicted by thermodynamics (Ponnamperuma, 1955). This sequence was experimentally verified by Takai *et al.* (1956, 1963a, b), and the results were reviewed by Takai and Kamura (1969) and Takai (1969). Air-dried soils were put in syringes with a cut end, immersed in water, and incubated. Changes in chemical and microbiological characteristics were followed. Figure 3 shows the pattern of chemical and microbiological changes in the soil during incubation. Takai (1978) summarized the sequence of reduction processes in relation to redox potential as shown in Table I. Reduction proceeds from the reactions with higher E_h7 (E_h at pH 7) to those with lower E_h7. In the convention used, the reactions at higher E_h are more oxidative.

Takai divided reduction processes into two major steps: facultatively anaerobic and obligately anaerobic ones. Takai hypothesized that facultatively anaerobic processes are coupled with the activity of aerobic or facultatively anaerobic microorganisms. In Fig. 3, the shifting period from facultatively anaerobic to obligately anaerobic processes is shown by arrows.

To understand the sequence of steps in anaerobic metabolism, H_2 evolution and organic acid production need special consideration. Although H_2 formation is listed at the bottom of Table 1, that does not mean it always takes place at the final stage of reduction. When an air-dried soil which is rich in decomposable organic matter or a soil amended with soluble carbohydrates is incubated, H_2 evolution and a concomitant drop in E_h, reaching to -400 mV, take place at an early stage of incubation (Yamane and Sato, 1964; Bell, 1969). The consump-

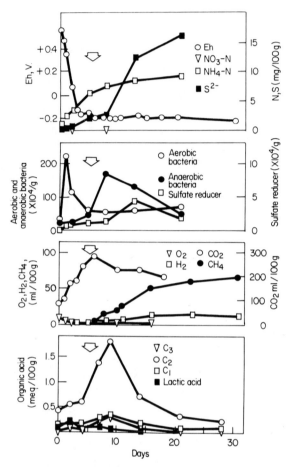

Figure 3. Course of microbial metabolism in waterlogged paddy soil during incubation in a syringe at 35°C. Before an arrow = the first step; after an arrow = the second step. From Takai and Kamura (1966).

tion of H_2 or the removal of H_2 by bubbling N_2 gas leads to an increase in E_h, and thereafter the sequential reductions as shown in Table I take place.

The H_2 production process is described at the bottom of Table I because the E_h of the soil is lowest when H_2 evolution occurs. Furthermore, the observed H_2 evolution must be considered as the balance between H_2 consumption and its evolution, because H_2 is consumed in an anaerobic paddy soil, as will be discussed later (Section 4.1.4). Nothing is known about rates of H_2 evolution relative to its consumption in connection with reductive processes.

The accumulation of organic acids must also be considered as the net result of their production and consumption. When a large amount of organic matter (green manure) is added to a submerged soil with a small amount of reducible iron, organic acids accumulate at an early stage (Takai, 1969).

Table I. Reduction Processes and Microbial Metabolism[a]

Reactions	E_h7 of the reactions (mV)	Ammonification	CO_2 evolution	Energy metabolism for acquiring energy	Anaerobiosis (hypothetical)
			Phase I		
Consumption of molecular O_2	+500 ~ +300 critical value +332[b]			Oxygen respiration	
Disappearance of nitrate and formation of N_2	+400 ~ +100 critical value +226[b]	Active	Active	E. coli type nitrate reduction and denitrification	Aerobic or facultatively anaerobic
Mn^{2+} formation	+400 ~ -100			Indirect reduction by microbial metabolites (Mn^{4+}, Mn^{3+})	
Fe^{2+} formation	+200 ~ -200 at pH 6 ~ 7			Direct or indirect reduction of Fe^{3+}	
			Phase II		
S^{-2} formation	0 ~ -200 critical value -150		Either slow, absent, or CO_2 consumed	Sulfate reduction	
		Slow			
CH_4 formation	-200 ~ -300			Methane fermentation	Strictly anaerobic
H_2 formation	-200 ~ -420			Fermentation	

[a]Modified from Takai (1978).
[b]Patrick (1960).

The sequential-reduction hypothesis was further confirmed by Patrick (1960), Patrick and Turner (1968), Connel and Patrick (1968), Bailey and Beauchamp (1971), and Patrick and Gotoh (1974). Takai *et al.* (1969a) observed a similar sequence of reduction processes after submerging planted pots. The addition of oxidized ions in the reactions that require higher E_h generally depresses the reduction in the reactions requiring lower E_h. Several mechanisms may be involved in the sequential reductions. This review will not deal with those, but in the following section, the microorganisms that take part in each reduction process are described.

4.1.1. Denitrification

In the submerged rice soil, nitrate is quickly denitrified, and almost all of the added nitrate is finally transformed to N_2 gas (Yamane, 1957). MacRae *et al.*

(1968) studied the loss of ^{15}N-labeled nitrate in six submerged Philippine soils. The loss of ^{15}N in 6 weeks ranged from 59 to 100%. The loss increased as organic matter levels in soil increased.

The sequence of products formed during denitrification in Senegal rice soils was studied by Garcia (1973). Most of the soils used by Garcia were acidic (pH less than 5.0). The maximum amount of N_2O and NO formed during denitrification was highly correlated with the soil acidity. The velocities of nitrate and N_2O reduction were correlated with the amount of organic matter. A higher N_2O and NO accumulation in soils at lower pH values was experimentally confirmed by controlling the pH of the submerged soil (Van Cleemput et al., 1975, 1976).

Garcia (1974, 1975a) proposed using the velocity of N_2O reduction as a method of evaluating denitrifying activity. The velocity of nitrate reduction was highly correlated with the velocity of N_2O reduction (Garcia, 1974) in Senegal soils. The hourly activity in nitrate reduction among 33 soils varied from 1.0 to 5.6 μg N/g. Assuming that 1 hectare has 10^9 g of actively denitrifying soil mass, the hourly rate would correspond to 1.0–5.6 kg N/ha. This rate is far higher than the nitrification rate predicted by Focht (1978) (see Section 3.1). Garcia et al. (1974) found a significant correlation between denitrifying capacity and soil organic matter content in 29 flooded rice soils in Senegal.

The numbers of denitrifying bacteria are usually estimated by the MPN method, which is based on gas production in nitrate-containing tubes. Gamble et al. (1977) reported the numbers of denitrifying bacteria in 19 soils, including 4 rice soils. The minimum and maximum numbers of denitrifiers in the 19 soils were 1.2×10^4 and 7.0×10^6/g soil. The MPN count of denitrifiers was very highly correlated with plate counts ($r = 0.69$). For all samples, the numerical relationship was denitrifiers < nitrite accumulators $\leq$ organisms capable of growth without nitrate. The average ratios for all samples among these groups were $0.20 \pm 0.23 : 0.81 \pm 0.23 : 1.0$.

Ishizawa and Toyoda (1964) obtained an average of 3×10^5 denitrifying bacteria/g soil from the plow layer of 18 Japanese paddy soils. Araragi and Tangcham (1975) surveyed 98 Thailand paddy soils and obtained an average of 6×10^5/g. In the studies of Garcia et al. (1974), the average was 2×10^3 (minimum of 1.4×10^2 to a maximum of 2.2×10^5).

Denitrification appears to be a property of a diverse group of gram-negative bacteria and some species of *Bacillus*. Gamble et al. (1977) reported on the numerical taxonomy of 146 denitrifying isolates from various sources, including agricultural soils, aquatic sediments, and poultry manure. The diversity of denitrifiers from a given sample was usually high, with at least two groups present. From the results of Gamble et al. (1977), it is evident that the isolates from rice-growing soils were also common in nonrice-growing soils, except that a rice-growing soil from Los Baños, Philippines yielded group-4 bacteria, resembling *Pseudomonas aeruginosa*.

Jordan *et al.* (1967) and Suleimanova (1971) described the generic composition of denitrifying bacteria isolated from flooded rice-growing soils. Garcia (1977a) studied mesophilic denitrifying bacteria in a Senegal rice-growing soil by culturing samples in a medium with high concentrations of potassium nitrate or nitrite (5 g/liter). The bacteria that tolerated nitrite and produced gas were more numerous than bacteria that produced gas only from nitrate. Among the nitrite-tolerant denitrifiers, spore formers were predominant. Some were nitrite dependent and could not reduce nitrate.

In some soils, thermophilic denitrifiers, which are grown at $60°C$, gave higher MPN counts than mesophilic denitrifiers. The thermophilic denitrifiers were spore formers and tolerated nitrite more or less well. The dominant thermophilic denitrifier was different from *Bacillus stearothermophilus* (Garcia, 1977b; Pichinoty *et al.*, 1977). Surprisingly, the isolates of denitrifiers obtained by Gamble *et al.* (1977) did not include *Bacillus*. The reason was not clear.

4.1.2. Manganese Reduction

Microbial activity is probably the major mechanism for Mn reduction in anaerobic rice-growing soils (Yoshida and Kamura, 1972a). All of the heterotrophic bacteria obtained from the aerobic counting plates showed some Mn-reducing activity. Thus, Mn reduction may be shared by a wide range of bacteria (Yoshida and Kamura, 1972b). Culture filtrates of Mn-reducing bacteria also have Mn-reducing activity, and MnO_2 that was separated by a cellophane tube from growing cells of Mn-reducing bacteria was reduced in the culture medium. Yoshida and Kamura (1972b), therefore, considered that Mn reduction may be brought about by some microbial metabolite. This contact-free reduction of MnO_2 decreased as the pH of the medium increased (Yoshida and Kamura, 1975a). The role of this acidity-induced reduction in a reduced paddy soil, which is near neutral, is doubtful. Trimble and Ehrlich (1968) had a similar view: acidity-induced Mn reduction is unlikely to take place in Mn nodules in sea sediments.

Trimble and Ehrlich (1968, 1970) isolated *Bacillus* and a coccus from Mn nodules and presented evidence that the MnO_2-reducing enzyme was induced by Mn^{2+} and was O_2 insensitive. Yoshida (1975) and Yoshida and Kamura (1975b) presented the view that Mn was reduced chemically by ferrous iron.

Much ambiguity remains in the reports on Mn-reduction mechanisms and the microorganisms in Mn reduction in flooded rice-growing soils.

4.1.3. Iron Reduction

Kamura and Takai (1961) presented evidence that Fe reduction in reduced rice-growing soils is brought about by microbial activities. They showed inhibition of Fe reduction by a mixture of antibiotics added to preincubated soil. The addition of acetate and alcohol to nonsterile soil enhanced Fe reduction, and the reduction was accompanied by the oxidation of substrates (Kamura et al., 1963). Asami and Takai (1970) showed that the addition of amorphous Fe_2O_3 to an anaerobic soil stimulated carbon dioxide evolution and the carbon dioxide thus stimulated was proportional to the amount of Fe^{2+} formed; this suggests that Fe reduction may be coupled to respiration.

From these data, it appears likely that Fe reduction is coupled with substrate oxidation, presumably through the electron-transport systems of facultative anaerobes. A series of studies by Ottow of Fe reduction in hydromorphic soils supports this hypothesis. The capacity for Fe reduction is widely distributed among microorganisms (Ottow, 1969a), including fungi, yeasts (Ottow, 1969b), facultatively anaerobic bacteria such as Pseudomonas, Micrococcus, Enterobacteriaceae, Corynebacteriaceae, and Bacillus, and obligate anaerobes like Clostridium (Ottow and Glathe, 1973).

The Fe-reducing isolates from the subsoil of a gley soil were studied in detail by Ottow and Glathe (1971). Among 71 isolates, 68 had nitrate-reducing activity, and 35 of them could denitrify. Ottow (1971) isolated and characterized Fe-reducing N_2-fixing saccharolytic clostridia from a gley soil. Based on a series of experiments, Ottow (1969a) presented the view that nitrate reductase was involved in iron reduction by facultative anaerobes. First, most of the Fe-reducing microorganisms (facultatively anaerobic bacteria, fungi, and yeasts) were simultaneously nitrate reductase inducible (Ottow, 1969c; Ottow and von Klopotek, 1969). Second, the addition of increasing amounts of nitrate and chlorate to a medium supporting growth of nitrate reductase-positive microorganisms depressed Fe reduction (Ottow, 1969c). In contrast, the Fe-reducing capacity was not changed by nitrate and chlorate in bacteria lacking nitrate reductase (Munch and Ottow, 1977).

To test that hypothesis, the effect of nitrate addition on Fe reduction by nitrate reductase-less mutants (nit^-) of some bacteria was studied (Ottow, 1970). In nit^- mutants, nitrate did not depress Fe reduction. Although the Fe-reducing capacity of the mutants was considerably less than that of the wild types, the capacity was not eliminated entirely. Further, the remaining Fe-reducing capacity could not be repressed by nitrate. To explain the difference in the effect of nitrate on Fe reduction, the involvement of at least two different and specific reductases—nitrate reductase and an unknown enzyme—was suggested. The former type of Fe reductase is distributed in nitrate-reducing and denitrifying, facultatively anaerobic microorganisms; the latter type is involved in Fe reduction by obligate anaerobes, which are insensitive to nitrate (Munch and Ottow, 1977).

In anaerobic, flooded rice-growing soils, nitrate greatly depressed Fe reduction (Ponnamperuma et al., 1956). However, in the presence of a large amount of nitrate, the Fe-reducing capacity was not eliminated entirely. This suggests a partial involvement of obligate anaerobes in Fe reduction.

4.1.4. Sulfate Reduction

As shown in Fig. 3, sulfide formation is accompanied by an increase in the number of sulfate reducers, indicating that sulfate reduction in a flooded rice soil is by sulfate reducers that require a low E_h condition. Furusaka and Hattori (1956) used the Warburg manometer to study H_2 uptake of a paddy-soil suspension. Endogenous H_2 uptake was increased by the addition of sulfite, thiosulfate, and sulfate, which are known as hydrogen acceptors for the dissimilatory sulfate reducer, *Desulfovibrio vulgaris* (Ishimoto *et al.*, 1954). The increment of H_2 uptake noted upon the addition of sulfite was taken as the rate of sulfate reduction in a field soil. The hourly activity in sulfite reduction was 3–5 $\mu l \, H_2/g$ soil in an alluvial paddy soil and more than 10 $\mu l \, H_2/g$ in a muck paddy soil. When sulfate-reducing activity was detected, the MPN count of sulfate reducers was 10^4–$10^5/g$, and *Desulfovibrio desulfuricans* (later reclassified as *D. vulgaris* by Postgate and Campbell, 1966) was isolated from the soil.

In wide-range surveys of the microflora in Japan, Ishizawa and Toyoda (1964) found that the number of sulfate reducers in a paddy was about 10^3–$10^4/g$ in the plow layer, and the number decreased greatly at lower soil depths; however, the number was higher than in a dryland field. Generally, the number was lower in the paddies in cooler regions, such as Hokkaido and Tohoku. Takai and Tezuka (1971) compared the population density of sulfate reducers in paddies and adjacent drylands at six sites in Japan. In the plow layer, the density of sulfate reducers was higher in the paddy (10^4–10^5 MPN/g) than in the dryland field (10–10^5). In the paddy, the density decreased sharply beneath the plow layer, but in the dryland field, the vertical distribution was more or less uniform among soil profiles.Takai and Tezuka (1971) pasteurized soil suspensions to test for the presence of spore-forming sulfate reducers (*Desulfotomaculum*). Sulfate reducers existing in the spore form predominated in dryland soil but not in paddy soil. Sulfate reducers in the dryland soils, or in drained paddy soils sampled after rice harvest, were resistant to air drying. The incubation of dryland soil for 1 year did not decrease the number of sulfate reducers, presumably due to the predominance of spore forms. Sulfate reducers grown in flooded soil were more susceptible to drying or to aerobic soil conditions than those in dryland soil, but 10% of the population in flooded soil survived during a 1-year incubation under aerobic conditions.

In an acid sulfate soil, sulfur reduction becomes marked enough to damage the rice plant when organic matter as organic amendments or exudates from rice seeds and roots are supplied (Jacq, 1973). Garcia *et al.* (1974) surveyed the population of sulfate reducers in acid sulfate rice soils (saline and nonsaline) in Senegal. The population density in plow layers of 29 soils ranged from 6×10^1 to $7 \times 10^5/g$ and was closely related to soil pH and sulfate concentration. The number of sulfate reducers increased at lower pH.

The distribution of sulfate reducers in paddy soils was studied in detail by

Wakao and Furusaka (1972, 1973) and Wakao *et al.* (1973), particularly the horizontal distribution of sulfate reducers in small areas. By counting sulfate-reducers by the anaerobic double-dish method in 100 samples from 4 cm^2, it was possible to analyze the microdistribution pattern by Morishita's (1959) Iδ index. In this case, the distribution pattern was neither homogenous nor random but contiguous. The larger the mean cell number, the greater the contiguity, indicating that growth of the sulfate-reducing bacteria brings about the contiguity. The authors pointed out that this contiguous distribution was related to the presence of organic debris (Wakao and Furusaka, 1976b). This contiguous contribution pattern coincides closely with the spotty deposit pattern of [^{35}S] sulfide, which is formed from uniformly distributed ^{35}S-labeled sulfate in the flooded soil (Furusaka, 1968).

The presence of microaggregates containing sulfate reducers was indicated by the fact that the number of sulfate reducers increased several times when soil was dispersed by sonic treatment. Furthermore, the change in the number of sulfate reducers in the supernatant fluid after repeated washing of soil particles and sedimentation of the washed suspension suggested the continuous release of sulfate reducers from the soil aggregates (Wakao and Furusaka, 1976a).

Lactate is known as the preferred substrate for most sulfate reducers. Cappenberg and Prins (1974) and Cappenberg (1974b) presented evidence, derived from studies using ^{14}C-labeled lactate and acetate and analogue inhibitors, that lactate was converted to acetate by sulfate reducers in lake sediments.

Kinetic studies on the utilization and formation of organic substrates have not been done in flooded rice soils.

As shown in Fig. 3, sulfate reduction is accompanied by the disappearance of acetate from flooded soil. It is not yet clear if this decrease in acetate concentration is exclusively due to methanogenic bacteria. Furusaka (1968) studied the effect of medium composition with lactate as the energy source and got higher MPN counts of sulfate reducers with the lactate medium. Consequently, the predominance of *D. vulgaris* type was assumed. Recently, Widdel and Pfennig (1977) described a new anaerobic spore-forming acetate-oxidizing sulfate reducer, *Desulfotomaculum acetoxidans*, which was isolated from anaerobic sediments. The presence of this type of bacterium or its prominence in flooded rice soils is not known. Jacq (1978) suggested that sulfur-reducing bacteria (Pfennig and Biebl, 1976) in Senegal rice soils caused blackening of sulfur-coated urea.

4.1.5. Methane Formation

The ecology of methanogenesis in anaerobic rice-growing soils has been studied less than in other submerged environments. Reviews on methanogenesis were written by Mah *et al.* (1977), Zeikus (1977), and Toerien and Hattingh (1969).

By extrapolating from the limited number of studies of methane-forming activities in flooded rice-growing soil, the activities being measured using anaerobic incubations, Koyama (1963) estimated annual production of methane in Japanese flooded soils as 112 liters/m^2. Methane formation in 28 Senegal rice-growing soils, measured at 37°C under anaerobic conditions, ranged from negligible to 25 ml/g each day (Garcia et al., 1974). In this case, the rate of methane formation had little correlation to the organic carbon content of the soils. In contrast, denitrifying capacity had a significant correlation to the organic carbon content. This contrast may suggest a difference in organic nutrient sources for the two reduction processes.

These data on methane formation, measured under anaerobic conditions, do not give a direct estimate of methane evolution form the flooded rice field because methane oxidation must be considered, as described in Section 3.2. The figures of Koyama (1963), therefore, appear to be an overestimate.

Temperature greatly affects methane production. At a temperature higher than 25°C, methanogenesis is greatly accelerated (Koyama, 1963; Yamane and Sato, 1967), and consequently, the accumulation of organic acids is diminshed. The diminished accumulation of organic acids appears to explain the minimal damage to wetland rice by organic amendments at higher temperature (Mitsui et al., 1959).

Takai (1970) reported the preferential formation of methane from the methyl ^{14}C of acetate in a flooded rice-growing soil. The formation of methane from carbon dioxide was of lesser importance. The addition of butyrate accelerated the reduction of carbon dioxide, presumably by the formation of H_2 by the oxidation of butyrate to acetate. In contrast to the reports of the inhibitory action of sulfate on methanogenesis in lake sediments (Cappenberg, 1974a,b; Winfrey and Zeikus, 1977), the addition of 571 ppm S as $(NH_4)_2SO_4$ in an anaerobically incubated soil did not depress methane formation (Bollag and Czlonkowski, 1973).

Little is known about the population of methanogenic bacteria in flooded rice-growing soils. Garcia et al. (1974) surveyed the number of methanogenic bacteria in Senegal rice soils and noted that the density varied from 10^2 to 10^6/g soil.

4.2. Nature of Aerobic and Anaerobic Bacteria

By using plate count methods, many researchers (De and Boss, 1938; Ch'en, 1963; Venkatesan and Rangaswami, 1965; Taha et al., 1967; Furusaka et al., 1969; Vostrov and Dolgikh, 1970; Shchapova, 1971; and Araragi and Tangcham, 1975) recognized that the bacteria predominated and fungi and actinomycetes were depressed during soil submergence. When soil was dried after rice harvest, the counts of actinomycetes and fungi increased but then decreased during the severe dryness later in the hot dry season (Venkatesan and Rangaswami, 1965; Rangaswami and Venkatesan, 1966).

Ishizawa and Toyoda (1964) took soil samples from 21 drained paddies after harvest and analyzed their microbial properties. By comparing the data obtained with those of dryland soils, they recognized that the characteristics of the microflora in the submerged paddy soil were maintained after the soil was drained. The density of anaerobic bacteria, measured using deep tubes of thioglycolate nutrient agar sealed with liquid paraffin, was higher in paddy-soil samples than in dryland-soil samples.

Furusaka and his associates have conducted extensive research (Furusaka, 1978; Hayashi et al., 1978) to clarify the floristic characteristics of the aerobic and anaerobic bacteria and to determine the environmental factors affecting microflora of paddy soils.

4.2.1. Aerobic Bacteria

Ushigoshi (1974) took soil samples before and several times after the flooding of a wetland rice field. Counts of aerobic bacteria on albumin agar increased sharply for 2-3 weeks after flooding and decreased to the level before flooding during the following weeks. About 80 colonies grown on albumin agar were picked each sampling time and classified to the generic level. Most of bacteria showed deep-layer growth in stab culture on semisolid agar containing cysteine. Only about 15-30% of the aerobic bacteria isolated from albumin agar grew on the surface of stab cultures. This indicates that most bacteria isolated aerobically are facultatively anaerobic or microaerophilic. The change in bacterial number occurred mainly among the facultative anaerobes. The predominance of facultative anaerobes in the bacterial flora may facilitate the smooth shift from aerobic metabolism to anaerobic respiration that occurs after submergence.

Ushigoshi (1974) proposed a tentative dichotomous scheme of classification applicable to the bacteria isolated from soils. This scheme was based on the characteristics of the isolated bacteria and the 7th edition of Bergey's Manual. The number of gram-negative bacteria almost equaled the number of gram-positive bacteria throughout the survey period. The gram-positive bacteria were grouped into five categories: bacilli, cocci, and three groups of coryneform bacteria. One-fourth of the isolates was coryneform bacteria. The classification of coryneforms is still incomplete. Ushigoshi classified them into three groups, based mainly on the type of pleomorphism of the cell during the life cycle of the organism.

Gram-negative bacteria were divided into nine groups. Because of the weak biochemical activities of most of the isolates on the usual media used for bacterial classification, most isolates fell into promiscuous groups that gave negative reactions in various tests (Leifson, Kovacs oxidase, indol, Voges–Proskauer, and methyl red tests for Enterobacteriaceae). One of the characteristics of the gram-negative bacterial flora was that *Pseudomonas* was a minor group in paddy soil. Some of the major groups were *Flavobacterium*, *Achromobacter*, *Acinetobacter*, *Alcaligenes*, and *Erwinia* (Hayashi et al., 1978).

4.2.2. Anaerobic Bacteria

Takeda and Furusaka (1970, 1975a,b) conducted extensive surveys of bacteria isolated from a paddy soil using anaerobic incubation procedures. Soil samples were taken before flooding, soon after flooding, about 2 months after flooding, and near harvest and soon after drainage. Anaerobic bacteria were counted on a glucose-peptone-yeast extract medium containing cysteine after an anaerobic incubation of 5 days. About 80 colonies were picked at each sampling time. Most of the bacteria isolated anaerobically could grow aerobically. The number of facultative anaerobes increased sharply after the submergence, followed by a decrease to the level found before flooding.

The number of strict anaerobes was always about 10% of the count of facultative anaerobes, and the seasonal changes in the former group were smaller. The numbers were about $0.7-2.0 \times 10^5/g$, and the strict anaerobes were mostly clostridia. The numbers of spores, including those of *Bacillus* and *Clostridium*, were almost constant during the surveys, except during a period of about 2-3 weeks after flooding, when the vegetative cells dominated.

Classification of the isolated facultative anaerobes and strict anaerobes was also performed. Among the facultative anaerobes, Enterobacteriaceae predominated during nonsubmerged periods, and polar flagellated bacteria (*Aeromonas hydrophila*) and coccoid bacteria predominated during flooding.

Bacillus and *Clostridium* numbers were almost constant irrespective of submergence, but the species composition changed before and after submergence, suggesting that these spore-forming bacteria were active in the paddy soil.

Takeda and Furusaka (1975a) made a further study of the biochemical activities of the clostridia isolated from paddy soil. All organisms, except *Clostridium tertium*, utilized solely amino acids as a source of energy. All of them produced acetic and butyric acids, and some of them also produced propionic acid. Isovaleric and isobutyric acids were also produced. These producers of branched fatty acids can carry out the Strickland reaction (oxidation-reduction reactions between two amino acids).

To get evidence that these proteolytic clostridia were active in flooded soil, Takeda and Furusaka (1975b) examined the formation of branched fatty acids in flooded soils. In a soil taken from a flooded rice field, isovaleric acid was detected, even though its amount was small. In a soil flooded with a peptone solution, the formation of isovaleric acid was greater. In addition to peptone, the addition of proline and leucine increased the isovaleric acid level.

All clostridia isolated from peptone-amended soil produced branched fatty acids, but the facultative anaerobes isolated from peptone-treated soil produced only acetic acid. In glucose-amended soil, branched fatty acids were not formed, and the numbers of clostridia were low.

Because acetic acid is the most abundant organic acid in flooded rice soils, and content of iso-acids is quite low, it is likely that facultative anaerobes play a major role in organic acid fermentation, but the bacterial flora active in this fermentation still remains to be characterized. Hiura *et al.* (1976, 1977) suggested that *Clostridium* may be involved preferentially in the mineralization of soil organic nitrogen. Their suggestion was based on the high correlation between the mineralized ammonia that was accelerated by soil disruption and the number of strict anaerobes.

By increasing the incubation period for anaerobic counts by the roll-tube method from 5 days to 2 weeks, Hayashi and Furusaka (unpublished) obtained counts similar (10^6–10^8/g) to those obtained when plates were incubated in air. The strictly anaerobic bacteria that appeared after prolonged incubation were mainly gram-positive rods of irregular shape. Most of these bacteria were identified as strains of *Propionibacterium*, and they produced propionate from lactate. Because about half of the lactate added to flooded soil was converted to propionate, these bacteria appear to be active in propionate fermentation in flooded soils.

It appears logical to assume that the dominance of facultative and strict anaerobes in flooded rice soils is due to the deficiency of O_2. Nagatsuka and Furusaka (1976) studied the effect of O_2 concentration on the bacterial flora in a paddy-soil suspension. A gas mixture, with varying O_2 content, was continuously passed through the stirred-soil suspension, and the O_2 concentration was monitored. The isolated bacteria were grouped into strict aerobes that grew only on the surface of stab cultures; facultative anaerobes that grew deep in the stab culture; and strict anaerobes that could not grow under O_2.

The predominance of strict anaerobes appeared only in the soil exposed to an O_2-free atmosphere. In this experiment, *Propionibacterium*, as mentioned above, was not taken into account. The relative composition of the above-mentioned three groups of bacteria in the soil that were found 10 days after incubation at 1–2 mm Hg was close to the composition found in the field. In other words, the bacterial composition of paddy soils is not reflected by data obtained in studies using entirely O_2-deficient conditions.

Greenwood and Goodman (1964) indicated that O_2 consumption by soil is retarded only at one-hundredth of the atmospheric O_2 tension (1.6 mm Hg). If this limit on O_2 consumption is applicable to the reduced layer of flooded rice soils, the result obtained by Nagatsuka and Furusaka might be explained. Oxygen content might not fall lower than 1–2 mm Hg because of the respiration of soil microorganisms. However, this explanation appears to be only one among many possible answers.

4.3. Other Microorganisms

4.3.1. Actinomycetes

Rangaswami and Venkatesan (1966) studied the actinomycete flora of a paddy during the submerged and dry periods. The actinomycetes—which are capable of utilizing a great variety of carbohydrates (including cellulose and pectin), liquefying gelatin, reducing nitrate, ammonifying organic nitrogen, and producing acids—are found more in wet than in dry soil. During the submerged period, more nonsporulating and fewer sporulating forms were found, whereas in the dry soil organisms sporulating more readily were found. It was observed that continued submergence greatly reduced the population of aerobic actinomycetes, and microaerophilic forms increased gradually. *Micromonospora* increased to as much as 20% of the total actinomycete isolates under submergence and decreased to less than 7% during the dry-soil period.

Ishizawa *et al.* (1969) studied the actinomycete flora of Japanese paddy soils. They concluded that *Micromonospora*, *Streptosporangium*, and *Streptomyces* that produce little or no aerial mycelium were found at a higher frequency in a wetland than in a dryland soil. These data are almost similar to those obtained by Rangaswami and Venkatesan (1966), although there are differences in minor points. The peculiarities of the actinomycete flora developed under submergence were more or less maintained in the dry period in Japanese paddy soils.

4.3.2. Fungi

The number of fungal propagules decreases sharply after soil submergence. Submergence is used to control soil-borne fungal pathogens (Stover *et al.*, 1953; Sewell, 1965). Other reports on the fungal flora of paddy soils were by Reynolds (1970), Lim (1967, 1972), Chen (1974), and Tu (1975). Sivasithamparam (1971) reported that most spores of seed-borne fungi cannot germinate well in flooded soils. The ecology of fungi related to soil-moisture conditions was reviewed by Griffin (1969, 1972), and the survival of pathogenic fungi in submerged soils is not described here.

Despite the depressed activity of fungi in submerged soils, fungi play a role in the initial stage of organic matter decomposition in paddy soil (Wada, 1971, 1974; Saito *et al.*, 1977b). However, little is known about the kind of fungi found during decomposition of organic matter in flooded paddy soils.

4.3.3. Cellulose Decomposers

Vostrov and Dolgikh (1970) reported that aerobic cellulose decomposers with yellow colonies became abundant during the decomposition of flax fiber in flooded soil, and they criticized the view that cellulose decomposition in a flooded soil is predominantly by anaerobes.

Saito *et al.* (1977a,b) modified Tribe's technique (Tribe, 1957) and used a cellophane membrane to observe microbial succession on cellulosic material. They divided the succession of cellulose decomposers into three phases: (1) after about 1 week of submerged incubation, decomposition by aerobic microorganisms, which include Phycomycete-like fungi and Cytophaga-like bacteria, takes place; (2) after about 2 weeks of submerged incubation, active decomposition takes place, the predominant bacteria resembling *Clostridium dissolvens* (*Bacillus cellulose dissolvens*); and (3) after about 1 month, aggregated colonies composed of many microorganisms appear, and sporulation by *Clostridium dissolvens* becomes evident.

4.3.4. N_2-Fixing Heterotrophs

Because N_2 fixation is sensitive to high O_2 tensions, reduced paddy soils appear to be a suitable medium for most anaerobic or microaerophilic N_2 fixers. Heterotrophic N_2 fixation in paddy soils was reviewed by Watanabe (1978), Yamaguchi (1979), and Matsuguchi (1979), and we shall deal with the topic briefly.

Azotobacter has been studied extensively. Yamagata (1924), Ishizawa and Toyoda (1964), and Ishizawa et al. (1975) made surveys on the distribution of N_2-fixing bacteria in Japanese paddy soils, and the numbers of *Azotobacter* were found to seldom exceed 10^4/g soil.

In Egypt and Iraq (Mahmoud et al., 1978; Hamdi et al., 1978), the population of *Azotobacter* was more than 10^5/g and sometimes up to 10^6/g. The high populations in these countries appear to be due to the alkalinity of the soils.

Derx (1950) and Becking (1961, 1978) made surveys of the distribution of *Beijerinckia* and concluded that this genus is a tropical type. Ishizawa et al. (1975) reported, however, quite high frequencies of *Beijerinckia* in Japanese wetland and dryland soils, and they raised the question of the tropical prevalence of *Beijerinckia*.

Little is known about the survival of aerobic N_2-fixing heterotrophic bacteria in anaerobic soils. Magdoff and Bouldin (1970) suggested that *Azotobacter* develops in the aerobic–anaerobic interface close to the soil surface by utilizing products of anaerobic metabolism and O_2. Further elucidation of their viewpoint is needed, however.

N_2-fixing Clostridia are more abundant than aerobic N_2 fixers. Takeda and Furusaka (1975a), however, showed that the dominant clostridia in flooded soils are proteolytic and not saccharolytic species such as *Clostridium pasteurianum*. N_2-fixing bacteria appear, however, at high population densities when an energy source is added to the flooded soil (Rice and Paul, 1972).

Yoneyama et al. (1977) showed that acetylene reduction occurred only in soil with E_h lower than -0.15 V when the flooded soil was mixed with straw. This indirectly suggested a possible contribution of strict anaerobes to N_2 fixation in the flooded soil. Because the energy supply in a flooded soil limits N_2 fixation and most of organic matter is supplied as organic debris (see Section 4.4), Wada et al. (unpublished data), in a study of the distribution of N_2 fixers on organic debris, found the predominance of N_2-fixing bacteria on the debris.

The number of N_2-fixing bacteria that are counted under aerobic or microaerophilic conditions is sometimes higher than that of *Azotobacter* or *Beijer-*

inckia (Ishizawa *et al.*, 1975; Watanabe *et al.*, 1978a). Other types of N_2 fixers may exist, but little is known about the predominant microaerophilic N_2-fixing bacteria of paddy soils.

4.4. Heterogeneous Distribution of Microorganisms and Role of Organic Debris

The characteristics of the microflora of reduced soil must be considered as the average of those of the microfloras at various microsites. Nishigaki *et al.* (1960) made multipoint measurements of soil E_h. In a highly productive paddy soil, the average soil $E_h 6$ (E_h at pH 6) was about 0.2 V, but the E_h values fluctuated widely from 0.0 to 0.5 V. In contrast, in a less productive paddy soil, the average $E_h 6$ was 0.1 V, and the E_h values fell between 0.0 and 0.2 V. The wide fluctuations in E_h values in highly productive soils indicated the heterogenous nature of the plowed reduced layer. The contiguous distribution of sulfate reducers, as discussed in Section 4.1.4, appears to present additional evidence of the heterogenous distribution. Wakao and Furusaka (1976b) assumed that the distribution of organic debris on which sulfate reducers grow determines the distribution pattern of sulfate reducers.

Wada and Kanazawa (1970) developed a technique to fractionate soil particles according to their size and density, and they found that about 30% of the organic matter in a paddy soil existed in particles larger than 37 μm. With the larger particles, organic debris in the separated particle was less decayed, was less contaminated with mineral particles, and had higher C/N ratios. Because less decayed, larger organic fragments are present in the larger soil aggregates, larger soil aggregates have higher ammonifying and Fe-reducing capacities, and higher microbial populations (Wada *et al.*, 1974). These observations led to the assumption that organic debris is an important site of microbial activities and creates heterogeneities in the microbial populations and activities in soil.

The decaying of organic debris, which appears to come from rice roots, stubble, aquatic weeds, and algae, was observed directly under the microscope (Wada, 1971, 1975). Sulfide formation, gas formation (Wada, 1974, 1976), and Mn oxidation (Wada *et al.*, 1978a) started from or near the organic debris.

4.5. Water Percolation and Microbial Activity

Most of the highly productive paddies in Japan are characterized by fairly good internal drainage (15–30 mm/day) and low water tables. To clarify the effect of water percolation on the chemistry and microbiology of flooded soil, Takai *et al.* (1968, 1969a, 1974) conducted laboratory experiments. Submerged soils with varying contents of readily decomposable organic matter were percolated either with water that was saturated with ambient air or with deoxygenated water. Comparisons were made with a nonpercolated (stagnant water) treatment.

Takai *et al.* (1974) assumed that changes in chemical and microbial properties brought about by percolation were determined by the amount of readily decomposable organic matter. If the soil contained an excess amount of readily decomposable organic matter, the supply of O_2 and removal of water-soluble substances brought about by percolation were conducive to enhanced microbial activities, increased formation of ferrous iron and ammonium, and decreased E_h of the soil. It was assumed that the removal of excess organic matter that inhibited microbial activity may enhance microbial transformations. However, if the soil contained a limited amount of readily decomposable organic matter, O_2 oxidized the upper part of the soil column, the limited amount of available organic matter was exhausted, and water-soluble substrates available to microorganisms were removed. These changes depressed microbial activities. Consequently, there was a decrease in the formation of ferrous iron and an increase in the E_h of the upper part of the soil.

The number of bacteria (plates incubated in air) was always higher in the percolated soil than in the nonpercolated one. The upper part of the soil always had a higher number of bacteria than the lower part. The stimulation of bacterial growth by percolation was more pronounced with the gram-negative bacteria than with the gram-positive or variable bacteria (Takai *et al.*, 1968, 1969b).

The removal of bacteria from the soil column was also increased by percolation. The increase in volume of percolating water brought about a greater leaching of gram-negative than gram-positive bacteria (Takai *et al.*, 1969b; Kagawa and Takai, 1969). They thought that the likeliest mechanism of preferential washing out of gram-negative bacteria was associated with their weaker adsorption to soil particles. The physiological properties of washed bacteria are also different from those of bacteria that remain at the lower part of the soil column. Most of the bacteria that were washed out were species of *Pseudomonas, Flavobacterium*, and *Caulobacter* (Kagawa and Takai, 1969). Based on these observations, Kagawa (1968) presented the view that bacteria that are capable of decomposing high-molecular-weight organic matter grow on organic matter in the soil particle–organic matter matrix and are firmly adsorbed on this matrix. Some gram-negative bacteria, which are incapable of decomposing the high-molecular-weight organic matter, are strongly adsorbed on soil particles and grow in water-filled pores from which they are easily washed out by percolating water. Sulfate reducers are also washed out, but they are most easily washed out during their logarithmic growth stage.

5. Subsoil

5.1. Microbial Enrichment as Affected by Leaching

Ishizawa and Toyoda (1964) noticed that more sulfate reducers and denitrifiers were found in subsoils of paddy soils in Japan than in subsoils of dryland soils.

As discussed in Section 4.5, bacteria, particularly gram-negative bacteria, at 10^6-10^4/ml and sulfate reducers at 10^4-10^2/ml were washed out from a soil column incubated in the laboratory. The washing out of bacteria was accompanied by the washing out of organic matter.

Based on field observations of the vertical distribution of microorganisms and laboratory experiments on the leaching of microbial mass and organic matter from the plow layer, Takai *et al.* (1970) drew attention to the presence of glistening, gray cutanic material that developed on the surface of prismatic peds in the subsoil.

Brewer (1960) defined cutanic formation as a modification of the texture, structure, or fabric at the natural surface of soil due to concentration of particular components or *in situ* modification of the plasma. The surface of peds in the subsoil was postulated as the site of microbial and organic enrichment. Takai *et al.* (1970) found that the number of bacteria, bacterial spores, gram-negative bacteria, and sulfate reducers was 10 times higher on the surface of the peds than inside the peds. The cutanic material was also higher in organic matter content than the material inside (Wada *et al.*, 1971).

The abundance of microorganisms on the surface of cutans is presumably due to the washing down of microbial cells *per se* from the plow layer and due to organic matter accumulation on the cutans.

5.2. Manganese Oxidation

Manganese oxidation takes place in the plow layer as well as in the oxidized subsoil. A wide variety of heterotrophic microorganisms can bring about Mn oxidation in a medium containing a certain amount and kind of organic matter (Bromfield, 1956). Motomura (1966) found that a medium with 0.1% peptone, 0.1% $MnSO_4$, and some inorganic salts was optimum for Mn oxidation when inoculated with a soil suspension. Using this medium, Motomura (1966) determined the most-probable-number of Mn oxidizers in soil layers having various Mn deposits. MPN values ranged from 10^3 to 10^5/g dry soil, and in most cases, they were higher in the surface of the plow layer than in the lower layers, whereas Mn deposits developed mostly at the second soil layer. Even in soils without Mn deposits *in situ*, large numbers of Mn oxidizers were observed. Therefore, the number of Mn-oxidizing bacteria is not a factor limiting Mn oxidation.

Wada *et al.* (1978a) presented the hypothesis that the initial stage of Mn oxidation takes place rapidly and results from activities of microorganisms and that the subsequent enlargement of Mn deposits occurs slowly by nonbiological means. To observe the formation of Mn deposits, they incubated (in air) films of agar composed of a suspension of reduced paddy soil and a Mn salt (Wada *et al.*, 1978b). When peptone was added to the agar film, Mn deposits developed around microbial colonies that were a distance away from soil particles. When the nutrient was not added to the agar film, Mn-oxidizing colonies developed around the organic debris, suggesting that organic debris is a site of supply of energy. From the colonies developed on the agar film with peptone, Mn-oxidizing bacteria were isolated. The isolates belonged to *Pseudomonas* and *Bacillus*. As found by Bromfield (1956), *Pseudomonas*-like bacteria oxidized Mn only after the stationary phase in batch culture.

The *in situ* role of Mn-oxidizing bacteria, which are seen in films of peptone-amended agar, has not been elucidated.

6. Root Zone of Rice

6.1. Physiological and Morphological Characteristics of Wetland Rice Roots

Van Raalte (1941) found that the rice plant has a system of air spaces through which atmospheric gas is transported from the aboveground parts to the root zone. By virtue of this system, the rhizosphere soil is oxidized, and rice roots are not suffocated. Green and Etherington (1977) observed that ferric oxide precipitates are formed not only on the surface of epidermal layers but in the intercellular space and middle lamella of cortex layers. The air transport system of rice plants can facilitate the transport of other gases such as N_2 (Yoshida and Broadbent, 1975), ethylene and acetylene (Lee and Watanabe, 1977), and methane (DeBont *et al.*, 1978) supplied to the top portion of the plant or evolved in the soil. Rice roots constitute a system through which gases are exchanged. Therefore, the rhizosphere, in addition to the surface of the paddy soil, must be considered as an air-liquid interface.

Wetland rice roots have less developed root hairs and are straighter than dryland rice roots (Kawata *et al.*, 1964a). Roots of wetland rice are more concentrated in the upper layers of soil than those of dryland rice. Organic acids formed in the reduced soil prevent the formation and reduce the longevity of root hairs (Kawata and Ishihara, 1961). Morphological and physiological differences of rice roots grown in flooded soil certainly affect the microflora in the rhizosphere.

6.2. Organic Substances Released from Rice Roots and "Reducing Power" of Rice Roots

Organic substances released from the rice root consist of root exudates, sloughed-off cells, detached mucilaginous layers, cell lysate, and cell contents released by the invasion of microorganisms. These substances can be differentiated conceptually, but it is difficult to separate them experimentally.

The release of organic substances from the roots is accelerated by the presence of microorganisms (Martin, 1977a; Barber and Lynch, 1977) and by root desiccation (Martin, 1977b). Experiments on exudate release under aseptic conditions in water culture have yielded data on only a fraction of the organic matter that is released under natural conditions. No data such as that obtained by Martin

(1977a) for dryland crops are available on the release of organic matter from the roots of wetland rice grown in nonsterile soil for a long period.

Experiments on root exudation have been reported only for young rice seedlings grown in aseptic hydroponic culture. Organic acids (Jacq, 1975; Boureau, 1977), amino acids (Andal *et al.*, 1956; MacRae and Castro, 1966; Jacq, 1975; Boureau, 1977), and carbohydrates (MacRae and Castro, 1966; Jacq, 1975; Boureau, 1977) were found in the root exudates. The excretion of enzymes from rice roots was reported by Okuda *et al.* (1964). Quantitative studies on the release of organic substances from wetland rice roots during the entire growth period are needed.

In addition to the oxidizing ability of wetland rice root, its capacity to reduce the rhizosphere was also recognized. Mitsui and Tensho (1951) drew attention to the reduction of nitrate to nitrite that took place in culture solutions, particularly when the rice plant approached the heading stage. They ascribed this capacity to "reducing power." The reduction of soil by wetland rice roots was more apparent in degraded paddy soils, which have a small amount of free iron oxide, than in normal soils (Mitsui and Tensho, 1952).

Okajima (1958) attributed the soil reduction by rice roots to physiological changes induced by nitrogen deficiency, which often occurs at the late stage of rice growth. Trolldenier (1973) reported that potassium deficiency at the middle stage of rice growth also brought about an anoxic condition in the water-culture medium, leading to lower E_h, bacterial multiplication, and iron reduction. Further, Trolldenier (1977a) found that the nitrogen and potassium deficiencies of rice grown in water, which was produced by the sudden cutoff of the nutrient supply, lowered the E_h of the culture solution more severely than the deficiency of each single nutrient.

Because the rice plant in pot experiments tends to be deficient in many elements at the late stage of growth, Okajima (1960) pointed out that the marked reduction of the soil, as effected by the rice plant, was unlikely to occur in the field. This "pot effect" must be taken into account in analyzing data obtained in pot experiments.

6.3. Characteristics of Microflora of Rice Rhizosphere

6.3.1. Comparison of the Rhizosphere of Wetland Rice and Dryland Crops

Mahmoud and Ibrahim (1970) compared the microflora of the rhizosphere soil of wetland rice grown in a lysimeter with the microflora in a flooded fallow soil. Bacteria grown aerobically—actinomycetes, *Azotobacter*, nitrifiers, and cellulose decomposers—showed a positive rhizosphere effect. On the other hand, bacterial spores, clostridia, sulfate reducers, and denitrifiers gave a negative rhizosphere effect. Mahmoud and Ibrahim (1970) presented the view that the rhizosphere effect is generally less pronounced in wetland rice than in other dryland crops. Ch'en (1963) advanced a similar view.

Kimura *et al.* (1977b,c) studied the rhizosphere effect of rice grown in flooded pots and noted that R/S values (counts of soil adhering to the rice root that was removed by forceps divided by counts of fallow soil) for many groups of microorganisms were lower than the values reported for dryland crops. Generic analysis of the isolates picked from the aerobic agar plates revealed the dominance of *Bacillus* and actinomycetes, but there was no dominance of gram-negative bacteria, which are known as rhizosphere inhabitants of dryland crops.

To test the hypothesis that the less pronounced rhizosphere effect in wetland rice is related to the moisture level of soil, Kimura *et al.* (1977d) compared the rhizosphere effect in rice seedlings grown in submerged and in dry-soil conditions. R/S values for total bacteria, gram-negative (crystal violent tolerant) bacteria, sulfate reducers, cellulose decomposers, and denitrifiers were higher for the rice grown under dry conditions than under flooded conditions.

Rangaswami and Venkatesan (1966) obtained higher R/S values for bacteria and actinomycetes in water-saturated soil than in unsaturated soil, but the number per unit weight of rhizosphere soil was the lowest in the water-saturated soil.

Peterson *et al.* (1965) discussed the less pronounced rhizosphere effect at high soil-moisture contents and speculated about the dilution of root exudates. The content of water-soluble carbohydrates and amino acids in the rhizosphere soil of rice was higher in flooded conditions than in dryland conditions (Kimura *et al.*, 1977a). Therefore, Kimura *et al.* (1977d) questioned the dilution by high moisture content.

Factors such as inefficient microbial synthesis under anaerobiosis, a presumably smaller amount of root exudates, smaller area of root surfaces as evidenced by less developed root hairs, and other such factors may explain the less pronounced rhizosphere effects of rice plants in flooded soils. Approaches based on quantitative analysis of organic matter release, bacterial growth kinetics, and electron microscopy are needed to clarify the nature of the rhizosphere microflora, as discussed by Bowen and Rovira (1976). Contrary to the data of Mahmoud and Ibrahim (1970), Kimura *et al.* (1977b) did not find a positive effect on nitrifiers. Kimura *et al.* (1977c) obtained a positive rhizosphere effect for heterotrophic Mn oxidizers, however.

At this time, no conclusive evidence can be assembled on the stimulation of aerobic chemoautotrophic bacteria in the oxidized zone of the rice rhizosphere. Joshi and Hollis (1977) presented the view that *Beggiatoa* in the rhizosphere of rice may be active in detoxifying hydrogen sulfide, but more information is needed on the ecology of *Beggiatoa* in flooded soils.

Garcia (1977a) presented data on the populations of denitrifiers in nine Senegal rice soils. The rhizosphere effect of nitrate-utilizing denitrifiers and nitrite-dependent denitrifiers was positive in some soils and negative in others. Therefore, the higher denitrifying capacity of rhizosphere soil (Garcia, 1975b) is considered to be determined by the energy supply, not by the denitrifying population.

6.3.2. Sulfate Reduction

In a paddy reclaimed from mangrove swamps and fluvial estuarine deposits in Senegal, sulfate reduction in the spermosphere or rhizosphere was cited as the cause of death of seeds or seedlings of rice (Jacq, 1973, 1975; Garcia et al., 1974). The sulfate-reducing bacteria sometimes reached a level of $10^8/g$ dry soil of the spermosphere and rhizosphere.

Exudates from seeds and roots provide energy for sulfate reduction. The presence of lactate in the exudate was confirmed (Jacq, 1975). After the seed and seedling damage, the surviving rice plant became tolerant of the sulfide damage, probably because of the development of the oxidizing ability of the root. Sulfide may be oxidized biologically, nonbiologically, or both.

After analyzing the statistical relationship between plant damage and the numbers of sulfate reducers and sulfide oxidizers (*Thiobacillus denitrificans*), Jacq and Roger (1978) advanced the hypothesis that damage to the rice plant is determined by the number of sulfate reducers relative to sulfide oxidizers in the rhizosphere. Nothing is known, however, about the cyclic oxidation and reduction of sulfur compounds in flooded soils.

6.3.3. N_2-Fixing Bacteria

The subject of N_2-fixing bacteria was reviewed by Dommergues and Rinaudo (1978), so the topic is dealt with only briefly here.

Sen (1929) suggested that N_2-fixing heterotrophs were present at the surfaces of wetland rice roots, but Sen's suggestion was long overlooked. By applying the sensitive acetylene-reduction technique, Rinaudo and Dommergues (1971) and Yoshida and Ancajas (1971) presented evidence that the wetland rice root has the ability to fix N_2 owing to bacteria living on its surface or in its tissue. Studies using $^{15}N_2$ gas have yielded direct evidence of N_2 fixation by wetland rice roots (Ito and Watanabe, unpublished data; Eskew et al., unpublished data). In both experiments, the rice plant was exposed to $^{15}N_2$ gas for a period of 7 or 10 days. Quantitative assessments of N_2 fixation associated with wetland rice during the entire growth cycle remain to be made. All of the acetylene-reduction assays (Watanabe et al., 1979), ^{15}N studies, and N balance studies (App et al., unpublished data) unequivocally showed that the N_2-fixing rate associated with rice roots would not be higher than 20% of the nitrogen requirement of the plant.

The presence of *Azotobacter* (Purushothaman et al., 1976), *Beijerinckia* (Döbereiner and Ruschel, 1961; Diem et al., 1978), *Azospirillum* (Kumari et al., 1976), Enterobacteriaceae (Watanabe et al., 1977), and *Achromobacter*-like bacteria (Watanabe and Barraquio, 1979) has been reported. The population of N_2-fixing *Clostridium* is smaller than that of aerobic N_2-fixing bacteria (Trolldenier, 1977b).

Watanabe et al. (1979) used the MPN method to determine the number of

N_2-fixing bacteria associated with wetland rice grown in the field. The soil dilution was inoculated into semisolid media containing both glucose and malate and supplemented with yeast extract. Positive tubes were detected by tests of acetylene reduction under aerobic conditions. The rhizosphere soil, rhizoplane, and histosphere (inner rhizoplane) were examined. More N_2-fixing rhizoplane bacteria were counted on the glucose medium than on the malate medium. Most of the bacteria in the malate medium resembled *Azospirillum*, but N_2-fixing bacteria on the glucose medium seemed to differ from any of the reported N_2-fixing bacteria. The colonies of aerobic heterotrophs on 0.1% tryptic soy plates inoculated with a dilution from rhizoplane and histosphere samples of rice variety IR26 were selected and tested for nitrogenase activity on yeast extract (0.1 g/liter) amended semisolid glucose medium. About 80% of the isolates gave positive results for N_2 fixation, but none of them could be grown on N-free medium. Various amino acids (0.1 g/liter) could substitute for yeast extract to derepress nitrogenase of the isolate. The bacteria require O_2 concentrations of 1% or less for N_2 fixation, and they are tentatively classified as *Achromobacter* (Watanabe and Barraquio, 1979).

Nothing is known, however, about the relative contribution of these "*Achromobacter*"-type bacteria and *Azospirillum* to N_2 fixation *in situ*.

There must be a close interaction of pectinolytic bacteria and N_2-fixing bacteria in the rhizosphere, because inoculation of sterile rice plants with N_2-fixing bacteria from rice roots and pectinolytic non-N_2-fixing bacteria greatly increased acetylene-reduction rates as compared to a single inoculation (Dommergues and Rinaudo, 1978). In studies of colonization by N_2-fixing bacteria of gnotobiotic rice plants, it was found that bacteria are embedded in mucigels around the root surface (Diem *et al.*, 1978). The surface sterilization of new roots killed N_2-fixing bacteria, but surface sterilization of basal and old roots did not kill N_2-fixing bacteria (Diem *et al.*, 1978), indicating colonization by N_2-fixing bacteria deep within old root tissues.

Nothing is known about either the organic substances needed to support N_2 fixation by associative bacteria or the sources of these substances.

6.4. Invasion of Rice Root Tissues by Microorganisms

Soil adjacent to the root (rhizosphere, in a strict sense), mucilaginous layers developed on the surface of the epidermis, intercellular spaces of the epidermis and cortex, and inner tissues of the epidermis and cortex are generally found to contain microbial colonies. These spaces and tissues provide more or less continuous media for microbial activities (Old and Nicolson, 1975).

Kawata *et al.* (1964b) observed by light microscopy surfaces or inner tissues of seemingly healthy rice roots in paddy soil were colonized by several kinds of microorganisms. A microorganism forming spherical sporangium-like globules was found only on the tips of root hairs. *Fusarium* species, yeasts, and bacteria were isolated from the roots. Miyashita *et al.* (1977) observed by light microscopy the process of microbial invasion into rice root tissues obtained from pot-grown plants. In young and vigorously growing roots, few microorganisms were found inside the root tissue. The first sites of the microbial invasion were cells of the epidermis. It was observed that an epidermal cell, which was full of microorganisms, was surrounded by microbe-free cells. After cell walls of the epidermis had disintegrated, the spreading of microorganisms in the epidermal cells occurred. Decomposition of the sclerenchyma epidermis layer was followed by the microbial invasion of inner layers (epidermis → exodermis → sclerenchyma). Bacteria were the predominant type of invading microorganisms. Sometimes fungal mycelia and spores were found. In the heavily rotted roots, actinomycetes were recognized. Protozoa were found mainly in the partially decomposed lateral roots. Black roots were already heavily colonized by many microorganisms.

7. Perspective and Conclusion

The current information available on the microbial ecology of flooded rice soils should contribute to the search for added information to improve cultural practices for increasing rice production. For this purpose, quantitative ecological information is likely to be the most valuable. The needed information includes: (1) knowledge of the nature of the system; (2) knowledge of energy and material flow (i.e., quantitative data on metabolism of the system); and (3) knowledge of the kinds and biomass of microorganisms occurring in the system.

Microbial ecology must add the following information for analyzing the ecosystem: (1) data on *in situ* physiological activities of microorganisms that take part in chemical, biochemical, and biological processes in the system; (2) physicochemical characteristics of microsites where microorganisms exist; and (3) factors that determine interactions among organisms and between microorganisms and soil components.

The principles for analysis of the microbial ecology of anaerobic digestion have been discussed by Toerien and Hattingh (1969). Quantitative descriptions of the metabolism of carbon, nitrogen, and other elements in flooded rice soils are still very limited. Ecologists who intend to describe *in situ* phenomena quantitatively should seek techniques that can be applied in rice fields. The techniques that have been applied to studies of other aquatic ecosystems and their sediments should be valuable in guiding ecological studies of flooded rice soils.

For describing material and energy metabolism in flooded rice soils, the interfaces between oxidized and reduced sites, such as the surface soil and rice rhizosphere, may be of great importance.

Determinations of the kinds and numbers of microorganisms in flooded rice soils are still dependent largely upon classical techniques. Slight modifications of such techniques often lead to discoveries of new types of microorganisms. The discovery of *Propionibacterium* (Hayashi and Furusaka, unpublished), N_2-fixing bacteria on rice roots (Watanabe and Barraquio, 1979), and nitrite-dependent denitrifiers (Garcia, 1975b) are examples. In addition to the modification of classical techniques such as platé and dilution-counting methods, the application of modern techniques such as immunofluorescence (Dommergues *et al.*, 1978) and microradiogram techniques (Hoppe, 1977) can undoubtedly add much ecological information. Applications of these quantitative and modern techniques may lead to the elucidation of *in situ* physiological activities of the resident microorganisms.

Many microbial ecologists believe that the physiological activities of microorganisms in natural environments cannot be determined simply from extrapolations based on knowledge of microbial behavior observed in laboratory-grown cultures. A model experimental approach, which simulates soil systems, can narrow the gap between our knowledge of microorganisms grown in natural ecosystems and those grown in the laboratory (Furusaka and Wakao, 1973; Hattori and Hattori, 1976).

Despite the tremendous importance of rice production in the fight against world hunger, knowledge of the role and potential of microorganisms in flooded rice soils is still very limited. Scientists who have an interest in various aspects of rice soils are not widely distributed, particularly in the developed world. To overcome these constraints, international research efforts should be organized immediately.

ACKNOWLEDGMENT. The authors are indebted to A. App and H. Wada for providing manuscripts of papers that are in press. We thank S. Hayashi, M. Saito, V. Jacq, and J. L. Garcia for collecting references and F. N. Ponnamperuma and P. A. Roger for valuable comments.

References

Alimagno, B. V., and Yoshida, T., 1977, *In situ* acetylene–ethylene assay of biological nitrogen fixation in lowland rice soils, *Plant Soil* 47:239–244.

Al-Kaisi, K. A., 1976, Contribution to the algal flora of rice fields of Southeastern Iraq, *Nova Hedwigia* 27:813–827.

Andal, R., Bhuvaneswari, K., and Subba Rao, N. S., 1956, Root exudates of paddy, *Nature (London)* 178:1063–1064.

Araragi, M., and Tangcham, B., 1975, Microflora related to the nitrogen cycle in paddy soils of Thailand, *Jpn. Agric. Res. Q.* 8:256–257.

Asami, T., and Takai, Y., 1970, Behavior of free iron oxide in paddy soils. IV. Relation between reduction of free iron oxide and formation of gases in paddy soils, *Nippon Dojohiryo Gaku Zasshi* 41:48–55 (in Japanese).

Bailey, L. D., and Beauchamp, E. G., 1971, Nitrate reduction and redox potentials measured with permanently and temporarily placed platinum electrode in saturated soils, *Can. J. Soil Sci.* 51:51–58.

Balandreau, J., Millier, C. R., and Dommergues, Y. R., 1974, Diurnal variations of nitrogenase activity in the field, *Appl. Microbiol.* **27**:662–672.

Barber, D. A., and Lynch, J. M., 1977, Microbial growth in the rhizosphere, *Soil Biol. Biochem.* **9**:305–308.

Becking, J. H., 1961, Studies on nitrogen-fixing bacteria of the genus *Beijerinckia*. I. Geographical and ecological distribution in soils, *Plant Soil* **14**:49–81.

Becking, J. H., 1978, *Beijerinckia* in irrigated rice soils, in: *Environmental Role of Nitrogen Fixing Blue Green Algae and Asymbiotic Bacteria*, Ecology Bulletin (U. Granhall, ed.), pp. 116–129, Stockholm.

Bell, R. G., 1969, Studies on the decomposition of organic matter in flooded soil, *Soil Biol. Biochem.* **1**:105–116.

Bollag, J. M., and Czlonkowski, S. T., 1973, Inhibition of methane formation in soil by various nitrogen-containing compounds, *Soil Biol. Biochem.* **5**:673–678.

Bouldin, D. R., 1968, Models for describing the diffusion of oxygen and other mobile constituents across the mud–water interface, *J. Ecol.* **56**:77–87.

Boureau, M., 1977, Application de la chromatographie en phase gazeuze a l'étude de l'exsudation racinaire du riz, *Cah. ORSTOM Ser. Biol.* **12**:75–81.

Bowen, G. D., and Rovira, A. D., 1976, Microbial colonization of plant roots, *Annu. Rev. Phytopathol.* **14**:121–144.

Brewer, R., 1960, Cutans: Their definition, recognition, and interpretation, *J. Soil Sci.* **11**:280–292.

Bromfield, S. M., 1956, Oxidation of manganese by soil microorganisms, *Aust. J. Biol. Sci.* **9**:238–252.

Bunt, J. S., 1961, Nitrogen-fixing blue-green algae in Australian rice soils, *Nature (London)* **192**:479–480.

Cappenberg, T. E., 1974a, Interrelations between sulfate-reducing and methane-producing bacteria in bottom deposits of a fresh-water lake. I. Field observations, *Antonie van Leeuwenhoek J. Microbiol. Serol.* **40**:285–296.

Cappenberg, T. E., 1974b, Interrelations between sulfate-reducing and methane-producing bacteria in bottom deposits of a fresh-water lake. II. Inhibition experiments, *Antonie van Leeuwenhoek J. Microbiol. Serol.* **40**:297–306.

Cappenberg, T. E., and Prins, R. A., 1974, Interrelations between sulfate-reducing and methane-producing bacteria in bottom deposits of a freshwater lake. III. Experiments with ^{14}C-labeled substrates, *Antonie van Leeuwenhoek J. Microbiol. Serol.* **40**:457–469.

Chen, H. K., and Chou, C., 1961, Investigation on nitrification and nitrifying organisms in rice field soils. I. Nitrification in rice field soils, *Acta Pedol. Sinica* **9**:56–64 (in Chinese).

Chen, T. W., 1974, The population of *Fusarium* spp. in paddy soil with special reference to *Fusarium moniliforme* and its pathogenicity to rice, *Proc. Natl. Sci. Coun. Taiwan* **7**:339–350 (in Chinese).

Ch'en, T. Y., 1963, Principal characteristics of rice root microflora, *Acta Microbiol. Sinica* **9**:186–192 (in Chinese).

Connel, W. E., and Patrick, W. H., Jr., 1968, Sulfate reduction in soil: Effects of redox potential and pH, *Science* **159**:86–87.

De, P. K., 1936, The problems of the nitrogen supply of rice. I. Fixation of nitrogen in the rice soils under waterlogged condition, *Indian J. Agric. Sci.* **6**:1237–1245.

De, P. K., and Boss, N. M., 1938, Study of the microbial conditions existing in rice soils, *Indian J. Agric. Sci.* **8**:487–498.

De Bont, J. A. M., and Mulder, E. G., 1976, Invalidity of acetylene reduction assay in alkane-utilizing, nitrogen-fixing bacteria, *Appl. Environ. Microbiol.* **31**:640–647.

De Bont, J. A. M., Lee, K. K., and Bouldin, D. R., 1978, Bacterial oxidation of methane in rice paddy, in: *Environmental Role of Nitrogen Fixing Blue Green Algae and Asymbiotic Bacteria*, Ecology Bulletin (U. Granhall, ed.), pp. 91–99, Stockholm.

Derx, H. G., 1950, Further researches on *Beijerinckia, Ann. Bogor.* 1:1–11.

Diem, H. G., Rougier, M., Hamad-Fares, I., Balandreau, J. P., and Dommergues, Y. R., 1978, Colonization of rice roots by diazotroph bacteria, in: *Environmental Role of Nitrogen Fixing Blue Green Algae and Asymbiotic Bacteria*, Ecology Bulletin (U. Granhall, ed.), pp. 305–311, Stockholm.

Döbereiner, J., and Ruschel, A. P., 1961, Inoculation to rice with N_2-fixing bacteria *Beijerinckia* Derx, *Rev. Bras. Biol.* 21:397–407.

Dommergues, Y. R., and Rinaudo, G., 1979, Factors affecting N_2-fixation in the rice rhizosphere, in: *Nitrogen and Rice*, pp. 241–260, The International Rice Research Institute, Los Baños, Philippines.

Dommergues, Y. R., Belser, L. W., and Schmidt, E. R., 1978, Limiting factors for microbial growth and activity in soil, in: *Advances in Microbial Ecology*, Vol. 2 (M. Alexander, ed.), pp. 49–104, Plenum Press, New York.

Focht, D. D., 1978, Microbial kinetics on nitrogen losses in paddy soils, in *Nitrogen and Rice*, pp. 119–139, The International Rice Research Institute, Los Baños, Philippines.

Focht, D. D., and Verstraete, W., 1977, Biochemical ecology of nitrification and denitrification, *Adv. Microb. Ecol.* 1:135–214.

Furusaka, C., 1968, Studies on the activity of sulfate reducers in paddy soil, *Tohoku Daigaku Nogaku Kenkyusho Iho* 19:101–184 (in Japanese).

Furusaka, C., 1978, Introduction to microbiology of a paddy soil, *Nippon Nogei Kagaku Kaishi* 52:R151–R158 (in Japanese).

Furusaka, C., and Hattori, T., 1956, Studies on the sulfate-reducing activity of several paddy soils, *Tohoku Daigaku Nogaku Kenkyusho Hokoku* 8:35–51 (in Japanese).

Furusaka, C., and Wakao, N., 1973, Distribution and chemical activity of sulfate reducing bacteria in a paddy soil, in: *Proceedings of Hydrogeochemistry*, Vol. 2 (E. Ingerson, ed.), pp. 422–435, Clark, Washington, D.C.

Furusaka, C., Hattori, T., Sato, K., Yamagishi, H., Hattori, R., Nioh, I., Nioh, T., and Nishio, M., 1969, Microbiological, chemical and physicochemical surveys of the paddy field soil, *Rep. Inst. Agric. Res. Tohoku Univ.* 20:89–101.

Gamble, T. N., Betlach, M. R., and Tiedje, J. M., 1977, Numerically dominant denitrifying bacteria from world soils, *Appl. Environ. Microbiol.* 33:926.

Garcia, J. L., 1973, Séquence des products formés au cours de la déntrification dans les sols de rizieres du Sénégal, *Ann. Microbiol. (Inst. Pasteur)* 124B:351–362.

Garcia, J. L., 1974, Réduction de l'oxyde nitreux dans les sols de rizieres du Sénégal: mesure de l'activité denitrifiante, *Soil Biol. Biochem.* 6:79–84.

Garcia, J. L., 1975a, Evaluation de la dénitrification dans les rizieres par la methode de réduction de N_2O, *Soil Biol. Biochem.* 7:251–256.

Garcia, J. L., 1975b, Effect rhizosphere du riz sur la dénitrification, *Soil Biol. Biochem.* 7:139–141.

Garcia, J. L., 1977a, Analyse de differents groups composant la microflore denitrificante des sols de rizieres du Senegal, *Ann. Microbiol. (Inst. Pasteur)* 128A:433–446.

Garcia, J. L., 1977b, Etude de la dénitrification chez une bacterie thermophile sporulee, *Ann. Microbiol. (Inst. Pasteur)* 128A:447–458.

Garcia, J. L., Raimbault, M., Jacq, V., Rinaudo, G., and Roger, P. A., 1974, Activités microbiennes dans les sols des rizieres du Sénégal: Relations avec les caracteristiques physicochimiques et influence de la rhizosphere, *Rev. Ecol. Biol. Sol.* 11:169–185.

Green, M. S., and Etherington, J. R., 1977, Oxidation of ferrous iron by rice (*Oryza sativa* L.) roots: A mechanism for waterlogging tolerance? *J. Exp. Bot.* 28:678–690.

Greenwood, D. J., and Goodman, D., 1964, Oxygen diffusion and aerobic respiration in soil spheres, *J. Sci. Food Agric.* 15:579–588.

Griffin, D. M., 1969, Soil water in the ecology of fungi, *Annu. Rev. Phytopathol.* **7**:289–310.

Griffin, D. M., 1972, *The Ecology of Soil Fungi*, p. 192, Chapman and Hall, London.

Gupta, B., 1966, Algal flora and its importance in the economy of rice fields, *Hydrobiologia* **28**:213–222.

Hamdi, Y. A., Yousef, A. N., Al-Azawi, S., Al-Tai, A., and Al-Baquari, M. S., 1978, Distribution of certain non-symbiotic nitrogen fixing organisms in Iraq soils, in: *Environmental Role of Nitrogen Fixing Blue Green Algae and Asymbiotic Bacteria*, Ecology Bulletin (U. Granhall, ed.), pp. 110–115, Stockholm.

Harrison, W. H., and Aiyer, P. A. S., 1915, The gases of swamp rice soils. II. Their utilization for the aeration of the roots of crops, *Mem. Dept. Agric. India Chem. Ser.* **4**:1–7.

Hattori, T., and Hattori, R., 1976, The physical environment in soil microbiology: An attempt to extend principles of microbiology to soil microorganisms, *Crit. Rev. Microbiol.* **4**:423–461.

Hayashi, S., Asatsuma, K., Nagatsuka, T., and Furusaka, C., 1978, Studies on bacteria in paddy soil, *Rep. Inst. Agric. Res. Tohoku Univ.* **29**:19–38.

Hirano, T., 1958, Studies on blue green algae. II. Studies on the formation of humus due to the growth of blue green algae, *Shikoku Nogyo Shikenjo Hokoku* **4**:63–74 (in Japanese).

Hiura, K., Hattori, T., and Furusaka, C., 1976, Bacteriological studies on the mineralization of organic nitrogen in paddy soils. I. Effect of mechanical disruption of soils on ammonification and bacterial number, *Soil. Sci. Plant Nutr. Tokyo* **22**:459–466.

Hiura, K., Sato, K., Hattori, T., and Furusaka, C., 1977, Bacteriological studies on the mineralization of soil organic nitrogen in paddy soils. II. The role of anaerobic isolates on nitrogen mineralization, *Soil Sci. Plant Nutr. Tokyo* **23**:201–205.

Hoppe, G., 1977, Analysis of actively metabolizing bacterial populations with autographic method, in: *Microbial Ecology of Brackish Water Environment* (G. Rheinheimer, ed.), pp. 179–197, Elsevier, Amsterdam.

Howeler, R. H., and Bouldin, D. R., 1971, The diffusion and consumption of oxygen in submerged soils, *Soil Sci. Soc. Am. Proc.* **35**:202–208.

Ichimura, S., 1954, Ecological studies on the plankton in paddy fields. I. Seasonal fluctuations in the standing crop and productivity of plankton, *Jpn. J. Bot.* **14**:269–279.

Ishimoto, M., Koyama, J., Omura, T., and Nagai, Y., 1954, Biochemical studies on sulfate-reducing bacteria. III. Sulfate reduction by cell suspension, *J. Biochem.* **41**:537–546.

Ishizawa, S., and Toyoda, H., 1964, Microflora of Japanese soils, *Nogyo Gijutsu Kenkyusho Hokoku* **14B**:204–284 (in Japanese).

Ishizawa, S., Araragi, M., and Suzuki, T., 1969, Actinomycete flora of Japanese soils. III. Actinomycete flora of paddy soils (a). On the basis of morphological, cultural and biochemical characters, *Soil Sci. Plant Nutr. Tokyo* **15**:104–112.

Ishizawa, S., Suzuki, T., and Araragi, M., 1975, Ecological study of free-living nitrogen fixers in paddy soil, in: *Nitrogen Fixation and Nitrogen Cycle* (H. Takahashi, ed.), pp. 41–50, University of Tokyo Press, Tokyo.

Jacq, V. A., 1973, Biological sulfate reduction in the spermatosphere and the rhizosphere of rice in some acid sulfate soils of Senegal, in: *Acid Sulfate Soil*, Vol. 2 (H. Dost, ed.), pp. 82–98, University of Wageningen, Holland.

Jaq, V., 1975, La sulfato-réduction en relation avec l'excretion racinaire, *Soc. Bot. Fr. Coll. Rhizosphere* **136**:136–146.

Jacq, V. A., 1977, Utilization du sulfur coated urea en riziere et production de sulfures toxiques, *Cah. ORSTOM Ser. Biol.* **13**:133–136.

Jacq, V. A., and Roger, P. A., 1978, Evaluation des risques de sulfato-réduction en rizière au moyen d'un critère microbiologique mesurable *in situ*, *Cah. ORSTOM Ser. Biol.* **13**:137–142.

Jordan, J. H., Jr., Patrick, W. H., Jr., and Willis, W. H., 1967, Nitrate reduction by bacteria isolated from waterlogged Crowley soil, *Soil Sci.* **104**:129-133.

Joshi, M. M., and Hollis, J. P., 1977, Interaction of *Beggiatoa* and rice plant: Detoxification of hydrogen sulfide in the rice rhizosphere, *Science* **195**:179-180.

Jutono, 1973, Blue-green algae in rice soils of Jogjakarta, Central Java, *Soil Biol. Biochem.* **5**:91-95.

Kagawa, H., 1968, Distribution of bacteria in paddy soils, *Tsuchi to Biseibutsu* **10**:1-8 (in Japanese).

Kagawa, H., and Takai, Y., 1969, Predominant aerobic bacteria in the water percolating through submerged paddy soil, *Nippon Dojohiryo Gaku Zasshi,* **40**:332-336 (in Japanese).

Kamura, T., and Takai, Y., 1961, The microbial reduction mechanisms of ferric iron in paddy soils (Part I), *Nippon Dojohiryo Gaku Zasshi* **32**:135-138 (in Japanese).

Kamura, T., Takai, Y., and Ishikawa, K., 1963, Microbial reduction mechanism of ferric iron in paddy soils (Part I), *Soil Sci. Plant Nutr. Tokyo* **9**:171-175.

Kawata, S., and Ishihara, K., 1961, Studies on the effects of some organic acids on the root hair formation in root of rice plants, *Nippon Sakumotsu Gakkai Kiji* **30**:72-78.

Kawata, S., Ishihara, K., and Shioya, T., 1964a, Studies on the root hairs of lowland rice plants in the upland field, *Nippon Sakumotsu Gakkai Kiji* **32**:250-253 (in Japanese).

Kawata, S., Ishihara, K., and Iizuka, K., 1964b, Some microorganisms on and in rice roots, *Nippon Sakumotsu Gakkai Kiji* **33**:164-167.

Kikuchi, E., Furusaka, C., and Kurihara, Y., 1975, Surveys of the fauna and flora in the water and soil of paddy fields, *Rep. Inst. Agric. Res. Tohoku Univ.* **26**:25-35.

Kimura, M., Wada, H., and Takai, Y., 1977a, Rhizosphere of the rice plant. I. Physico-chemical features of the rhizosphere (1), *Nippon Dojohiryo Gaku Zasshi* **48**:85-90 (in Japanese).

Kimura, M., Wada, H., and Takai, Y., 1977b, Rhizosphere of the rice plant. II. Microbiological features of the rhizosphere (1), *Nippon Dojohiryo Gaku Zasshi* **48**:91-95 (in Japanese).

Kimura, M., Wada, H., and Takai, Y., 1977c, Rhizosphere of the rice plant. III. Microbiological features of the rhizosphere (2), *Nippon Dojohiryo Gaku Zasshi* **48**:111-114 (in Japanese).

Kimura, M., Wada, H., and Takai, Y., 1977d, Rhizosphere of rice plants. V. Rhizosphere effects of rice at the nursery stage growing under upland condition, *Nippon Dojohiryo Gaku Zasshi* **48**:540-545 (in Japanese).

Kobayashi, M., and Haque, M. Z., 1971, Contribution to nitrogen fixation and soil fertility by photosynthetic bacteria, *Plant Soil, Spec. Vol.* **443**:456.

Kobayashi, M., Katayama, T., and Okuda, A., 1965a, Nitrogen fixation in mixed culture of photosynthetic bacteria (*R. capsulatus*) with other heterotrophic bacteria (3), Association with *B. subtilis, Soil Sci. Plant Nutr. Tokyo* **11**:74-77.

Kobayashi, M., Katayama, T., and Okuda, A., 1965b, Nitrogen fixation in mixed culture of photosynthetic bacteria (*R. capsulatus*) with other heterotropic bacteria, (8), Effect of light on the mixed culture of *R. capsulatus* with *B. megaterium, Soil Sci. Plant Nutr. Tokyo* **11**:200-203.

Kobayashi, M., Takahashi, E., and Kawaguchi, K., 1967, Distribution of nitrogen-fixing microorganisms in paddy soils of Southeast Asia, *Soil Sci.* **104**:113-118.

Kobo, K., and Uehara, H., 1943, Fertility increase of paddy soil during submergence, *Nippon Dojohiryo Gaku Zasshi* **17**:344-346 (in Japanese).

Koyama, T., 1963, Gaseous metabolism in lake sediments and paddy soils and the production of atmospheric methane and hydrogen, *J. Geophys. Res.* **68**:3971-3973.

Kumari, L. M., Kavimandan, S. K., and Subba Rao, N. S., 1976, Occurrence of nitrogen-fixing *Spirillum* in roots of rice, sorghum, maize and other plants, *Indian J. Exp. Biol.* **14**:638-639.

Kurasawa, H., 1956, The weekly succession in the standing crop of plankton and zoobenthos in the paddy field (Part I), *Shigen Kagaku Kenkyusho Iho* **41/42**:86-98 (in Japanese).

Kurasawa, H., 1957, The weekly succession in the standing crop of plankton and zoobenthos in the paddy field (Part II), *Shigen Kagaku Kenkyusho Iho* **45**:73-83 (in Japanese).

Lee, K. K., and Watanabe, I., 1977, Problems of acetylene reduction technique applied to water-saturated paddy soils, *Appl. Environ. Microbiol.* **34**:654-660.

Lim, G., 1967, *Fusarium* populations in rice field soils, *Phytopathology* **57**:1152-1153.

Lim, G., 1972, *Fusarium* in paddy soils of West Malaysia, *Plant Soil* **36**:47-51.

MacRae, I. C., and Castro, T. F., 1966, Carbohydrates and amino acids in the root exudates of rice seedlings, *Phyton* **23**:95-100.

MacRae, I. C., Ancajas, R. R., and Salandanan, S., 1968, The fate of nitrate nitrogen in some tropical soils following submergence, *Soil Sci.* **105**:327-334.

Magdoff, F. R., and Bouldin, D.R., 1970, Nitrogen fixation in submerged soil–sand–energy material media and the aerobic–anaerobic interface, *Plant Soil* **33**:49-61.

Mague, T. H., 1977, Ecological aspects of dinitrogen fixation by blue-green algae, in: *Agronomy and Ecology*, Vol. 4 (R. W. F. Hardy and A. H. Gibson, eds.), pp. 85-140, John Wiley & Sons, New York.

Mah, R. A., Ward, D. M., Baresi, L., and Glass, T. L., 1977, Biogenesis of methane, *Annu. Rev. Microbiol.* **31**:309-341.

Mahmoud, S. A. Z., and Ibrahim, A. N., 1970, Studies on the rhizosphere microflora of rice, *Acta Agron. Acad. Sci. Hung.* **19**:71-78.

Mahmoud, S. A. Z., El-Sawy, M., Ishae, Y.Z., and El-Safty, M. M., 1978, The effect of salinity and alkalinity on the distribution and capacity of N_2-fixation by *Azotobacter* in Egyptian soils, in: *Environmental Role of Nitrogen Fixing Blue Green Algae and Asymbiotic Bacteria*, Ecology Bulletin (U. Granhall, ed.), pp. 99-109, Stockholm.

Martin, J. K., 1977a, Factors influencing the loss of organic carbon from wheat roots, *Soil Biol. Biochem.* **9**:1-7.

Martin, J. K., 1977b, Effect of soil moisture on the release of organic carbon from wheat roots, *Soil Biol. Biochem.* **9**:303-304.

Matsuguchi, T., 1979, Factors affecting heterotrophic nitrogen fixation in submerged rice soils, in: *Nitrogen and Rice*, pp. 207-222, The International Rice Research Institute, Los Baños, Philippines.

Mitsui, S., and Tensho, K., 1951, The reducing power of the roots of growing plants as revealed by nitrite formation in the nutrient solution, *Nippon Dojohiryo Gaku Zasshi* **22**:301-307 (in Japanese).

Mitsui, S., and Tensho, K., 1952, The mechanism of nitrite formation by metabolizing plant roots, *Nippon Dojohiryo Gaku Zasshi* **23**:5-8 (in Japanese).

Mitsui, S., Kumazawa, K., and Mukai, N., 1959, The growth of the rice plant in poorly drained soil as affected by the accumulation of volatile organic acids (Part I), *Nippon Dojohiryo Gaku Zasshi* **30**:345-348 (in Japanese).

Miyashita, M., Wada, H., and Takai, Y., 1977, Decomposition of the root of the rice plant. II. Invasion by microorganisms of the rice root, *Nippon Dojohiryo Gaku Zasshi* **48**:558-563 (in Japanese).

Morishita, M., 1959, Measuring of the dispersion of individuals and analysis of the individual patterns, *Mem. Fac. Sci. Kyushu Univ. Ser. E. Biol.* **2**:215-253.

Motomura, S., 1966, Estimation of the number of manganese-oxidizing bacteria in paddy soils: Studies on the oxidative sediments in paddy soils (Part II), *Nippon Dojohiryo Gaku Zasshi* **37**:263-268 (in Japanese).

Munch, J. C., and Ottow, J. C. J., 1977, Model experiments about mechanism of bacterial iron-reduction in waterlogged soils, *Z. Pflanzenernaehr. Dueng. Bodenkd.* **140**:549-562.

Nagatsuka, T., and Furusaka, C., 1976, Effect of oxygen tension on bacterial number in a soil suspension, *Soil Sci. Plant Nutr. Tokyo* **22**:287-294.

Nishigaki, S., Shibuya, M., and Hanaoka, I., 1960, The method of measurement of soil Eh in relation to plant growth, in: *Manual of Studying Crops*, Vol. II (N. Yamada, ed.), pp. 497-540, *Nogyo Gijutsu Kyokai, Tokyo* (in Japanese).

Okajima, H., 1958, On the relationship between the nitrogen deficiency of the rice plant roots and the reduction of the medium, *Nippon Dojohiryo Gaku Zasshi* **29**:175-180 (in Japanese).

Okajima, H., 1960, Studies on the physiological function of the root system in rice plant, viewed from the nitrogen nutrition, *Tohoku Daigaku Nogaku Kenkyusho Iho* **12**:1-49 (in Japanese).

Okuda, A., Yamaguchi, M., and Kamata, S., 1957, Nitrogen-fixing microorganisms in paddy soils. III. Distribution of non-sulfur purple bacteria in paddy soils, *Soil Sci. Plant Nutr. Tokyo* **2**:131-133.

Okuda, A., Yamaguchi, M., and Kobayashi, M., 1960, Nitrogen fixation in mixed culture of photosynthetic bacteria (*Rhodopseudomonas capsulatus* species) with other heterotrophic bacteria (Part I), *Soil Sci. Plant Nutr. Tokyo* **6**:35-39.

Okuda, A., Yamaguchi, M., and Yong-Gil, S., 1964, Effect of organic substances on the growth of higher plants. I. A new device of sterile water culture apparatus and excretion of some enzymes by rice plant root, *Nippon Dojohiryo Gaku Zasshi* **35**:311-314 (in Japanese).

Old, K. M., and Nicolson, T. H.. 1975, Electron microscopical studies of the microflora of roots of sand dune grasses, *New Phytol.* **74**:51-58.

Ottow, J. C. G., 1969a, Mechanism of iron-reduction by nitrate reductase inducible aerobic microorganisms, *Naturwissenschaften* **56**:371-372.

Ottow, J. C. G., 1969b, The distribution and differentiation of iron-reducing bacteria in gley soils, *Zentralbl. Bakteriol. Parasitenkde Hyg. Abt. 2* **123**:601-615.

Ottow, J. C. G., 1969c, Effect of nitrate, chlorate, sulfate, iron oxide form and growth conditions on the extent of bacterial reduction of iron, *Z. Pflanzenernaehr. Dueng. Bodenkd.* **124**:238-253.

Ottow, J. C. G., 1970, Selection, characterization and iron-reducing capacity of nitrate reductaseless (nit⁻) mutants from iron-reducing bacteria, *Z. Allg. Mikrobiol.* **10**:55-62.

Ottow, J. C. G., 1971, Iron reduction and gley formation by nitrogen-fixing clostridia, *Oecologia (Berlin)* **6**:164-175.

Ottow, J. C. G., and Glathe, H., 1971, Isolation and identification of iron-reducing bacteria from gley soils, *Soil Biol. Biochem.* **3**:43-55.

Ottow, J. C. G., and Glathe, H., 1973. Pedochemie and Pedomikrobiologie hydromorphor, *Bodenkd. Chem. Erde* **32**:1-37.

Ottow, J. C. G., and von Klopotek, A., 1969, Enzymatic reduction of iron oxide by fungi, *Appl. Microbiol.* **18**:41-43.

Pandey, D. C., 1965, A study of the algae from paddy soils of Ballia and Ghazipur districts of Uttar Pradesh, India. I. Cultural and ecological consideration, *Nova Hedwigia* **9**:229-334.

Pantastico, J. B., and Suayan, Z. A., 1973, Algal succession in the rice-fields of College and Bay, Laguna, *Philipp. Agric.* **57**:313-326.

Patrick, W. H., 1960, Nitrate reduction rates in a submerged soil as affected by redox potential, *Trans. 7th Int. Cong. Soil Sci.* **2**:494-500.

Patrick, W. H., and Gotoh, S., 1974, The role of oxygen in nitrogen loss from flooded soils, *Soil Sci.* **118**:78-81.

Patrick, W. H., and Reddy, K. R., 1977, Fertilizer nitrogen reactions in flooded soil, in: *Proceedings of the International Symposium, Soil Environment and Fertilizer Management in Intensive Agriculture*, pp. 275-280, Japanese Society of Soil Science, Tokyo.

Patrick, W. H., and Turner, F. T., 1968, Effect of redox potential on manganese transformation in waterlogged soil, *Nature (London)* 220:476–478.

Patrick, W. H., and Tusneem, M. E., 1972, Nitrogen loss from flooded soil, *Ecology* 53:732–737.

Pearsall, W. H., and Mortimer, C. H., 1939, Oxidation–reduction potentials in waterlogged soils, natural water and muds, *J. Ecol.* 27:485–501.

Peterson, E. A., Rouatt, J. W., and Katznelson, H., 1965, Microorganisms in the root zone in relation to soil moisture, *Can. J. Microbiol.* 11:483–489.

Pfennig, N., and Biebl, H., 1976, *Desulforomonas acetoxidans*, gen. nov. and sp. nov., a new anaerobic, sulfur-reducing, acetate-oxidizing bacterium, *Arch. Microbiol.* 110:3–12.

Pichinoty, F., Mandel, M., Greenway, B., and Garcia, J. L., 1977, Isolation and properties of a denitrifying bacterium related to *Pseudomonas lemoignei*, *Int. J. Syst. Bacteriol.* 27:346–348.

Ponnamperuma, F. N., 1955, The Chemistry of Submerged Soils in Relation to the Growth and Yield of Rice, Ph.D. Thesis, Cornell University, Ithaca, New York.

Ponnamperuma, F. N., 1972, Chemistry of submerged soils, *Adv. Agron.* 24:29–96.

Ponnamperuma, F. N., Bradfield, R., and Peech, M., 1956, The chemistry of submerged soils in relation to the growth of rice, *Proc. 6th Int. Soil Sci. Soc.* R4:503.

Postgate, J. R., and Campbell, L. L., 1966, Classification of *Desulfovibrio* species, the non-sporulating sulfate-reducing bacteria, *Bacteriol. Rev.* 30:732–738.

Purushothaman, D., Oblisami, G., and Balasun, C. S., 1976, Nitrogen fixation by *Azotobacter* in rice rhizosphere, *Madras Agric. J.* 63:595–599.

Rangaswami, G., and Narayanaswami, R., 1965, Studies on the microbial population of irrigation water in rice field, *Int. Rice Comm. News.* 14:35–42.

Rangaswami, G., and Venkatesan, R., 1966, *Microorganisms in Paddy Soil*, p. 192, Annamalai University, Madras, India.

Reddy, K. R., and Patrick, W. H., Jr., 1976, A method for sectioning saturated soil cores, *Soil Sci. Soc. Am. J.* 40:611–612.

Reddy, K. R., Patrick, W. H., Jr., and Philipps, R. E., 1976, Ammonium diffusion as a factor in nitrogen loss from flooded soils, *Soil Sci. Soc. Am. J.* 40:528–533.

Reynaud, P. A., and Roger, P. A., 1977, Milieu selectifs pour la numeration des algues eucaryotes, procaryotes et fixatrices d' azote, *Rev. Ecol. Biol. Sol* 14:421–428.

Reynaud, P. A., and Roger, P. A., 1978a, N_2 fixing algal biomass in Senegal rice fields. in: *Environmental Role of Nitrogen Fixing Blue Green Algae and Asymbiotic Bacteria*, Ecology Bulletin (U. Granhall, ed.), pp. 148–157, Stockholm.

Reynaud, P. A., and Roger, P. A., 1978b, Vertical distribution of algae and acetylene-reducing activity in an algae mat on a sandy waterlogged tropical soil, in: *Limitations and Potentials for Biological Nitrogen Fixation in the Tropics* (J. Döbereiner, ed.), pp. 346–347, Plenum Press, New York.

Reynolds, D. R., 1970, Fungi isolated from rice paddy soil at Central Exp. Station, UP College of Agriculture, *Philipp. Agric.* 54:55–59.

Rice, W. A., and Paul, E. A., 1972, The organisms and biological process involved in symbiotic nitrogen fixation in waterlogged soil amended with straw, *Can. J. Microbiol.* 18:715–723.

Rinaudo, G., and Dommergues, Y., 1971, Valedité de l'estimation de la fixation biologique de l'azote dans la rhizosphere par la methode de reduction de l'acetylene, *Ann. Inst. Pasteur (Paris)* 121:93–99.

Roger, P. A., and Reynaud, P. A., 1976, Dynamique de la population algael an cours d'un cycle de culture dans une rizière sahelienne, *Rev. Ecol. Biol. Sol* 13:545–560.

Roger, P. A., and Reynaud, P. A., 1977, La biomass algae dans les rizieres du Senegal: importance relative des Cyanophycées fixatrices de N_2, *Rev. Ecol. Biol. Sol* 14:519–530.

Roger, P. A., and Reynaud, P. A., 1978, Ecology of blue-green algae in paddy fields, in:

Nitrogen and Rice, pp. 287–310, The International Rice Research Institute, Los Baños, Philippines.

Saito, M., and Watanabe, I., 1978, Organic matter production in rice field flood water, *Soil Sci. Plant Nutr. Tokyo* 24:427–444.

Saito, M.,Wada, H., and Takai, Y., 1977a, Microbial ecology of cellulose decomposition in paddy soils. I. Modification of Tribe's cellophane film method and staining methods for the observation of microorganisms growing on cellulose material, *Nippon Dojohiryo Gaku Zasshi* 48:313–317 (in Japanese).

Saito, M., Wada, H., and Takai, Y., 1977b, Microbial ecology of cellulose decomposition in paddy soils. II. Succession of microorganisms growing on cellulose material, *Nippon Dojohiryo Gaku Zasshi* 48:318–322.

Sen, M. A., 1929, Is bacterial association a factor in nitrogen assimilation by rice plants?, *Agric. J. India* 24:229–231.

Sewell, G. W. F., 1965, The effect of altered physical condition of soil on biological control, in: *Ecology of Soil-borne Plant Pathogens*–Prelude to Biological Control (K. F. Baker *et al.*, eds.), pp. 479–494, University of California Press, Berkeley.

Shchapova, L. N., 1971, Microflora of rice soils of Primore, *Mikrobiologiya* 40:702–706.

Shioiri, M., 1941, On denitrification in the paddy field, *Kagaku* 11:24–29 (in Japanese).

Singh, R. N., 1961, Role of Blue Green Algae in Nitrogen Economy of Indian Agriculture, p. 175, Indian Council for Agricultural Research, New Delhi.

Sivasithamparam, K., 1971, Survival of seed borne fungi in submerged soils, *Trop. Agric. Ceylon* 127:85–92.

Sørensen, J., 1978, Occurrence of nitric and nitrous oxides in a coastal marine sediment, *Appl. Environ. Microbiol.* 36:809–813.

Stover, R. H., Thornton, N. C., and Dunlap, V. C., 1953, Flood-fallowing for eradication of *Fusarium oxysporum f. cubense*: I. Effect of flooding on fungus flora of clay loam soils in Ulua valley, Honduras, *Soil Sci.* 76:225–238.

Suleimanova, S. I., 1971, Denitrifying bacteria in the soils of Kazakhstan rice paddies, *Izv. Akad. Nauk Kaz. SSR Ser. Biol.* 9:31–35.

Suzuki, T., 1967, Characteristics of microorganisms in paddy field soils, *Jpn. Agric. Res. Q.* 2:8–11.

Taha, S. M., Mahmoud, S. A. Z., and Ibrahim, A. N., 1967, Microbiological and chemical properties of paddy soil, *Plant Soil* 26:33–48.

Takai, Y., 1969, The mechanism of reduction in paddy soil, *Jpn. Agric. Res. Q.* 4:20–23.

Takai, Y., 1970, The mechanism of methane fermentation in flooded paddy soil, *Soil Sci. Plant Nutr. Tokyo* 16:238–244.

Takai, Y., 1978, Reduction mechanism of paddy soils, in: *Suidendojogaku* (K. Kawaguchi, ed.), pp. 23–55, Kodansha, Tokyo (in Japanese).

Takai, Y., and Kamura, T., 1969, The mechanism of reduction in waterlogged paddy soil, *Folia Microbiol.* 11:304–313.

Takai, Y., and Tezuka, C., 1971, Sulfate-reducing bacteria in paddy and upland soils, *Nippon Dojohiryo Gaku Zasshi* 42:145–151.

Takai, Y., and Uehara, Y., 1973, Nitrification and denitrification in the surface layer of submerged soils. I. Oxidation–reduction condition, nitrogen transformation and bacterial flora in the surface and deeper layers of submerged soils, *Nippon Dojohiryo Gaku Zasshi* 44:463–470.

Takai, Y., Koyama, T., and Kamura, T., 1956, Microbial metabolism in reduction process of paddy soil (Part I), *Soil Plant Food Tokyo* 2:63–66.

Takai, Y., Koyama, T., and Kamura, T., 1963a, Microbial metabolism in reduction process of paddy soils. III. Effect of iron and organic matter on the reduction process (2), *Soil Sci. Plant Nutr. Tokyo* 9:207–211.

Takai, Y., Koyama, T., and Kamura, T., 1963b, Microbial metabolism in reduction process of paddy soils. II. Effect of iron and organic matter on the reduction process (1), *Soil Sci. Plant Nutr. Tokyo* 9:176-180.

Takai, Y., Kagawa, H., and Kobo, K., 1968, Movement of bacteria by water percolation in submerged paddy soils (Part I), *Nippon Dojohiryo Gaku Zasshi* 39:219-223.

Takai, Y., Koyama, T., and Kamura, T., 1969a, Effects of rice plant roots and percolating water on the reduction process of flooded paddy soil in pot. V. Microbial metabolism in reduction process of paddy soils, *Nippon Dojohiryo Gaku Zasshi* 40:15-19 (in Japanese).

Takai, Y., Kagawa, H., and Kobo, K., 1969b, Movement of bacteria by water percolation in submerged paddy soils. II. On the behavior of aerobic Gram-negative bacteria, *Nippon Dojohiryo Gaku Zasshi* 40:358-363 (in Japanese).

Takai, Y., Shimazu, T., Yoshida, H., Kagawa, H., Kondo, H., and Wada, H., 1970, Enrich-ment of bacteria on the surface, a cutanic material, of prismatic peds developed in paddy subsoil (Part I), *Nippon Dojohiryo Gaku Zasshi* 41:401-405 (in Japanese).

Takai, Y., Wada, H., Kagawa, H., and Kobo, K., 1974, Microbial mechanism of effects of water percolation on E_h, iron, and nitrogen transformation in the submerged paddy soils, *Soil Sci. Plant Nutr. Tokyo* 20:33-45.

Takeda, K., and Furusaka, C., 1970, Studies on the bacteria isolated anaerobically from paddy field soil, *Rep. Inst. Agric. Res. Tohoku Univ.* 21:1-22.

Takeda, K., and Furusaka, C., 1975a, Studies on the bacteria isolated anaerobically from paddy field soil. III. Production of fatty acids and ammonia by *Clostridium* species, *Soil Sci. Plant Nutr. Tokyo* 21:113-118.

Takeda, K., and Furusaka, C., 1975b, Studies on the bacteria isolated anaerobically from paddy field soil. IV. Model experiments on the production of branched-chain fatty acids, *Soil Sci. Plant Nutr. Tokyo* 21:119-127.

Toerien, D. F., and Hattingh, W. H. J., 1969, Anaerobic digestion. I. The microbiology of anaerobic digestion, *Water Res.* 3:385-408.

Tribe, H. T., 1957, Ecology of microorganisms in soils as observed during their development upon buried cellulose film, in: *Microbial Ecology* (R. E. O. Williams, ed.), pp. 287-298, Cambridge University Press, Cambridge.

Trimble, R. B., and Ehrlich, H. L., 1968, Bacteriology of manganese nodules. III. Reduction of MnO_2 by two strains of nodule bacteria, *Appl. Microbiol.* 16:695-702.

Trimble, R. B., and Ehrlich, H. L., 1970, Bacteriology of manganese nodules. IV. Induction of an MnO_2-reductase system in a marine *Bacillus*, *Appl. Microbiol.* 19:966-972.

Trolldenier, G., 1973, Secondary effects of potassium and nitrogen nutrition of rice: Change in microbial activity and iron reduction in the rhizosphere, *Plant Soil* 38:267-279.

Trolldenier, G., 1977a, Mineral nutrition and reduction processes in the rhizosphere of rice, *Plant Soil* 47:193-202.

Trolldenier, G., 1977b, Influence of some environmental factors on nitrogen fixation in the rhizosphere of rice, *Plant Soil* 47:203-217.

Tu, C. C., 1975, Fusarium wilt suppressive soil in Taiwan and the existence of *Fusarium oxysporum f. sp. lini* in rice root, *Natl. Sci. Coun. Monogr. (Taiwan)* 3(9):20-36.

Uehara, Y., Wada, E., and Takai, Y., 1978, Nitrification and denitrification in the surface layers of submerged soils, *Proc. 11th Int. Soil Sci. Soc. Cong., Edmonton, Canada* 1: 299-300.

Van Cleemput, O., Patrick, W. H., Jr., and McIlhenny, R. C., 1975, Formation of chemical and biological denitrification products in flooded soil at controlled pH and redox po-tential, *Soil Biol. Biochem.* 7:329-332.

Van Cleemput, O., Patrick, W. H., Jr., and McIlhenny, R. C., 1976, Nitrite decomposition in flooded soil under different pH and redox potential conditions, *Soil Sci. Soc. Am. J.* 40:55-60.

Van Raalte, M H., 1941, On the oxygen supply of rice roots, *Ann. Bot. Garden Buitenzorg* 50:99–114.

Venkataraman, G. S., 1975, The role of blue-green algae in tropical rice cultivation. in: *Nitrogen Fixation by Free-living Microorganisms* (W. D. P. Stewart, ed.), pp. 207–218, Cambridge University Press, London.

Venkatesan, R., and Rangaswami, G., 1965, Studies on the microbial populations of paddy soil as influenced by moisture percentage and rice crop, *Indian J. Exp. Bot.* 3:30–31.

Vostrov, I. S., and Dolgikh, Y. R.. 1970, Microflora of submerged soils of the rice field, *Izv. Akad. Naukkaz. SSR Ser. Biol.* 1:64–69 (in Russian).

Wada, H., 1971, Stereoscopic observation of paddy soils (Part I), *Nippon Dojohiryo Gaku Zasshi* 42:421–428 (in Japanese).

Wada, H., 1974, Chemical and biological changes at micro-site of submerged soils. I. Generation and disappearance of bubbles and biological reduction of TTC, *Nippon Dojohiryo Gaku Zasshi* 45:435–440 (in Japanese).

Wada, H., 1975, Micropedological approach to the study of dynamics of paddy soils, *Jpn. Agric. Res. Q.* 9:24–29.

Wada, H., 1976, Chemical and biological changes at micro-sites of submerged soil. II. Formation of ferrous sulfide and reduction of NTB and TB, *Nippon Dojohiryo Gaku Zasshi* 47:109–113 (in Japanese).

Wada, H., and Kanazawa, S., 1970, Method of fractionation of soil organic matter according to its size and density (Part I), *Nippon Dojohiryo Gaku Zasshi* 41:273–280 (in Japanese).

Wada, H., Yoshida, H., and Takai, Y., 1971, Cutans developed in subsurface horizons of clayey paddy soils (Part I), *Nippon Dojohiryo Gaku Zasshi* 42:12–17 (in Japanese).

Wada, H., Ishii, H., and Takai, Y., 1974, Distribution pattern of substances and microorganisms in connection with aggregates of different sizes of paddy soil. II. Content of chemical substances and population of microorganisms, *Nippon Dojohiryo Gaku Zasshi* 45:208–212 (in Japanese).

Wada, H., Seiyarosakol, A., Kimura, M., and Takai, Y., 1978a, The process of manganese deposition in paddy soils. I. A hypothesis and its verification, *Soil Sci. Plant Nutr. Tokyo* 24:55–62.

Wada, H., Seiyarosakol, A., Kimura, M., and Takai, Y., 1978b, The process of manganese deposition in paddy soils. II. Microorganisms responsible for manganese deposition, *Soil Sci. Plant Nutr. Tokyo* 24:319–325.

Wakao, N., and Furusaka, C., 1972, A new agar plate method for the quantitative study of sulfate-reducing bacteria in soil, *Soil Sci. Plant Nutr. Tokyo* 18:39–44.

Wakao, N., and Furusaka, C., 1973, Distribution of sulfate-reducing bacteria in paddy-field soil, *Soil Sci. Plant Nutr. Tokyo* 19:47–52.

Wakao, N., and Furusaka, C., 1976a, Presence of microaggregates containing sulfate-reducing bacteria in a paddy-field soil, *Soil Biol. Biochem.* 8:157–159.

Wakao, N., and Furusaka, C., 1976b, Influence of organic matter on the distribution of sulfate-reducing bacteria in a paddy-field soil, *Soil Sci. Plant Nutr. Tokyo* 22:203.

Wakao, H., Hattori, T., and Furusaka, C., 1973, Study on the distribution patterns of sulfate reducing bacteria in a paddy-field soil by $I\delta$-index, *Soil Sci. Plant Nutr. Tokyo* 19:201–203.

Watanabe, I., 1974, A statistical study on the relationship between *Nitrosomonas* population and nitrifying activity of soil, *Nippon Dojohiryo Gaku Zasshi* 45:279–284 (in Japanese).

Watanabe, I., 1978, Biological nitrogen fixation in rice soils, in: *Soil and Rice*, pp. 465–478, The International Rice Research Institute, Los Baños, Philippines.

Watanabe, I., and Barraquio, W. L., 1979, Low levels of fixed nitrogen required for isolation of free-living N_2-fixing organisms from rice roots, *Nature (London)* 277:565.

Watanabe, I., and Cholitkul, W., 1979, Field studies on nitrogen fixation in paddy soils. in: *Nitrogen and Rice*, pp. 223-239, The International Rice Research Institute, Los Baños, Philippines.

Watanabe, I., Lee, K. K., Alimagno, B. V., Sato, M., Del Rosario, D. C., and De Guzman, M. R., 1977, Biological Nitrogen Fixation in Paddy Fields Studied by *in situ* Acetylene-Reduction Assays, IRRI Research Paper Series, No. 3, p. 9, The International Rice Research Institute, Los Baños, Philippines.

Watanabe, I., Lee, K. K., and Alimagno, B. V., 1978a, Seasonal change of N_2-fixing rate in rice field assayed by *in situ* acetylene reduction technique. I. Experiments in long term fertility plots. *Soil Sci. Plant Nutr. Tokyo* **24**:1-14.

Watanabe, I., Lee, K. K., and De Guzman, M. R., 1978b, Seasonal change of N_2-fixing rate in rice field assayed by *in situ* acetylene reduction technique. II. Estimate of nitrogen fixation associated with rice plants, *Soil Sci. Plant Nutr. Tokyo* **24**:465-472.

Watanabe, I., Barraquio, W., De Guzman, M. R., and Cabrera, D. A., 1979, Nitrogen-fixing (acetylene reduction) activity and population of aerobic heterotrophic nitrogen-fixing bacteria associated with wetland rice, *Appl. Environ. Microbiol.* **37**:813-819.

Widdel, F., and Pfennig, N., 1977, A new anaerobic, sporing, acetate-oxidizing, sulfate-reducing bacterium, *Desulfotomaculum* (*emend.*) *acetoxidans*, *Arch. Microbiol.* **112**: 119-122.

Winfrey, M. R., and Zeikus, J. G., 1977, Effect of sulfate on carbon and electron flow during microbial methanogenesis in freshwater sediments, *Appl. Environ Microbiol.* **33**: 275-281.

Yamagata, U., 1924, On the distribution of *Azotobacter* in relation to the reaction of soils in Japan, *Nippon Nogei Kagaku Kaishi* **1**:85-126 (in Japanese).

Yamaguchi, M., 1979, Biological nitrogen fixation in flooded rice field in: *Nitrogen and Rice*, pp. 193-204, The International Rice Research Institute, Los Baños, Philippines.

Yamane, I., 1957, Nitrate reduction and denitrification in flooded soils, *Soil Plant Food Tokyo* **3**:100-107.

Yamane, I., and Sato, K., 1964, Decomposition of glucose and gas formation in flooded soil, *Soil Sci. Plant Nutr. Tokyo* **10**:127-133.

Yamane, I., and Sato, K., 1967, Effect of temperature on the decomposition of organic substances in flooded soil, *Soil Sci. Plant Nutr. Tokyo* **13**:94-100.

Yoneyama, T., Lee, K. K., and Yoshida, T., 1977, Decomposition of rice residues in tropical soils. IV. The effect of rice straw on nitrogen fixation by heterotrophic bacteria in some Philippine soils, *Soil Sci. Plant Nutr. Tokyo* **23**:287-295.

Yoshida, K., 1975, The reduction mechanism of manganese in paddy soils. VIII. The role of ferrous iron in manganese reduction in waterlogged paddy soils, *Nippon Dojohiryo Gaku Zasshi* **46**:458-461 (in Japanese).

Yoshida, T., 1975, Microbial metabolism of flooded soils, in: *Soil Biochemistry*, Vol. 3 (E. A. Paul and A. D. MacLaren, eds.), pp. 83-122, Marcel Dekker, New York.

Yoshida, T., and Ancajas, R. R., 1971, Nitrogen fixation by bacteria in the root zone of rice, *Soil Sci. Soc. Am. Proc.* **35**:156-157.

Yoshida, K., and Kamura, T., 1972a, The reduction mechanisms of manganese in paddy soils. II. Role of microorganisms in the reduction process of manganese, *Nippon Dojohiryo Gaku Zasshi* **43**:447-450 (in Japanese).

Yoshida, K., and Kamura, T., 1972b, The reduction mechanism of manganese in paddy soils. III. Manganese-reducing microorganisms in soils and the reduction mechanisms of manganese in the culture solution, *Nippon Dojohiryo Gaku Zasshi* **43**:451-455 (in Japanese).

Yoshida, T., and Broadbent, F. E., 1975, Movement of atmospheric nitrogen in rice plants, *Soil Sci.* **120**:288-291.

Yoshida, K., and Kamura, T., 1975a, The reduction mechanism of manganese in paddy soils. VI. The reaction conditions of manganese reduction by metabolic products of microorganisms, *Nippon Dojohiryo Gaku Zasshi* **46**:377–381 (in Japanese).

Yoshida, K., and Kamura, T., 1975b, The reduction mechanism of manganese in paddy soils. VII. Model experiments on the role of ferrous iron in manganese reduction, *Nippon Dojohiryo Gaku Zasshi* **46**:382–388 (in Japanese).

Zeikus, J. G., 1977, The biology of methanogenic bacteria, *Bacteriol. Rev.* **41**:514–541.

5

Microbial Oxidation of Organic Matter of Histosols

ROBERT L. TATE III

1. Introduction

Although deposits of organic soils (Histosols) are found more extensively in the cold, moist portions of the northern hemisphere, especially in regions of past glaciation, this type of soil occurs throughout the world. About 60% of the deposits are located in the USSR (Farnham and Finney, 1965). Of the approximately 55–77 million hectares located within the U.S., nearly 41 million hectares are located in Alaska (Davis and Lucas, 1959). The largest known tract of Histosols in the world is located in the Florida Everglades. This region consists of approximately 8×10^5 ha of peat and muck soils. Prior to drainage for agricultural development, the Everglades was approximately 65 km $\times$ 160 km in area and extended from the base of Lake Okeechobee south through the center of the Florida peninsula and intersected with the ocean at the tip of Florida (Stephens, 1969).

Although large portions of the peat deposits throughout the world remain undeveloped, drainage of the soils as has occurred in the Everglades has primarily allowed for the agricultural development of the soils. Histosols are rich soils for the growth of truck crops, sugarcane, and sod as well as for several minor crops such as mint, cranberries, and blueberries. Drainage of the soil results in the process common to all drained organic soils which spells the ultimate demise of agriculture on Histosols. This process is soil subsidence. When the soils are flooded, the organic matter is protected by, among other factors, the anaerobic condi-

ROBERT L. TATE III • University of Florida, Agricultural Research and Education Center, Belle Glade, Florida 33430. This work is Florida Agricultural Experiment Stations Journal Series No. 1933.

tions imposed by the high water content. When the soils are drained, the soil organic matter again becomes available to the microbial community; thus, microbial oxidation of the carbon reserves is again initiated. Since a major portion of the soil mass consists of organic matter, conversion of the insoluble carbon to gaseous and water-soluble products results in a loss of soil mass and a parallel loss in soil elevation.

Because of the moisture and organic matter contents of Histosols, these soils provide a unique and varied environment for the microbe. As opposed to mineral soils which, in general, contain less than 5 or 6% organic matter (Buckman and Brady, 1969), organic soils are defined to be soils containing greater than 20 to 30% soil organic matter (the percentage varies depending upon the clay content of the mineral fraction) and having a minimum depth of the high organic matter soil of at least 30 cm (McCollum *et al.*, 1976; Farnham and Finney, 1965). The high organic matter content of these soils results from the accumulation of partially decayed plant and animal remains under waterlogged conditions. A portion of the diversity in the Histosols is provided by the variation in moisture from the flooded state of the peat bogs and marshes to the drained conditions typical of the agricultural soils. Given and Dickinson (1975) recently reviewed the microbiology of flooded peats. Accordingly, this review will concentrate on the microbial transformations of drained peats and mucks.

The microbiological changes during the transition from the flooded to the drained state, the biological and physical factors limiting the microbial oxidation of the soil organic matter in drained Histosols, and the biochemical reactions involved in the oxidation of the soil organic matter will be examined. Since the majority of the Histosols which have been drained were developed for agriculture, the bulk of the data describing microbial oxidation of the soil organic matter is necessarily biased in favor of that type of soil which will provide adequate growth of a crop. Thus, those limiting factors and reactions which shall be discussed will be related to soils which are of a more neutral pH and are in a well-drained state for proper crop development. The objective of this presentation is not to exhaustively review the literature describing microbial transformations of organic soils, but rather it is to present an idea of the state of the art of research in this area and to provide a general description of the microbial properties of this interesting environment.

1.1. Subsidence—The Process

When peats are first drained, the most dramatic—and, in many of the early studies, totally unexpected—occurrence (see review of Powers, 1932) is the subsidence of the Histosol. When the soils are drained, as was indicated above, both the protective effect of the water on the organic matter and the natural buoyancy resulting from the high water content are lost. Therefore, the soil elevation declines (subsides). Shortly after drainage, the majority of the loss of the soil elevation results from physical factors such as the loss of buoyancy, compaction, and shrinkage due to drying; but after this initial phase is completed, the role of the microbial community becomes paramount. It has been estimated that the microbial oxidation of the soil organic matter contributes 58 to 73% of the loss of surface elevation in the Everglades' mucks as they currently exist (Volk, 1972). These values were obtained by a comparison of carbon dioxide evolution rates with known subsidence rates. Schothorst (1977), after comparing bulk densities of the organic matter of low-moor peat soils in the western Netherlands at different levels of the soil profile, concluded that over the last 1000 years 15% of the subsidence could be attributed to compaction. He therefore ascribes 85% of the subsidence to the oxidation of the organic matter. The magnitude of the effect of the microbial oxidation on the soil elevation is dramatically demonstrated in Fig. 1. The pole depicted, which is located in the Everglades at the Agricultural Research and Education Center, Belle Glade, Florida, was sunk to bedrock so that the surface of the pole was at ground level on April 3, 1924 (Stephens, 1969; Thomas, 1965). The picture was taken in April 1979. In the 55 years between placement and the current observation, 152 cm of soil was lost. Chains were placed around the pole so that compaction by traffic was avoided. Currently the soils of the Everglades are subsiding at 3 cm/year. Comparable

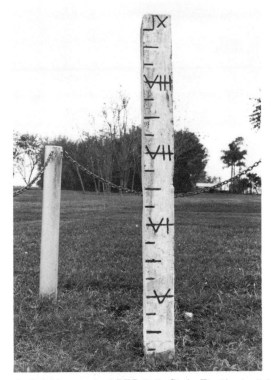

Figure 1. Subsidence pole, AREC, Belle Glade, Florida, April 1979.

rates have been recorded elsewhere: 2.07 cm/year in southern Quebec (Millette, 1976), 1.75 cm/year in the western Netherlands (Schothorst, 1977), and 7.6 cm/year in the California delta (Weir, 1950). The rate in California was augmented by the frequent burning of the surface layers of the muck for weed and parasite control. As was reflected in these rates, subsidence rates are decreased in temperate zones by the freezing of the soil during the winter months. The significance of this loss of rich soil is emphasized by comparison of the subsidence rate with the time required for deposition of peat. Peat accumulates at a rate of approximately 30 cm/500 years (Broadbent, 1960; Dawson, 1956). Thus, the muck lost each year in the Everglades required approximately 50 years to accumulate.

Currently, the best means of controlling this loss of soil is by returning a portion of the soil profile to its original flooded state. Several workers (Clayton, 1938; Neller, 1944; Ellis and Morris, 1945; Stephens, 1969) have demonstrated that the subsidence rate of muck soils is linearly proportional to the depth of the water table. Thus, a doubling of the depth to the water table (from the surface of the soil) results in a doubling of the subsidence rate. Since this relationship of water table and subsidence rate was derived from manipulation of the water tables at depths of 30 to 60 cm, considering that the microbial flora are the primary causal agents of subsidence, some interesting relationships between the microbial activity and position in the soil profile are implied (see Section 2.2).

1.2. Products of Subsidence

Peat accumulation represents a break in the carbon and nitrogen cycles. Removal of the inhibition to microbial oxidation of the soil organic matter by draining the swampy soil could be described as repairing the break. The impact of such a repair on the global carbon cycle is substantial when the production of carbon dioxide by the microbial community is considered. This can be exemplified by consideration of the carbon-oxidation rates of Pahokee muck in the Florida Everglades. The soil consists of approximately 45% carbon (Terry, 1980). With an average bulk density in the soil profile above the water table of 0.25 g/cm^3 soil, approximately 1.1×10^4 kg C/ha/year is produced based on a subsidence rate of 1 cm/year. Since the average subsidence rate is currently 3 cm/year, about 3.3×10^4 kg C/ha/year is returned to the atmosphere. About 4×10^5 ha of these soils in the region are under cultivation so that this subsidence process results in an annual production of approximately 1.3×10^{10} kg C/year—a sizable return of fixed carbon to the atmosphere in the south Florida region alone. Comparison of these values with the carbon production by other microbial soil communities further emphasizes the importance of the Histosols to the carbon cycle (Table I). The above carbon evolution rate is equivalent to 3300 g C/m^2/year. Soils from tropical forests are estimated to evolve 405 to 2117 g C/m^2/year (Schlesinger, 1977). This latter figure includes root respiration of the plants growing on the soil, whereas that for the muck includes simply the

Table I. Carbon Dioxide Evolution from
Soil Ecosystems

Ecosystem	g C/m^2/yr
Pahokee muck	3300
Tropical forest	405-2117[a]
Temperate forest	171-1414[a]
Tropical savanna	515-785[a]
Temperate grassland	150-452[a]
Desert	22[a]

[a]After Schlesinger (1977).

carbon mineralized from the soil organic matter. Greater contrast in the figures is obtained by comparison of the carbon yield from temperate grasslands. These soils produce 150-452 g C/m^2/year.

A more immediate impact of the mineralization of the organic matter of drained Histosols results from the entry of the mineralized nitrogen into the lakes and streams. The same soil described above, Pahokee muck, at a subsidence rate of 3 cm/year will produce 1400 kg N/ha/year. Although most of the nitrogen mineralized is denitrified (Terry and Tate, 1980), some of the nitrogen leaches into the regional waterways (Florida Sugar Cane League, 1978). Similar nitrogen problems have been reported for muck soils in Israel (Avnimelech, 1971) and New York (Duxbury and Peverly, 1978).

2. Microbial Activities in Newly Drained Peats

Prior to drainage, the microbial community of the swamp or marsh is in equilibrium with the limitations imposed by the chemical and physical environments. Although aerobic and anaerobic bacteria are found throughout the soil profile (Waksman and Purvis, 1932), the aerobic microbial activity is low partly because of the O_2-limiting conditions imposed by the high moisture content of the soils. This minimal activity results in the slow accumulation of partially decayed plant and animal remains, which is characteristic of this ecosystem. After drainage, the aerobic microbial community, which could be described as quiescent, is free of many of the growth constraints. The limitations imposed by O_2 deprivation are lost because of the decreased moisture content of the soil. Since most of the soils are drained for agricultural development, one of the first procedures following draining is to adjust the soil pH through the addition of limestone, if necessary, and to add any micro- or macronutrients needed for crop production. Thus, the microbe is faced with an environment of partially humified plant remains, proper aeration, and a pH generally found to be conducive to growth. Consequently, the aerobic microflora is stimulated, and soil organic matter is depleted rather than accumulated. It is the transition in the microbial community from the predominantly anaerobic metabolism of the swamp to the active aerobic catabolism of the drained cultivated Histosols that is the subject of this section. Literature will be assembled describing (a) alterations in the microbial populations following draining of the swamp and (b) the chemical nature of the modifications of the soil organic matter during the transition from a swamp to a drained Histosol.

2.1. Microbial Population

When peat lands are initially drained, the microorganisms are faced with a veritable smorgasbord of potential carbon sources. Among these are the following: the partially humified plant remains (including polysaccharides, proteins, and lignin); the humified substituents common to soil organic matter, humic and fulvic acids and humin; and the various cellular components of the microbial community (Waksman and Stevens, 1929; Davis and Lucas, 1959; Given and Dickinson, 1975). Since peats are formed from a wide variety of plant constituents under many environmental conditions (Davis and Lucas, 1959), the physical and chemical environmental traits of newly drained peats are varied. For example, Given and Dickinson (1975) report that the pH values of peats range from 2.8 to 7.8. Thus, after initial draining, even in this high-carbon environment, the microbial activity can be limited by factors such as the soil pH. Since generally the soils are drained for agricultural use, one of the first procedures implemented in the newly drained field is the adjustment of the soil pH through the addition of limestone to a value conducive for crop growth. Thus, much of this variability is diminished. As a result, considerable data have accumulated in the literature describing the effect of these environmental modifications on the microbial activity. This has created a bias in favor of those soils of the more neutral pH range conducive to good crop growth.

Dramatic effects of pH adjustment have been recorded. An increase in nitrifier populations and in total soil respiration (as measured by carbon dioxide production) in peat after adjustment of the pH from 3.8 to between 6 and 7 has been reported (Ivarson, 1977). This worker also observed a decrease in fungal populations, which suggests that fungi became less important in the peat microbial community as the pH rose. Similar variations in respiration and nitrification in various limed peats were reported by Sunk (1929) and Vandecaveye (1932). The latter work described the effect of fertilization of freshly drained peats. Pereverzev and Golovko (1968) also found that several soil enzymatic activities (urease, catalase, invertase, and protease) were stimulated by the increase in soil pH resulting from liming. These data reflect the general increase in microbial activity provided by removal of the limitations of pH on microbial growth.

Early studies with flooded peats suggested that these soils were sterile, but it is now apparent that the peat, when initially drained, contains bacteria throughout most of the soil profile. Waksman and Purvis (1932), in studies reported nearly 50 years ago, demonstrated aerobic bacteria to a depth of 360 cm (maximum depth tested) in muck samples collected in Maine and to the same depth in Florida peats. This work also showed that the aerobic microbial population density decreased with increasing soil depth, while the anaerobic bacterial numbers increased, as would currently be predicted. This work answered most of the debate concerning the question of the existence of bacteria deep in the profile of flooded peats. This question has been answered repeatedly by many different techniques in the intervening 50 years since the study of Waksman and Purvis (Visser, 1964; Holding *et al.*, 1965; Dickinson *et al.*, 1974; Kavanaugh and Herlihy, 1975; Wheatley *et al.*, 1976; Casagrande and Park, 1978). The methods ranged from viable counts through reflectance microscopy to muramic acid analysis, but the conclusion remains the same—bacteria exist deep in peat profiles prior to drainage. The question becomes, how many of the organisms found in the profile are metabolically active? Wheatley *et al.* (1976) examined the ATP content of peat samples from the profile of a raised bog and suggested that, although bacteria were found in the profile, many of them were inactive. It can reasonably be concluded from these data that peat is an active medium for microbial metabolism while flooded but that, as would be expected, anaerobic microorganisms come to the forefront.

Examination of aerobic microbial activity in swamps and marshes suggests that this activity would be localized primarily in the surface layers of the soil receiving O_2 through diffusion. Also, aerobic populations would be stimulated during the annual wet and dry cycles, when the surface of the water table ranges from several inches above the soil surface to a depth within the soil profile. Such a phenomenon is common in the Florida Everglades, where the bulk of the rain falls during the summer months. This would explain the existence of humified plant remains deep in the soil profile, where anaerobic conditions would be expected to protect the decaying plant remains.

Several workers have examined the changes in microbial numbers after draining of peats. One very useful population that provides a good indication of the changes in the microenvironment after draining is the obligately aerobic autotrophic nitrogen oxidizers, the nitrifiers. These organisms occur in very low numbers in undrained peat. This was documented by Herlihy (1973) in his comparison of nitrifier populations in drained, cultivated, uncultivated, and undrained peats. In the surface soils of undrained peat, as measured by the most-probable-number (MPN) method, they detected no *Nitrosomonas* and populations of 1.3×10^3 *Nitrobacter*/g peat. As the peats were drained and the soils became aerobic, the *Nitrosomonas* and *Nitrobacter* population densities increased to 2.3×10^6 and 2.4×10^8/g peat, respectively. Again, as in the above reports of the effect of liming on peats, further studies by these workers (Herlihy, 1973) with wood fens in which drained but uncultivated soils were compared with those that had been drained, limed, and fertilized to provide for crop growth substantiated the stimulatory affect of pH adjustment on acid peat microflora. An approximate 10-fold increase in the nitrifiers was noted in the cultivated field. Mishustin *et al.* (1974), in a similar study of the changes in microbial populations following draining of a peat soil, noted increases in saprophytic bacteria, nitrifiers, denitrifiers, and cellulose oxidizers following draining. The changes were quite marked in the top 15 cm of the soil profile, where the maximum effect of the increased aeration would be expected.

Several conclusions concerning the overall microbiological changes occurring in peats following draining can be made from these data. Although the microbes are found throughout the soil profile prior to draining of the swampy soil, it is unlikely that the aerobic contingent of the microflora is active. As the soils are drained, a general stimulation of the microflora occurs as the water content is diminished. It is dangerous, though, to conclude that this increased activity is proportional to the increases in microbial populations measured either by direct-observation techniques or through viable-count methods. The time has come when enzyme activity or oxidation of various ^{14}C-labeled substrates of high specific activity must be used as an indication of microbial activity *in situ* (Tate, 1979a).

2.2. Changes of Microbial Activity in Soil Profile

One of the more dramatic changes that occurs once the peats are drained is the increased aerobic metabolism throughout the soil profile. As was demonstrated by Broadbent (1960), minute concentrations of O_2 are necessary to stimulate aerobic catabolism of soil organic matter of peat soils. In a series of laboratory experiments, he measured carbon dioxide evolution from peat incubated under various O_2 concentrations ranging from atmospheric levels of 21% to 0.1% by volume. Even at the minimum O_2 concentration tested, only a 30% decline in the soil respiration occurred. Thus, oxidation of the soil organic matter of peats can occur at exceedingly low O_2 concentrations. This conclusion was supported by the observations of Pereverzev and Golovko (1968) when they examined the enzymatic activity of drained and undrained peats. In the swampy soil, the enzymatic activity, which would be expected to be proportional to the microbial activity, decreased rapidly with depth. The decrease in activity was much less severe in the drained soil. Mandrovskaya (1971) found that the lower water content in a peat profile resulted in augmented dehydrogenase and catalase in the rhizosphere of maize growing on newly drained peat. An interesting observation of these workers was that the enzymatic activity of the rhizosphere was different in the first years of development of the peat. As the peat developed, dehydrogenase activities increased and catalase decreased.

Tate (1979a) recently reported the use of high-specific-activity ^{14}C-labeled radioisotopes to examine the changes in the microbial activity in various substrata of Pahokee muck of the Everglades. This technique proved useful in determination of carbon metabolism *in situ* in that the carbonaceous substrates could be amended to the muck samples without significantly increasing the concentration of the substrate in the soil samples. Thus, the induction of the soil microflora capable of oxidation of the test substrate was avoided. The microbial oxidation rate for succinate, an indicator of general aerobic respiration, salicylate, a model compound for aromatic-ring decomposition, and acetate diminished as the depth of soil increased from the surface to the 60- to 70-cm depth (the maximum depth of the soils in the field under study). Aromatic-ring oxidation was most sensitive to the increased moisture and decreased O_2 deep in the soil profile, as would be expected from the direct participation of O_2 in the catabolism of aromatic compounds (Gibson, 1968). At the end of the wet season in Florida (May through September), a 90% decrease in aromatic-ring oxidation was recorded between the surface and the 60- to 70-cm depth. The oxidation of acetate and succinate decreased approximately 75% in the same distance. Thus, in contrast to the undrained state, in which O_2 would be limiting throughout the profile, and thus aerobic carbon oxidation minimal, the aerobic catabolism of the soil organic matter in the muck soils occurred to the water table.

The same isotopic method was used to determine the effect of depth of muck on the proportion of the overall soil respiration contributed by the oxidation of various carbon sources commonly found in soil organic matter (Tate, 1980). Succinate was again used as an indicator of overall soil respiration, salicylate for aromatic-ring oxidation, glucose and amino acids for polysaccharides and protein monomers, and acetate for short-chain-acid oxidation. Little effect of depth of soil was observed. Although the contribution of various substrates to the overall metabolic rate of the soil varied somewhat with time, the changes in the oxidation of the various substrates varied little with depth of soil. The greatest effect was observed with aromatic-ring oxidation. The contribution of aromatic-ring oxidation to the overall soil respiration decreased 40% between the surface and the 60- to 70-cm depth at the end of the wet season. Considerable variation was noted in the oxidation of glucose, but that variation may have reflected availability of the substrate rather than O_2 limitation.

These observations on the extent of aerobic metabolism in the soil profile are supported by physical changes in the soil organic matter. As the soil organic matter is oxidized, the bulk density of the soil increases as does the ash content. The effect of the oxidation of the organic matter on the bulk density was documented by Yerdokimova et al. (1976) in peats of the Ukrainian poles'ye. They observed that the bulk density in the surface layers (0–10 cm) increased from 0.197 to 0.283 g/cm^3 over the first 60 years of soil use after drainage. Comparable increases were measured to a depth of 30 cm. The density of the 40- to 50-cm stratum increased from 0.11 to 0.13 g/cm^3 soil. The increases in bulk density were accompanied by the increased humification of the organic matter. As the organic matter is mineralized to carbon dioxide, ammonium, and water, the mineral matter is left in the soil profile. Broadbent (1960) used this accumulation of ash content of the soils as an indicator of the amount of microbial oxidation of the soil organic matter. After making the assumptions that the ash content of the soil had been deposited uniformly in the soil profile prior to drainage and that comparable losses of mineral matter due to leaching in the drainage waters occurred throughout the profile, he measured the ash content of the drained soil to a depth of 60 cm, the average watertable depth. His calculations suggested that approximately 66% decomposition of the top 7.5 cm of the soil had occurred. Between 28- and 38-cm depth, this percentage decreased to 49%. Little decrease in the percentage was again recorded to the 48- to 60-cm depth, where it decreased to 22.5%. Thus, active microbial oxidation had occurred at least to a depth of 60 cm in the profile. Similar results were obtained by Lupinovich (1968).

As is obvious from each of the methods detailed above, the degree of microbial activity decreases with depth of soil. This is an interesting observation considering that the subsidence rate of drained peats is linearly proportional to the depth of the water table. This included depths of 60–70 cm where it was demonstrated above that a considerable decrease in the oxidation of carbonaceous substrates had occurred. Tate (1979a) recently reported an explanation of this apparent discrepancy between the observation of microbial activity, the greatest contributor to soil subsidence, and field measurement of subsidence. The explanation is found by comparing bulk-density changes in the soil profile with decreases in microbial activity. In Pahokee muck of the Everglades from the surface of the soil to the underlying limestone layer, there is approximately a twofold decrease in the bulk density (g soil/cm^3) (Duxbury and Tate, 1979). Thus, in the deeper layers of the soil, the oxidation of one gram of soil organic matter would have a twofold effect on changes of soil volume as would be observed in the soil surface. Thus, only one-half the microbial activity would have to occur to provide the same loss of soil volume as would be noted in surface-soil samples. Thus, a variation in the physical structure of the soil is compensating for a decrease in the biological activity to produce the nearly linear effect of watertable variation on soil-subsidence rates that has been observed. Considering that the bulk density will increase in proportion to the microbial activity as the muck is oxidized, it is reasonable to assume that this relationship between bulk density and microbial activity can be observed elsewhere in peat and muck soils in which the linear relationship between subsidence and water table has been recorded.

2.3. Biomass Variation in the Soil Profile

Few data exist recording the variation of biomass within the profile of a cultivated Histosol, but the predictability of the subsidence rate, the relationships between carbon dioxide evolution and biomass (Wagner, 1975), and the variation of microbial activity with depth of soil (Tate, 1979a) allow for the estimation of the aerobic biomass within the soil profile. These values, since they are derived from bare (fallow) soil, allow for the calculation of the biomass necessary for the oxidation of the soil organic matter alone. As was indicated in Section 1.2, approximately 3.3×10^4 kg C/ha/year is produced by the microbial oxi-

Table II. Estimated Biomass within a Pahokee
Muck Profile

Depth (cm)	mg $C/cm^3/hr$	Active biomass (mg/cm^3)
0–10	0.0075	0.23
30–40	0.0054	0.16
60–70	0.0037	0.11

dation of Pahokee muck in the Everglades region. The source of this carbon dioxide within the soil profile can be determined using the microbial activity data of Tate (1979a). As is shown in Table II, the carbon evolution rate is estimated to range from 0.0075 to 0.0037 mg $C/cm^3/hr$ between depths of 0–10 and 60–70 cm, respectively. Wagner (1975) reports that 0.033 mg C/hr is necessary to yield 1 mg of active biomass. Thus, the biomass in the profile of a fallow (bare) Pahokee muck is estimated to vary between 0.23 and 0.11 mg/cm^3 soil. Two assumptions that will result in this estimate being high are (a) that all of the carbon mineralized from the soil organic matter is lost as carbon dioxide and that none is lost in drainage waters and (b) that microbial oxidation of the soil organic matter is the sole source of carbon dioxide. Sufficient fulvic acids and BOD levels are found in the drainage waters to indicate that a portion of the carbon is lost from the soil by this route (Florida Sugar Cane League, 1978). The second assumption is trivial in that the vast majority of the carbon produced results from biological activity rather than spontaneous chemical reactions. Although the estimates are likely overestimates of the actual aerobic biomass of the system, they do provide an indication of the high aerobic biomass deep in the profile of Pahokee muck.

2.4. Chemical Nature of the Soil Organic Matter Following the Transition Period

The foregoing studies document the loss of soil elevation, the production of gaseous and water-soluble products, and the increased microbial activity accompanying draining of peat deposits. The effect of the lowered water table is equally dramatic on the organic matter remaining in the soil following the transition period. As the microbial community develops in the drained peat, humification of the organic matter increases. Volk and Schnitzer (1973) compared cultivated Histosols from the Everglades by chemical and spectroscopic methods to determine the changes in the humic acids as subsidence proceeds. They found an increase in carboxyl groups, phenolic hydroxyls, quinones, and ketonic groups and a decrease in aliphatic and alcoholic hydroxyl groups. A decrease in the molecular complexity of the humic fraction was suggested by the E_4/E_6 ratios and possibly by the measurement of free radicals. They postulated that the reactions involved in the oxidation of the soil organic matter were oxidative, which led first to a degradation of the aliphatic structures prior to degradation of the more stable aromatic groups. Greater oxidation of the surface soils was demonstrated by decreased aliphatic structures over those detected deeper in the soil profile (Volk and Zelazny, 1973). Similar results on peat soils were reported by Plotkina et al. (1975). These data support the contention that as the microbial population is established in the newly drained organic soil, the readily oxidizable hydrocarbon chains and carbohydrates are degraded first. Thus, the product of this microbial activity is a more oxidized, more greatly humified product. For a review of the structure of the amorphous, alkali-soluble component characteristic of the majority of the soil humic matter, the humic acids, see Felbeck (1971), Kononova (1966), Haider et al. (1975), Schnitzer and Khan (1972), or Grant (1977).

When the soils are first drained, we can postulate that because of the accumulation of partially humified material (Given and Dickinson, 1975), the microbes are not carbon limited, but as the quantity of this easily oxidizable substrate is decreased, as was indicated by the spectroscopic data of Volk and Schnitzer (1973), the microbe must oxidize the more resistant humified products. The effect of the carbon limitations imposed by this sort of substrate was demonstrated by Waksman and Purvis (1932) in their study of macronutrient limitations of drained peats. They observed the production of carbon dioxide in unamended and phosphate- or nitrogen-amended peat samples. They concluded that the limitation to microbial growth rate was not a shortage of nitrogen or phosphorus, but rather a source of oxidizable carbon. Treatment of the peat with chemicals such as toluene to solubilize some of the carbon increased the microbial respiration (carbon dioxide evolution). Thus, although a wide variety of humified carbon sources was present in the soil organic matter, availability of nutrients or the ability of the microbes to solubilize the nutrients from the soil humic matter provided a limiting factor to the growth of the microbes. Soluble carbon, not total carbon, apparently provides the best indication of the capability of the microbes to grow in this complex system. Similarly, Terry and Tate (1980) noted a stimulation of denitrification potential by the addition of readily decomposable substrates, such as crop residues, to Pahokee muck.

The effect of the humification processes on microbial oxidation of carbonaceous substrates has been elegantly demonstrated by Martin and co·workers. They have studied the effect of humification on the oxidation of algal cells (Verma and Martin, 1976), fungal cells (Martin et al., 1959), and various aromatic compounds (Haider and Martin, 1975; Martin and Haider, 1976; Haider et al., 1977) as well as proteins, peptides and amino acids (Verma et al., 1975), and amino sugars (Bondietti et al., 1972). When ^{14}C-labeled whole cells, cell walls, protoplasm, and extracellular polysaccharides from algae were added to soil, after 22 weeks between 61 and 81% of the added carbon had been evolved as carbon dioxide (Verma and Martin, 1976). Complexing of the cell wall and cytoplasmic components with model humic acid-type phenolic polymers resulted in a reduction of the decomposition of the cell walls by 40% and the cytoplasm by 70%. Over half of the residual activity remained in the 0.5% sodium hydroxide-extracted soil. Thus, as shall be discussed later, the humification process results in the accumulation of a product that is more oxidized than the original parent material and enriched with more biodegradation-resistant aromatic-ring structures.

A physical manifestation of this humification process is observed in the structure of the organic matter. When drained, the organic material retains some of the original plant structure of the parent plant tissue. At that time, the material is referred to as peat. As the organic matter becomes more humified, the plant structure is gradually lost. When the plant structure has been lost from the soil organic matter, the product is muck.

3. Carbon Metabolism in Well-Humified, Cultivated Histosols

The product of draining, liming, and fertilizing peat deposits, following the transition period previously discussed, is a well-humified, agriculturally rich peat or muck soil. Even though a unique environment is provided by the flooded, swampy soils, as discussed by Given and Dickinson (1975), drained cultivated Histosols provide an equally interesting environment, especially for the study of the oxidation of soil organic matter. By definition these soils can contain large concentrations of mineral matter, such as clays, but frequently, such as occurs in the Everglades deposits, the content of clays and other minerals is minimal. Although the augmented concentration of colloidal organic matter would be expected to result in greater reaction rates for the biological oxidative processes involved in carbon transformations in peat soils even in the presence of clay, the absence of clays in some peat deposits provides an ideal environment for the study of the mechanism of oxidation of soil organic matter and of the physical and chemical factors affecting soil organic matter oxidation. Accordingly, this section will consist of an analysis of the microbial oxidation of the organic matter of Histosols to provide not only a better understanding of the processes involved in subsidence of Histosols themselves, but also an overall understanding of the microbial oxidation of soil organic matter in general.

3.1. The Substrate

The greatest handicap to the soil biochemist interested in the microbial catabolism of the colloidal soil organic matter complex involves the lack of an understanding of the complete structure of the substrate. Except in very general terms, such information is essentially nonexistent for the more complex

substrates such as humic and fulvic acids. Soil organic matter can be divided into
(a) readily degradable compounds composed essentially of recently incorporated
plant and animal remains plus the components of the growing and dying microbial populations and (b) the more degradation-resistant components made up
primarily of the soil humic acids. Humic acids are the major component of soil
organic matter and thus would be of greatest importance in consideration of
the reactions involved in soil subsidence. The biodegradation-resistant compounds of soil organic matter are classified primarily as humic acids, fulvic
acids, and humin. The delineation depends upon the solubility of the substituents in acid and base. Humin is soluble in neither acid nor base, and humic
acids are soluble in alkaline solutions but are precipitated when the solutions
are made acidic. Fulvic acids are soluble in both acid and base. A complete description of the current ideas of the structure of these compounds is given by
Kononova (1966), Schnitzer and Khan (1972), and Felbeck (1971). For the
purposes of this review, humic and fulvic acids will be considered to be random
polymers of aromatic compounds.

Several theories have been proposed to account for the formation of humic
acids (Felbeck, 1971), but two of the most probable involve the microbial and
chemical polymerization of the precursor molecules. In all likelihood, the final
product is a result of the combination of the two mechanisms. Several fungi and
bacteria produce humic acid-like polymers while growing on simple carbon compounds (Haider and Martin, 1967; Haider and Martin, 1970; Martin et al., 1972;
Huntjens, 1972; Bailly and Nkundikije-Dessaux, 1975; Kosinkiewicz, 1977).
Comparison of the chemical structure of these polymers reveals several similarities to humic acids found in soil organic matter. Martin *et al.* (1967) compared
the degradability of a humic acid-like compound produced by *Epicoccum
nigrum* with the humic acid fraction extracted from leonardite and found the
structure and the degradability of the two compounds to be quite similar.
Humic-acid-like polymers labeled with ^{14}C from *Eurotium echinulatum* and
Aspergillus glaucus were nearly as resistant to degradation in soil as were soil
and peat humic acids (Linhares and Martin, 1978). After 12 weeks' incubation
at $22°C$, the proportion of the labeled carbon evolved as carbon dioxide ranged
from 4 to 6% for *E. echinulatum* and 4 to 13% for *A. glaucus.* Huntjens (1972)
compared the amino acid composition of a streptomycete humic acid with the
humic acids isolated from arable land and from soil collected from a 10-year-old
pasture and found the compositions to be quite similar.

In each case cited above, the microbial-produced humic acid was found to
resemble soil humic acids. The fungal products are similar to soil humic acids
(Martin *et al.*, 1974), but considerable differences still exist (Schnitzer *et al.*,
1973). It must be remembered that the microbial humic acids were produced
in pure culture under controlled conditions, and thus it should not be expected
that they would be identical to those found in nature. The important conclusion is that several soil microbes are capable of producing substances resembling
soil humic acids. Since the soil contains a very diverse array of microbial species,

it would be expected that these compounds, if produced in soil, would be modified by both the microbial populations and the various spontaneous chemical reactions known to occur. The ultimate conclusion is that the product, whatever the source, presents a complicated structure to the biochemist interested in its catabolism.

3.2. Catabolism of Humic Acids

There have been several recent reports on the role of microorganisms in the oxidation of soil humic acids. The capability to oxidize these complex aromatic polymers appears to be widespread in the microbial world in that those organisms reported to catabolize humic acid include several bacterial species, especially pseudomonads (Huntjens, 1972; Taha et al., 1973; Andriiuk et al., 1973; Mal'tseva et al., 1975), yeast (Shiskova et al., 1972), actinomycetes (Steinbrenner and Mundstock, 1975; Ibrahim and Ibrahim, 1977; Sidorenko et al., 1978) and several fungal species (Paul and Mathur, 1967; Biederbeck and Paul, 1971; Ruocco and Burton, 1978). Nocardia corallina decomposed 23.4% of the humic acid added to a mineral medium as the sole carbon source (Sidorenko et al., 1978). Evidence that the bacterium was using the humic acid as a substrate for growth included a 15-fold increase in the number of cells. An increase in the H/C ratio of the cultural solids suggested that the bacteria were oxidizing the aromatic rings. Infrared analysis of the product demonstrated increased methyl and CH_2 groups after 6 months of growth. Utilization of the humic acid nitrogen has been demonstrated with Pseudomonas sp. (Huntjens, 1972) and with several soil fungi (Biederbeck and Paul, 1971). As much as 76% of the nitrogen mineralized by the fungi originated from the acid-hydrolyzable amino acid fraction of the humates. Penicillium frequentans was shown to attack the aromatic humate components and, thus, to utilize the primarily non-acid-hydrolyzable nitrogen forms.

Although humic and fulvic acids are composed of undefined polymers of aromatic compounds, studies of simple model aromatic compounds such as ferulic acid, a component of lignin (Martin and Haider, 1975; Turner and Rice, 1975), phthalic acid (Englehardt and Wollnöfer, 1978), vanillic acid (Crawford and Olson, 1978), and other simple aromatics in culture have provided information which can be used to speculate about the nature and properties of the oxidative reactions in the soil environment. Probably, the most outstanding feature common to all aerobic catabolic pathways for aromatic compounds is the role of the oxygenases. These enzymes hydroxylate aromatic rings through the direct incorporation of O_2. The ubiquity and details of these reactions have been adequately reviewed (Hayaishi and Nozaki, 1969; Gibson, 1968; Dagley, 1971; Gunsalus et al., 1975) and will not be delved into here. This mechanism of hydroxylation results in aromatic ring oxidation in the soil environment being particularly sensitive to the depletion of O_2. This was suggested by studies of aromatic-ring oxidation in a Pahokee muck profile (Tate, 1979a) and in flooded fields of the same muck (Tate, 1979b). As the concentration of O_2 diminished, either through the increase in water content of the soil via flooding or through the limitations of diffusion along with increased moisture resulting from increased depth of soil, aromatic-ring oxidation decreased. For example, when the aromatic-ring oxidation decreased approximately 90% in the soil profile [comparison of rates from surface soil (0–10 cm) and soil from a depth of 60 to 70 cm] general carbon catabolism as demonstrated by succinate oxidation was diminished only by 65%.

A limiting factor in the oxidation of aromatic-ring polymers of soil organic matter is the solubilization of the substrate. Humic acids are insoluble in water. Thus, the microbe must either produce soluble monomers of the substrate or devise mechanisms of oxidation of a portion of the insoluble substrate *in situ*. To accomplish the latter, the microbe could oxidize a portion of the ring structure while the ring is still attached to the polymer. This is exemplified by the oxidation of naphthalene (Gibson, 1968) and biphenyl (Catelani *et al.*, 1973). With the oxidation of polynuclear substrates such as naphthalene or anthracene, the rings are oxidized separately. For example, the initial oxidation product in naphthalene oxidation is 1,2-dihydroxynaphthalene. Generally, in the catabolism of polynuclear compounds, a dihydroxylated polyaromatic compound is the first product (Gibson, 1968). Production of this compound is followed by cleavage of the hydroxylated ring. Similarly, in the oxidation of biphenyl, a structure that could be related more closely to humic acids than naphthalene, the rings are oxidized in sequence. In this case, biphenyl is initially hydroxylated to 2,3-dihydroxybiphenyl (Catelani *et al.*, 1973). Support for the role of mechanisms of this type in the oxidation of humic and fulvic acids was provided by Volk and Schnitzer (1973) when they demonstrated that as the soil organic matter of histosols was oxidized, the number of oxygen-containing substituents was increased. The importance of catechol oxygenase-type enzymes in humic and fulvic acid oxidation was suggested by the observation that fulvic acids serve as inducers of this type of enzyme (DeHaan, 1976). DeHaan (1977) also reports that addition of benzoic acid to water from Tjeukemeer (The Netherlands) stimulated fulvic acid oxidation.

Any enzymatic involvement in the catabolism of humic and fulvic acids would necessarily be limited by the capability of the enzymes to approach the polymer. This steric hindrance of the enzymes was demonstrated by Alexander and Lustigman (1966) with simple aromatic monomers when they demonstrated the importance of location and nature of sidechain substituents in determining the biodegradability of aromatic compounds. They noted that such substituents as chloro and nitro groups retarded oxidation and that *meta* isomers were more resistant to biodegradation, although some *ortho* isomers were more resistant. There is no reason to suspect that such restrictions on biodegradation would not be even more important in the oxidation of the more complex humic acids.

Two mechanisms, one biological and one chemical, can be proposed for the actual cleavage of aromatic subunits from the humic acid polymer. The biological mechanism would involve the energetically problematic reaction of etherases to cleave the polymers. Evidence for such an enzyme system was reported by Paul and Mathur (1967). With the chemical mechanism, the microbial population would not be directly involved in the cleavage of the polymer, but the random chemical cleavage of the ether bonds linking the aromatic substituents of humic acids would provide the monomers for further microbial oxidation. Perhaps, if the latter mechanism was operative in humic acid catabolism, the microbes would play a role in the solubilization by creating an environment, but not an enzyme, which would make the spontaneous chemical cleavage more probable. The answer to the nature of the role of the microbes in the solubilization of humic acids awaits further research.

There have been several recent reports describing the anaerobic oxidation of aromatic compounds (Ferry and Wolfe, 1976; Evans, 1977; Balba and Evans, 1977; Healy and Young, 1978). Basically the anaerobic process can be classified into three groups: light-driven reduction of the aromatic ring, reduction through nitrate respiration, and reduction through methanogenesis (Evans, 1977). These reactions, especially the nitrate respiration pathway, in that nitrates are known to occur in high concentrations in histosols (Avnimelech, 1971; Tate, 1976; Terry and Tate, 1980), may occur in flooded organic soils. But, since under flooded conditions complex aromatic compounds—such as humic acids—accumulate, anaerobic catabolism of aromatic structures must be limited in histosols. Anaerobic catabolism of aromatic compounds could occur, for example, if a soil that contained a pool of simple aromatic compounds produced under aerobic conditions, plus high nitrate concentrations, was suddenly flooded. It can be predicted that anaerobic aromatic-ring catabolism would occur shortly after flooding.

3.3. Readily Oxidizable Substrates

Substrates such as proteins and polysaccharides, which are considered to be easily oxidizable substrates for the microbial community, are found in soil in two general pools. One consists of the recently incorporated carbon substrates such as plant and animal debris, plus the growing and dying microbial cells, whereas the other consists of those proteins and polysaccharides which have been humified or incorporated into the soil humic acids. Because of the ease with which the noncomplexed substrates are oxidized, it can be concluded that these compounds supply the majority of the carbon respired by the soil microbial community receiving an influx of carbonaceous substrate. For comparison of oxidation rates, Martin et al. (1967) reported that in the majority of the fungal humic acids tested, 5-7% of the carbon yielded was carbon dioxide after 8 weeks incubation in a soil environment, whereas Verma et al. (1975) noted that after 4 to 12 weeks, 71–95% of the ^{14}C-labeled proteinaceous substrate had been converted to carbon dioxide when added to soil. Rates similar to those for protein substrate were recorded for amino sugars (Bondietti et al., 1972), algal cells (Verma and Martin, 1976), and fungal cells (Wolf and Martin, 1976). Oxidation of the fungal cellular matter was complicated by the interaction with fungal melanins (Martin et al., 1959). The process involved in the oxidation of these readily oxidizable substrates has been recently reviewed by Haider et al. (1975).

The fate of the readily oxidizable carbon in soil is twofold. The majority of the carbon, as indicated above, is converted to carbon dioxide, but a small portion is incorporated into microbial cells and into soil humic and fulvic acid fractions. For example, when acetate is added to soil, all of the original substrate has been oxidized after 6 days, but only 70% is accounted for as carbon dioxide (Sorensen and Paul, 1974). That which remains with the soil is found in polysaccharides, amino acids, and in an insoluble residue. Similarly, Chesire et al. (1973) found that some of the ^{14}C-labeled plant polysaccharides were incorporated directly into the soil humic acids after 224 days. These workers incubated ^{14}C-labeled rye straw in soil. Fractionation of the soil–straw mixture, after 224 days' incubation, revealed that the specific activity of glucose was higher in the humin than the fulvic acid fraction. This would be expected if the remaining ^{14}C were still in the form of unchanged plant material.

3.4. Soil Enzymes

One of the most useful tools, and to date the least exploited, in the study of soil organic matter oxidation is soil enzymes. Past studies of soil enzymes have been primarily involved with those activities predictive of soil properties of agronomic importance. For example, dehydrogenase has been used as a general indicator of microbial activity (Skujins, 1973; Casida, 1977), whereas general enzmatic activity has been used to indicate the cultivation state of the soil (Konovalova, 1970). Considerable effort has been applied to understanding the dynamics of urease, an agronomically important enzyme in the soil environment. A more basic approach was taken when enzymatic activity was used as an indicator of soil organic matter structure (Mathur, 1971; Mathur and Morley, 1975). Two reviews provide a good summary of the research on soil enzymes (Skujins, 1967; Kiss et al., 1975). Although little effort has been put forth directly in the study of soil enzymes involved in the oxidation of soil organic matter, the methods involving agronomically important enzymes can be applied to the more basic questions of soil organic matter oxidation.

Probably the greatest limitation to soil enzymology in the past has been the inability to separate the enzyme from the remainder of the soil organic matter. Recently, it has been reported that enzymes involved in the oxidation of tryptophan (Chalvignac and Mayaudon, 1971) and urease (Burns et al., 1972; Nannipieri et al., 1975; Ceccanti et al., 1978) can be partially purified from the soil organic matter. Generally, the procedure involves extraction with solutions such as sodium carbonate or pyrophosphate (although Burns *et al.* (1972) were able to extract urease with water) and concentration of the extract. Ceccanti *et al.* (1978) concentrated urease by exhaustive ultrafiltration of the soil extract against 0.1 M pyrophosphate at pH 7.1. They reported increases in the specific activity by this method of 6.9 to 18-fold over that of the soil extract. Such preparations will be useful in determining the variability of the types of isoenzymes found in the soil organic matter as well as in the study of enzyme kinetics.

Simple studies on the kinetics of a soil enzyme *in situ* are useful for elucidation of the nature of the soil–enzyme interactions. Urease is known to exist in soil in two forms: associated directly with the soil microorganisms or free from the microorganisms and adsorbed on the soil colloids (Paulson and Kurtz, 1969). The Michaelis constant (K_m) of urease in soil varies with fluctuations in the microbial populations (Paulson and Kurtz, 1970). Thus, Paulson and Kurtz concluded that the variations in the K_m were in propagation to the urease existing in the living cells and that exoenzyme complexed to the soil organic matter. They measured K_m values of 0.057 M for the microbial form as opposed to 0.252 M for that adsorbed on the soil organic matter. The difference between the two K_m values stresses the inhibitory effect of humic acid on an enzymatic activity.

3.4.1. Soil Proteases

Recent studies of soil proteolytic activity have the greatest bearing on the soil organic matter oxidation question in that these enzymes would be active in soil organic matter hydrolysis. The studies of these enzymes have involved an examination of the amino acids released during hydrolysis of soil organic matter, studies of proteases extracted from the soil organic matter, and an examination of the relationship of proteolytic activity to humic acids.

Sowden (1970) examined the release of amino acids from soil organic matter by various proteolytic enzymes. Pronase was found to hydrolyze 2–10% of the aspartic acid plus asparagine, threonine, serine, glutamic acid plus glutamine, glycine, lysine, and histidine in some of the fractions of soil organic matter. This enzyme also hydrolyzed 15–35% of the alanine, valine, isoleucine, leucine, tyrosine, phenylalanine, and arginine. There was no release of proline, ornithine, or ammonia, but when leucine aminopeptidase was added to the pronase hydrolysate, 15% of the proline was hydrolyzed. Brisbane *et al.* (1972) noted that pronase and thermolysin released about one-sixth of the acid-hydrolyzable amino acids from humic acids. These studies demonstrated that the proteins of soil organic matter were available to the action of proteases.

Natural soil proteases have been extracted from soil organic matter (Ladd, 1972; Mayaudon *et al.*, 1975). Ladd found that peptides and proteins were hydrolyzed by buffer extracts of dry soils. Tris-borate buffer extracts of a Mount Gambier soil contained proteolytic activity accounting for approximately 50–

75% of the activity of the unextracted soil. Although enzymatic activity could be extracted with water, these extracts were less active than those prepared with buffer. The substrate specificity of the proteases in the extracts varied from that of the proteases of the soil prior to extraction. The specificity of the extracted proteases more closely resembled that of proteases associated with plant fragments in the soil.

The effect of humic acids on the proteolytic activity is quite varied (Ladd and Butler, 1969a). Neutralized solutions of soil humic acids were found to inhibit carboxypeptidase A, chymotrypsin A, pronase, and trypsin activities. The same solutions were stimulatory to papain, ficin, subtilopeptidase A, and thermolysin activities and had no effect on phaseolain and tyrosinase activities. The stimulation of papain and ficin was not due to the formation of metal ion–humic acid complexes. Polycondensates derived from p-benzoquinone and catechol had a similar affect on proteolytic activity. Crop history of the soil and preincubation of the humic acids with pronase had little effect on the inhibitory action (Ladd and Butler, 1969b). Methylation of the humic acids prevented both the inhibitory and the stimulatory effect of the humic acids. Variation in the degree of methylation of the humic acids suggests that carboxyl groups but not phenolic groups were responsible for the effect of humic acids on the enzymatic activity (Butler and Ladd, 1969). Further studies where the protease was acetylated suggest that the humic acids are binding to the protease by a cation-exchange mechanism which links the amino groups of the protein to the carboxyl groups of the humic acids (Ladd and Butler, 1971).

4. Factors Limiting Soil Organic Matter Oxidation

Because of the economic value of drained, cultivated Histosols, considerable effort has been put forth in search of a mechanism to mitigate subsidence of these soils. Since after the initial transition period the majority of the subsidence of Histosols, at least in temperate climates, results from microbial interactions with the soil organic matter, these studies of subsidence rates under various environmental conditions provide a resource of literature describing both field and laboratory studies of the effect of parameters such as temperature, soil amendment with clay or carbonaceous substrates, moisture, etc. on the respiration of the microbial community of Histosols.

4.1. Moisture and Oxygen

These two factors are discussed together because they are intimately related. As the moisture content of the soil increases, the soil pores will be filled with free water and the content of air will decrease (Buckman and Brady, 1969). Thus, in many cases, the effects of augmented moisture on the microbial community are merely expressions of limited O_2. Soil moisture is the prime factor limiting microbial respiration in Histosols. It has been known for over 30 years that as the water-table depth in the soil decreased, the moisture content of that portion of the soil which had previously been above the water table increased, and the subsidence rate decreased (Neller, 1944; Allison, 1956; Stephens, 1969; Zimenko, 1972). That this effect of the water table is on the microbial population and not just an effect of the buoyancy or some other physical property of the increased soil water content was demonstrated by Volk (1972) when he measured the carbon dioxide evolution rates from soil columns of various Histosols collected from throughout the Everglades under varying water tables. As was observed in the field with measurements of the subsidence rate, the carbon dioxide evolution was proportional to the depth of the water table.

The effect of variation of soil moisture concentrations on microbial respiration has been recorded. Waksman and Purvis (1932) incubated peat samples collected in the Everglades at several moisture levels and measured the carbon dioxide produced (Table III). Maximal carbon dioxide production was observed between 52.8 and 71.3% moisture (wet weight basis). Similarly, Revinskaya and Filipshonova (1973) found the highest mineralization of the organic matter at 60% moisture. A practical effect of exploiting the minimal moisture level was reported by Vavulo et al. (1974) when they studied sprinkle irrigation of peat bog soils of Polesie. Their objective was to encourage the mineralization of the organic matter so as to increase the soil fertility. Ammonifying, nitrifying, and denitrifying bacterial populations were enhanced by the moisture additions.

A succession of wet and dry cycles increases the mineralization of the organic matter over maintenance of a constant moisture level. Sorensen (1974) and Birch and Friend (1961) both report the increased oxidation of humic material in mineral soils with wet/dry cycles of various lengths. Terry (1980) found that wet/dry cycles of 28 days (3 cycles) increased nitrogen mineralization of Pahokee muck by 39% over that held at a constant moisture. Similarly, Broadbent (1960) observed increased mineralization of Egbert muck and Staten peaty muck collected from the California delta and incubated under varying wet/dry cycles. Greatest oxidation was observed for a 10-day cycle as opposed to 20-, 30-, or 60-day cycles. Similarly, Reddy and Patrick (1975) found the highest organic matter oxidation rates in soils with greatest number of anaerobic/aerobic cycles.

Maximum inhibition of microbial respiration could be expected from returning the muck or peat land to the flooded state existing prior to drainage. Consideration of the linear relationship between subsidence rate and depth of the water table predicts that the subsidence or microbial oxidation of the soil organic matter would be negligible in a flooded soil. Aerobic oxidation of the soil organic matter is apparently not completely stopped by flooding in that Knipling et al. (1970) measured carbon dioxide evolution from a flooded muck and concluded that soil decomposition continued under high moisture conditions. Examination of the carbon oxidation rates in soils collected from flooded

Table III. Effect of Soil Moisture on
Microbial Activity[a]

Moisture[b] (%)	CO_2 evolved (mg)
81.5	820
80.2	1183
71.3	2966
52.8	286
33.3	775

[a]After Waksman and Purvis (1932).
[b]Wet weight basis.

Pahokee muck with the use of high-specific-activity radioisotopes suggests that aerobic carbon oxidation occurs at about 25% of the preflood rate (Tate, 1979b). Also, lake sediments and stream sediments frequently contain an oxidized layer on the surface although the soils below the surface are anaerobic. The flooded muck should behave analogously.

Tate (1979b) demonstrated that while aerobic oxidation is continuing under flooded conditions, the overall effect of the diminished O_2 concentration resulting from flooding is a modification of the types of carbon compounds oxidized by the microbial community. Microbial respiration in the top 10 cm of a flooded Pahokee muck field, as measured by ^{14}C-succinate oxidation, decreased 65% during the first 10 days of flooding. Little change in this activity was noted during the remainder of the flood period. Aromatic-ring degradation, because of the direct participation of O_2 in the ring-cleavage reactions, was the most sensitive to the flooded state. A 90% decrease in the oxidation rate for salicylate, the model aromatic test compound used, occurred during flooding. Catabolic rates for glucose, amino acids, and acetate decreased approximately 50, 50, and 65%, respectively. Comparison of the proportion of the overall soil respiration contributed by the various carbon sources studied indicated that aromatic-ring oxidation became a minor participant in the overall respiration, whereas glucose- and amino-acid-oxidation contributions increased during flooding. Thus, although aerobic oxidation was continuing under the flooded state, subsidence was essentially stopped due to the inhibition of aromatic-ring (the major component of soil organic matter) oxidation.

4.2. Temperature

Once the organic soil has been drained, the second most important factor after O_2 status determining the oxidation rate of the organic matter would have to be soil temperature. Since a major portion of the observed subsidence results from biological catabolism of the organic matter, classical biochemical principles relating to the temperature effects on substrate catalysis would hold. Diminished activity would be expected under freezing conditions of the winter in the cooler regions and increased activity would be found as the soil temperature increased. Stephens and Stewart (1976) recently developed a model for determination of the effect of temperature on soil subsidence which explains the varying subsidence rates found in various climates. As reviewed by Stephens and Stewart, Russian soil scientists felt that compaction and desiccation were the major factors in subsidence whereas only 14% of the soil loss resulted from biological factors. The Everglades data suggest that the microbial contribution would be greater than 50% in the subtropical region. By comparing the effect of temperature on biological oxidation rates in the soils (Q_{10}) and the temperatures in the various regions, Stephens and Stewart proposed that discrepancies in the microbial contribution to soil subsidence resulted from the effect of the soil temperature on the microbial activity. At the colder temperatures of the soils examined in the USSR, microbial processes were reduced to the point at which physical factors became more important, whereas in the Everglades, microbial populations were stimulated to become the greatest cause of soil subsidence. Basically, the conclusion can be reduced to a single statement: as the temperature of the soil declines, so does the microbial oxidation of the soil organic matter.

4.3. Organic Amendments

One of the earliest means of mediating colloidal organic matter oxidation in soils was the use of green manures. This procedure involves incorporation of crop residues into the soil following harvest. Several effects on the microbial community of the soil can be proposed. Probably one of the most frequently investigated of these effects is the "priming effect" as described by Bingeman et al. (1953). This phenomenon is a stimulation of the oxidation of the native soil organic matter by the incorporation of exogenous carbonaceous substrates. This process, which would actually result in a stimulation of subsidence, was noted to occur in Histosols by Chew et al. (1978) in their study of Malaysian peat and by Terry (1980) with Pahokee muck from the Everglades. In both studies, small quantities of the peat or muck were amended with a carbonaceous substrate and the nitrogen mineralization assayed. Chew and associates used incorporation of the nitrogen into the crop as an indication of the quantity of nitrogen mineralized, whereas Terry examined the question with [15]N. In both studies, greater quantities of mineral nitrogen were detected in the soil than could be accounted for by complete oxidation of the exogenous nutrients. Hayes and Mortensen (1963) examined this problem from a different view. They prepared natural-soil cores of Rifle peat profiles and incubated them in the laboratory for 102 days prior to the addition of ^{14}C-labeled rye grass. Carbon released as carbon dioxide and the soluble carbon in the drainage waters from the cores were assayed. In contrast to the experiments described above, no priming effect was observed. The authors proposed that the priming effect was more likely an artifact of the incubation of the soils in small samples rather than being a real effect found in natural systems.

A second question involved in the use of carbon amendments to control soil subsidence is whether sufficient plant debris could be added to the soil to counteract the losses of soil organic matter due to microbial oxidation. Some early studies (Bingeman *et al.*, 1953; Stotzky and Mortensen, 1957, 1958) suggest a net gain of organic matter in soil amended with plant remains, but a few calculations reveal the futility of this approach for the reversal of soil subsidence. In the Everglades region (as indicated in Section 1.2), 3.3×10^4 kg C/ha/year is estimated to be mineralized during the microbial oxidation of 3 cm of muck. Generally, when plant residues (such as those entering the soil community through crops) are added to soil, approximately 70% of the carbon is oxidized to carbon dioxide during the first year. Of the remaining 30% found in the soil, 10% will have been incorporated into microbial cells (Jenkinson, 1971). Thus, should it be desired to counteract the subsidence by incorporation of plant material, approximately 1.1×10^5 kg C/ha/year would have to be added to the soil. Although this is an aesthetic solution to the problem, it lacks practicality in the real situation.

Other carbonaceous amendments that have been added to muck soils with no measurable effect or practical value in reducing subsidence of organic soils are various pesticides (Stotzky *et al.*, 1956). These workers tried a combination of herbicides, fumigants, and insecticides with little effect on the mineralization rate. It is reasonable to assume that with the necessary interactions between the microbes of the rhizosphere and the growing plant, any chemical treatment that would significantly reduce subsidence through eradication of the microbes would also diminish the growth of the crop.

4.4. Mineral Additives

Whereas the mineral content of Histosols can be quite low, such as occurs in the Everglades peats and mucks, by definition, Histosols can contain quite large concentrations of minerals such as sands and clays. These materials are of interest in that their presence in the soil matrix could retard the oxidation of the soil organic matter and thus serve as environmentally safe compounds to prevent the oxidation of this valuable natural resource.

The principal property of clay which would affect the degradation of humic matter lies in its adsorption of the various soil organic components. Among the products that have been demonstrated to be adsorbed by clays are fulvic acids (Kodema and Schnitzer, 1974), humic acids (Greenland, 1971; Guckert *et al.*, 1975), soil polysaccharides (Guckert *et al.*, 1975), enzymes (McLaren and Packer, 1970; Haska, 1975), and bacteria (Marshall, 1969). The effect of this adsorption of the microbial carbon source or the cell itself on the clay varies with the soil type and the environmental conditions. Martin *et al.* (1976) examined the effect of montmorillonite on growth of several bacterial species. They found that the growth, glucose consumption, and carbon dioxide evolution were all accelerated.

More efficient oxidation of glucose by the microbes was suggested. Kunc and Stotzky (1974) observed variation in heterotrophic activity (carbon dioxide evolution or consumption) after amendment of soil which had been augmented with various clay minerals and a wide array of carbon sources. The initial rate and/or extent of oxidation of the carbon substrate was found to be increased, decreased, or unaffected by addition of various clay minerals. All predictable reactions were observed. Thus, the effect of clay addition on oxidation of soil organic matter is quite variable, depending upon the nature of the clay added.

The effect of clay amendments is further complicated by the incubation conditions of the cultures. Lynch and Cotnoir (1956) observed that bentonite inhibited decomposition of many substrates. In other studies (Martin and Haider, 1971), bentonite increased humic acid formation in sand, whereas in a liquid culture of mixed soil flora, degradation of humic acid was stimulated by bentonite. Thus, the effect of clay amendment on the microbial oxidation of carbon appears to be complicated by effects of different clays, environmental variables, etc. Apparently, the only way of predicting the mitigation effect of clay amendment on soil subsidence is to add clay to the natural environment and to measure the effect.

Pessi (1960) reported that admixture of strips of cultivated peat bog with 300 m^3 clay/ha resulted in a stimulation of the oxidation rate of the soil organic matter. After 35 years of cultivation, there was an average greater subsidence of 12.5 cm in the clayed soil vs that not receiving clay. The differences were attributed partly to increased humification, which was observed to occur under the conditions of higher temperatures that were observed in the clayed soils.

Similarly, the mixture of peat soils with 400 m^3 sand/ha decreased the number of bacteria per soil unit but increased nitrifier, spore-forming bacteria, and mineral nitrogen-utilizing bacteria. Protease and urease activities were enhanced, while catalase, polyphenoloxidase, and peroxidase activities and carbon dioxide evolution were diminished (Zimenko et al., 1974).

The value of admixture of mineral material with organic soils to retard subsidence is uncertain. Certainly, the economic factors, if the mineral material had to be transported over long distances, would be prohibitive for such practices. The data reported above suggest that even if it were economically feasible, admixture of mineral matter with organic soils to retard subsidence at concentrations short of those diluting the organic matter to the levels found in mineral soils would be useless.

5. Concluding Remarks

The foregoing presentation reveals the unique properties of Histosols; yet, it also provides a basis for a discussion of the importance of this soil type in research concerning the nature of microbial interactions with soil organic matter. Drained Histosols provide a valuable resource for the study of soil organic matter oxidation in that (a) they contain high concentrations of organic matter not unlike that found at considerably lower concentrations in mineral soils and (b) frequently this organic matter is deposited in the absence of clay minerals so that the complications in the biochemical studies created by clay adsorption are avoided. Thus, samples of the soil are useful for the electron microscopist interested in examining the physical attachment or relationship of the microbial cell to the soil organic matrix and the biochemist wishing to study the production of enzymes involved in the oxidation of the soil organic matter.

An ecosystem, such as a Histosol, can be studied from two viewpoints. The system can be dissected into its component parts and the individual parts studied in the hope of being able to assemble a description of the processes involved in the function of the whole ecosystem, or the system can be examined as a whole by observing the effect of environmental modification on the overall metabolism of the entire system. As can be concluded from the above presentation, most of the studies of drained Histosols have involved isolation and characterization of the individual microbial species comprising the soil community. This information is interesting and in some cases useful in understanding the microbial community and its relationship with the soil organic matter, but because of the extreme metabolic diversity of the individual heterotrophic microbes, the information to be gained by such studies is limited. Mere possession of a physiological trait by a heterotrophic bacterium does not insure that the trait will be expressed in the soil community. Methods such as the use of fluorescent antibodies prepared for specific members of the microbial community, the use of high-specific-activity ^{14}C-labeled substrates, or even the simple measurement of biomass must now be employed to expand our meager knowledge of this interesting ecosystem. For example, data describing the overall metabolic properties of the soil community under a variety of environmental stresses could be combined with autecological studies of specific heterotrophs prossessing a metabolic trait in question to provide a greater understanding of the role of specific carbon-oxidation reactions in the soil. We must conclude that there is a wide array of modern biochemical and microbiological methods available to the soil microbiologist that were nonexistent for past studies. The soil microbiologist must now meet the challenge to provide a more complete understanding of the microbial interactions of the soil ecosystem.

ACKNOWLEDGMENTS. The author is extremely grateful to Ann C. Tate and Richard E. Terry for their helpful discussions during the preparation of this review and their critical reading of the final manuscript.

References

Alexander, M., and Lustigman, B. K., 1966, Effect of chemical structure on microbial degradation of substituted benzenes, *J. Agric. Food Chem.* **14**:410–413.
Allison, R. V., 1956, The influence of drainage and cultivation on subsidence of organic

soils under conditions of Everglades reclamation, *Soil Crop Sci. Soc. Fla. Proc.* **16**:21–31.

Andriiuk, K. I., Hordienko, S. O., Havrysh, I. N., Konotop, H. I., and Martynenko, V. A., 1973, Decomposition of peat humic acids by associative cultures of microorganisms, *Mikrobiol. Zh. Kiev* **35**:554–559 (in Russian).

Avnimelech, Y., 1971, Nitrate transformations in peat, *Soil Sci.* **111**:113–118.

Bailly, J. R., and Nkundikije-Desseaux, V., 1975, Sur la formation de substances noires a partir de phenols simples par des microorganisms de sol, *Plant Soil* **43**:235–258.

Balba, M. T., and Evans, W. C., 1977, The methanogenic fermentation of aromatic substrates, *Biochem. Soc. Trans.* **5**:302–304.

Biederbeck, V. O., and Paul, E. A., 1971, Fungal degradation of soil humic nitrogen, *Agron. Abstr.* **1971**:80.

Bingeman, C. W., Varner, J. E., and Martin, W. P., 1953, The effect of the addition of organic materials on the decomposition of an organic soil, *Soil Sci. Soc. Am. Proc.* **17**:34–38.

Birch, H. F., and Friend, M. T., 1961, Resistance of humus to decomposition, *Nature (London)* **191**:731–732.

Bondietti, E., Martin, J. P., and Haider, K., 1972, Stabilization of amino sugar units in humic-type polymers, *Soil Sci. Soc. Am. Proc.* **36**:597–602.

Brisbane, P. G., Amato, M., and Ladd, J. N., 1972, Gas chromatographic analysis of amino acids from the action of proteolytic enzymes on soil humic acids, *Soil Biol. Biochem.* **4**:51–61.

Broadbent, F. E., 1960, Factors influencing the decomposition of organic soils of the California delta, *Hilgardia* **29**:587–612.

Buckman, H. O., and Brady, N. C., 1969, *The Nature and Properties of Soils*, Macmillan, New York.

Burns, R. G., Pukite, A. H., and McLaren, A. D., 1972, Concerning the location and persistence of soil urease, *Soil Sci. Soc. Am. Proc.* **36**:308–311.

Butler, J. H. A., and Ladd, J. N., 1969, The effect of methylation of humic acids on their influence on proteolytic enzyme activity, *Aust. J. Soil Res.* **7**:263–268.

Casagrande, D. J., and Park, K., 1978, Muramic acid levels in bog soils from the Okefenokee swamp (Georgia), *Soil Sci.* **125**:181–183.

Casida, L. E., Jr., 1977, Microbial metabolic activity in soil as measured by dehydrogenase determinations, *Appl. Environ. Microbiol.* **34**:630–636.

Catelani, D., Colombi, A., Sorlini, C., and Treccani, V., 1973, Metabolism of biphenyl, *Biochem. J.* **134**:1063–1066.

Ceccanti, B., Nannipieri, P., Cervelli, S., and Sequi, P., 1978, Fractionation of humus-urease complexes, *Soil Biol. Biochem.* **10**:39–45.

Chalvignac, M. A., and Mayaudon, T. J., 1971, Extraction and study of soil enzymes metabolizing tryptophan, *Plant Soil* **34**:25–31.

Cheshire, M. V., Mundie, C. M., and Shepherd, H., 1973, The origin of soil polysaccharide: Transformation of sugars during the decomposition in soil of plant material labeled with ^{14}C, *J. Soil Sci.* **24**:54–68.

Chew, W. Y., Williams, J. C. N., and Ramli, K., 1978, The effect of green manuring on the availability to plants of nitrogen in Malaysian peat, *Soil Biol. Biochem.* **10**:151–153.

Clayton, B. S., 1938, Subsidence of Florida peat soil, *Trans. Int. Soc. Soil Sci. Commun. Zurich* **6B**:840–843.

Crawford, R. L., and Olson, P. P., 1978, Microbial catabolism of vanillate: Decarboxylation to guaiacol. *Appl. Environ. Microbiol.* **36**:539–543.

Dagley, S., 1971, Catabolism of aromatic compounds by microorganisms, *Adv. Microbiol. Phys.* **6**:1–46.

Davis, J. F., and Lucus, R. E., 1959, Organic Soils, Their Formation, Distribution, Utilization and Management, Special Bulletin 425, Michigan Agricultural Experimental Station, East Lansing.

Dawson, J. E., 1956, Organic soils, *Adv. Agron.* 8:377–401.

DeHaan, H., 1976, Evidence for the induction of catechol-1, 2-oxygenase by fulvic acid, *Plant Soil* 45:129–136.

DeHaan, H., 1977, Effect of benzoate on microbial decomposition of fulvic acid in Tjeukemeer (The Netherlands), *Limnol. Oceanogr.* 22:38–44.

Dickinson, C. H., Wallace, B., and Given, P. H., 1974, Microbial activity in Florida Everglades peat, *New Phytol.* 73:107–113.

Duxbury, J. M., and Peverly, J. H., 1978, Nitrogen and phosphorus losses from organic soils, *J. Environ. Qual.* 7:566–570.

Duxbury, J. M., and Tate, R. L., III, 1979, Enzymatic activities in cultivated Histosols, *Soil Sci. Soc. Am. J.*, submitted.

Ellis, N. K., and Morris, R. E., 1945, Preliminary observations on the relation of yield of crops grown on organic soils with controlled water table and the area of aeration in soil and subsidence of the soil, *Soil Sci. Soc. Am. Proc.* 10:282–283.

Engelhardt, G., and Wollnöfer, P. R., 1978, Metabolism of di- and mono-*n*-butyl phthalate by soil bacteria, *Appl. Environ. Microbiol.* 35:243–246.

Evans, W., 1977, Biochemistry of the bacterial catabolism of aromatic compounds in anaerobic environments, *Nature (London)* 270:17–22.

Farnham, R. S., and Finney, H. R., 1965, Classification of organic soils, *Adv. Agron.* 17: 115–162.

Felbeck, G. T., Jr., 1971, Chemical and biological characterization of humic matter, in: *Soil Biochemistry*, Vol. 2 (A. D. McLaren and J. Skujins, eds.), pp. 36–57, Marcel Dekker, New York.

Ferry, J. G., and Wolfe, R. S., 1976, Anaerobic degradation of benzoate to methane by a microbial consortium, *Arch. Microbiol.* 107:33–40.

Florida Sugar Cane League, 1978, Water Quality Studies in the Everglades Area of Florida, The Florida Sugar Cane League, Clewiston, Florida.

Gibson, D. T., 1968, Microbial degradation of aromatic compounds, *Science* 161:1093–1097.

Given, P. H., and Dickinson, C. H., 1975, Biochemistry and microbiology of peats, in: *Soil Biochemistry*, Vol. 3 (E. A. Paul and A. D. McLaren, eds.), pp. 123–212, Marcel Dekker, New York.

Grant, D., 1977, Chemical structure of humic substances, *Nature (London)* 270:709–710.

Greenland, D. J., 1971, Interactions between humic and fulvic acids and clays, *Soil Sci.* 111: 34–41.

Guckert, A., Valla, M., and Jacquin, F., 1975, Adsorption of humic acids and soil polysaccharides on montmorillonite, *Sov. Soil Sci.* 7:89–95.

Gunsalus, I. C., Pederson, T. C., and Sligar, S. G., 1975, Oxygen-catalyzed biological hydroxylations, *Annu. Rev. Biochem.* 44:377–407.

Haider, K., and Martin, J. P., 1967, Synthesis and transformation of phenolic compounds by *Epicoccum nigrum* in relation to humic acid formation, *Soil Sci. Soc. Am. Proc.* 31: 766–772.

Haider, K., and Martin, J. P., 1970, Humic acid-type phenolic polymers from *Aspergillus sydowi* culture medium, *Stachybotrys* spp. cells and autoxidized phenol mixtures, *Soil Biol. Biochem.* 2:145–156.

Haider, K., and Martin, J. P., 1975, Decomposition of specifically carbon-14 labeled benzoic and cinnamic acid derivatives in soil, *Soil Sci. Soc. Am. Proc.* 39:657–662.

Haider, K., Martin, J. P. and Filip, Z., 1975, Humus biochemistry, in: *Soil Biochemistry*, Vol. 4 (E. A. Paul and A. D. McLaren, eds.), pp. 195–244, Marcel Dekker, New York.

Haider, K., Martin, J. P., and Rietz, E., 1977, Decomposition in soil of ^{14}C-labeled coumaryl alcohols: Free and linked in dehydropolymer and plant lignins and model humic acids, *Soil Sci. Soc. Am. J.* 41:556–562.

Haska, G., 1975, Influence of clay minerals on sorption of bacteriolytic enzymes, *Microbiol. Ecol.* 1:234-245.

Hayaishi, O., and Nozaki, M., 1969, Nature and mechanisms of oxygenases, *Science* 164: 389-396.

Hayes, M. H. B., and Mortensen, J. L., 1963, Role of biological oxidation and organic matter solubilization in the subsidence of rifle peat, *Soil Sci. Soc. Am. Proc.* 27:666-668.

Healy, J. B., Jr., and Young, L. Y., 1978, Catechol and phenol degradation by a methanogenic population of bacteria, *Appl. Environ. Microbiol.* 35:216-218.

Herlihy, M., 1973, Distribution of nitrifying and heterotrophic microorganisms in cutover peats, *Soil Biol. Biochem.* 5:621-628.

Holding, A. J., Franklin, D. A., and Watling, R., 1965, Microflora of peat-podzol transitions, *J. Soil. Sci.* 16:44-59.

Huntjens, J. L. M., 1972, Amino acid composition of humic acid-like polymers produced by Streptomycetes and of humic acids from pastures and arable land, *Soil Biol. Biochem.* 4:339-345.

Ibrahim, A. N., and Ibrahim, I. A., 1977, Biodegradation of soil humus by streptomycetes, *Agrokem. Talajtan.* 26:415-423.

Ivarson, K. C., 1977, Changes in decomposition rate, microbial population and carbohydrate content of an acid peat bog after liming and reclamation, *Can. J. Soil Sci.* 57:129-137.

Jenkinson, D. S., 1971, Studies on the decomposition of C^{14} labeled organic matter in soil, *Soil Sci.* 111:64-70.

Kavanagh, T., and Herlihy, M., 1975, Microbiological aspects, *Appl. Bot.* 3:39-49.

Kiss, S., Dragan-Bularda, M., and Radulescu, D., 1975, Biological significance of enzymes accumulated in soil, *Adv. Agron.* 27:25-87.

Kodema, H., and Schnitzer, M., 1974, Adsorption of fulvic acid by nonexpanding clay minerals, *Trans. 10th Int. Cong. Soil Sci.* 2:51-56.

Kononova, M. M., 1966, *Soil Organic Matter: Its Nature, Its Role in Soil Formation and Soil Fertility*, Pergamon Press, New York.

Konovalova, A. S., 1970, Enzymatic activity as a diagnostic criterion of virgin and cultivated sod podzolic soils, *Sov. Soil Sci.* 1970:506-507.

Kosinkiewicz, B., 1977, Humic like substance of bacterial origin. I. Some aspects of the formation and nature of humic-like substances produced by *Pseudomonas*, *Acta Microbiol. Pol.* 26:377-386.

Knipling, E. B., Schroder, V. N., and Duncan, W. G., 1970, CO_2 evolution from Florida organic soils, *Soil Crop Sci. Soc. Fla. Proc.* 30:320-326.

Kunc, F., and Stotzky, G., 1974, Effect of clay minerals on heterotrophic microbial activity in soil, *Soil Sci.* 118:186-195.

Ladd, J. N., 1972, Properties of proteolytic enzymes extracted from soil, *Soil Biol. Biochem.* 4:227-237.

Ladd, J. N., and Butler, J. H. A., 1969a, Inhibition and stimulation of proteolytic enzyme activities by soil humic acids, *Aust. J. Soil Res.* 7:253-261.

Ladd, J. N., and Butler, J. H. A., 1969b, Inhibitory effect of soil humic compounds on the proteolytic enzyme pronase, *Aust. J. Soil Res.* 7:241-251.

Ladd, J. N., and Butler, J. H. A., 1971, Inhibition by soil humic acids of native and acetylated proteolytic enzymes, *Soil Biol. Biochem.* 3:157-160.

Linhares, L. F., and Martin, J. P., 1978, Decomposition in soil of the humic acid-types polymers (melanins) of *Eurotium echinulatum*, *Aspergillus glaucus* and other fungi, *Soil Sci. Soc. Am. J.* 42:738-743.

Lupinovich, I. S., 1968, Change in the physiobiochemical properties of peat-bog soils under influence of melioration and agricultural use, *Sov. Soil Sci.* 1968:755-767.

Lynch, D. L., and Cotnoir, L. J., Jr., 1956, The influence of clay minerals on the breakdown of certain organic substrates, *Soil Sci. Soc. Am. Proc.* 20:367-370.

Mal'tseva, N. N., Gordienko, S. A., and Izzheurova, V. V., 1975, Use of humic acids by oligonitrophilous microorganisms, in: *Tr. S'ezda Mikrobiol. Ukr. 4th* (D. G. Zatula, ed.), pp. 61–62, Naukova Dumka, Kiev, USSR (in Russian).

Mandrovskaya, N. M., 1971, Enzyme activity of peat soil during the first years of its development, *Fiz. Biokhim Kul't Rast* 3:176–179 (in Russian).

Marshall, K. C., 1969, Sorption of illite and montmorillonite to rhizobia, *J. Gen. Microbiol.* 56:301–306.

Martin, J. P., and Haider, K., 1971, Microbial activity in relation to soil humus formation, *Soil Sci.* 111:54–63.

Martin, J. P., and Haider, K., 1975, Decomposition of specifically ^{14}C labeled ferulic acid: Free and linked into model humic acid-type polymers, *Agron. Abstr.* 1975:128.

Martin, J. P., and Haider, K., 1976. Decomposition of specifically carbon-14 labeled ferulic acid: Free and linked into model humic acid-type polymers, *Soil Sci. Soc. Am. J.* 40:377–380.

Martin, J. P., Ervin, J. O., and Shepherd, R. A., 1959, Decomposition and aggregating effect on fungus cell material in soil, *Soil Sci. Soc. Am. Proc.* 23:217–220.

Martin, J. P., Richards, S. J., and Haider, K., 1967, Properties and decomposition and binding action in soil of "humic acid" synthesized by *Epicoccum nigrum*, *Soil Sci. Soc. Am. Proc.* 31:657–662.

Martin, J. P., Haider, K., and Wolf, D., 1972, Synthesis of phenols and phenolic polymers by *Hendersonula toruloidea* in relation to humic acid formation, *Soil Sci. Soc. Am. Proc.* 36:311–315.

Martin, J. P., Haider, K., and Saiz-Jimenez, C., 1974, Sodium amalgam reductive degradation of fungal and model phenolic polymers, soil humic acids, and simple phenolic compounds, *Soil Sci. Soc. Am. Proc.* 38:760–764.

Martin, J. P., Filip, A., and Haider, K., 1976, Effect of montmorillonite and humate on growth and metabolic activity of some actinomycetes, *Soil Biol. Biochem.* 8:409–413.

Mathur, S. P., 1971, Characterization of soil humus through enzymatic degradation, *Soil Sci.* 111:147–157.

Mathur, S. P., and Morley, H. V., 1975, Biodegradation approach for investigating pesticide incorporation into soil humus, *Soil Sci.* 120:238–240.

Mayaudon, J., Batistic, L., and Sarkar, J. M., 1975, Properties of proteolytically active extracts of fresh soil, *Soil Biol. Biochem.* 7:281–286.

McCollum, S. H., Carlisle, V. W., and Volk, B. G., 1976, Historical and current classification of organic soils in the Florida Everglades, *Soil Crop Sci. Soc. Fla. Proc.* 35:173–177.

McLaren, A. D., and Packer, L., 1970, Some aspects of enzyme reactions in heterogenous systems, *Adv. Enzymol.* 33:245–308.

Millette, J. A., 1976, Subsidence of an organic soil in southwestern Quebec, *Can. J. Soil Sci.* 56:499–500.

Mishustin, E. N., Tepper, E. Z., Pushkareva, T. V., and Golovanova, L. F., 1974, Mineralization of organic matter during the development of peat-bog soil, *Pochvovedenie* 1974: 78–83 (in Russian).

Nannipieri, P., Cervelli, S., and Pedrazzini, F., 1975, Extraction of enzymically active organic matter from soil, *Experientia* 31:513–515.

Neller, J. R., 1944, Influence of cropping, rainfall, and water table on nitrates in Everglades peat, *Soil. Sci.* 57:275–280.

Paul, E. A., and Mathur, S. P., 1967, Cleavage of humic acids by *Penicillium frequentans*, *Plant Soil* 27:297–299.

Paulson, K. N., and Kurtz, L. T., 1969, Locus of urease activity in soil, *Soil Sci. Soc. Am. Proc.* 33:897–901.

Paulson, K. N., and Kurtz, L. T., 1970, Michaelis constant of soil urease, *Soil Sci. Soc. Am. Proc.* 34:70–72.

Pereverzev, V. N., and Golovko, E. A., 1968, Effect of cultivation on physiochemical properties and biological activity of peat-bog soils, *Sov. Soil Sci.* 1968:359-367.

Pessi, Y., 1960, The effects of claying upon the settling of the soil surface on cultivated sphagnum bogs, *Maataloust Aikak.* 32:5-7.

Plotkina, Y. M., Yerkevich, E. A., and Falyushin, P. L., 1975, Effect of agricultural development of peat-bog soils on humic acid, *Agrokhimiya* 10:95-100.

Powers, W. L., 1932, Subsidence and durability of peaty lands, *Agric. Eng.* 13:71-72.

Reddy, K. R., and Patrick, W. H., Jr., 1975, Effect of alternate aerobic and anaerobic conditions on redox potential, organic matter decomposition and nitrogen loss in a flooded soil, *Soil Biol. Biochem.* 7:87-94.

Revinskaya, L. S., and Filipshonova, L. I., 1973, Effect of water and temperature conditions on the activity of microorganisms promoting the mineralization of nitrogen- and carbon-containing organic compounds of peat, *Ispol'z. Microorg. Ikh. Metab. Nar. Khoz.* 1973:133-138 (in Russian).

Ruocco, J. J., and Barton, L. L., 1978, Energy-driven uptake of humic acids by *Asperigillus niger, Can. J. Microbiol.* 24:533-536.

Schlesinger, W. H., 1977, Carbon balance in terestrial detritus, *Annu. Rev. Ecol. Syst.* 8:51-81.

Schnitzer, M., and Khan, S. U., 1972, *Humic Substances in the Environment*, Marcel Dekker, New York.

Schnitzer, M., Ortiz de Serra, M. I., and Ivarson, K., 1973, The chemistry of fungal humic acid-like polymers and of soil humic acids, *Soil Sci. Soc. Am. Proc.* 37:229-236.

Schothorst, C. J., 1977, Subsidence of low moor peat soils in the western Netherlands, *Geoderma* 17:265-291.

Shishkova, Z., Kalnins, A., Gailitis, Y., Popov, S., and Krastins, V., 1972, Utilization of humin substances by yeast, *Proc. 4th Int. Peat Cong. 1972* 4:239-249.

Sidorenko, O. D., Aristarkhova, V. I., and Chernikov, V. A., 1978, Changes in the composition and properties of humic acids after treatment with Nocardia microorganisms, *Izv. Akad. Nauk Kaz. SSSR Ser. Biol.* 2:195-202 (in Russian).

Skujins, J. J., 1967, Enzymes in soil, in: *Soil Biochemistry*, Vol. 1 (A. D. McLaren and G. A. Peterson, eds.), pp. 371-414, Marcel Dekker, New York.

Skujins, J. J., 1973 Dehydrogenase: An indicator of biological activity in soil, *Bull. Ecol. Res. Commun. NFR Statens Naturvetensk. Forskningsrad* 17:235-241.

Sorensen, L. H., 1974, Rate of decomposition of organic matter in soil as influenced by repeated air drying–rewetting and repeated additions of organic matter, *Soil Biol. Biochem.* 6:287-292.

Sorensen, L. H., and Paul, E. A., 1974, Transformation of acetate carbon into carbohydrate and amino acid metabolites during decomposition in soil, *Soil Biol. Biochem.* 3:173-180.

Sowden, F. J., 1970, Action of proteolytic enzymes on soil organic matter, *Can. J. Soil Sci.* 50:233-241.

Steinbrenner, K., and Mundstock, I., 1975, Untersuchungen zum Bituminstaff Abbau durch Nokardien, *Arch. Acker Pflanzenbau Bodenkd.* 19:243-255.

Stephens, J. C., 1969, Peat and muck drainage problems, *J. Irr. Drainage Div. Am. Soc. Civil Eng.* 95:285-305.

Stephens, J. C., and Stewart, E. H., 1976, Effect of climate on organic soil subsidence, in: *Proceedings, Anaheim Symposium*, December 1976, International Association of Hydrological Sciences, Publication No. 121 pp. 647-655, Washington, D.C.

Stotzky, G., and Mortensen, J. L., 1957, Effect of crop residues and nitrogen additions on decomposition of an Ohio muck soil, *Soil Sci.* 83:165-174.

Stotzky, G., and Mortensen, J. L., 1958, Effect of addition level and maturity of rye tissue on the decomposition of a muck soil, *Soil Sci. Soc. Am. Proc.* 22:521-524.

Stotzky, G., Martin, W. P., and Mortensen, J. L., 1956, Certain effects of crop residues and fumigant applications on the decomposition of an Ohio muck soil, *Soil Sci. Soc. Am. Proc.* **20**:392–396.

Sunk, I. V., 1929, Microbiological activities in the soil of an upland bog in eastern North Carolina, *Soil Sci.* **27**:283–303.

Taha, S. M., Zayed, M. N., and Zohdy, L., 1973, Studies in humic acids decomposing bacteria in soil II. Isolation and identification of microorganisms, *Z. Bakteriol. Parasiteinkd. Infektionskr.* **128**:168–172.

Tate, R. L., III, 1976, Nitrification in Everglades Histosols: A potential role in soil subsidence, in *Proceedings, Anaheim Symposium,* December 1976, International Association of Hydrological Sciences, Publication No. 121, pp. 657–663, Washington, D. C.

Tate, R. L., III, 1979a, Microbial activity in organic soils as affected by soil depth and crop, *Appl. Environ. Microbiol.* **37**:1085–1090.

Tate, R. L., III, 1979b, Effect of flooding on microbial activities in organic soil: Carbon metabolism, *Soil Sci.* **128**:267–273.

Tate, R. L., III, 1980, Effect of several environmental parameters on carbon metabolism in Histosols, *Microbial. Ecol.* **5**:329–336.

Terry, R. E., 1980, Nitrogen mineralization in Histosols, *Soil Sci. Soc. Am. J.* (in press).

Terry, R. E., and Tate, R. L., III, 1980, Denitrification as a pathway for nitrate removal from organic soils, *Soil Sci.* **129**:162–166.

Thomas, F. H., 1965, Subsidence of peat muck soils in Florida and other parts of the United States—A review, *Soil Crop Sci. Soc. Fla. Proc.* **25**:153–160.

Turner, J. A., and Rice, E. L., 1975, Microbial decomposition of ferulic acid in soil, *J. Chem. Ecol.* **1**:41–58.

Vandecaveye, S. C., 1932, Effect of stable manure and certain fertilizers on the microbiological activities in virgin peat, *Soil Sci.* **33**:279–299.

Vavulo, F. P., Vorob'eva, E. N., and Plotkina, N. N., 1974, Effect of sprinkling on the seasonal dynamics of the biological activity of shallow bog-peat soils, *Fiz. Khim. Geokhim. Mickrobiol. Protsessy Melior irovannykh Pochv Poles'ya* **1974**:131–147 (in Russian).

Verma, L., and Martin, J. P., 1976, Decomposition of algal cells and components and their stabilization through complexing with model humic acid-type phenolic polymers, *Soil Biol. Biochem.* **8**:85–90.

Verma, L., Martin, J. P., and Haider, K., 1975, Decomposition of carbon-14-labeled proteins, peptides and amino acids; free and complexed with humic polymers, *Soil Sci. Soc. Am. Proc.* **39**:279–284.

Visser, S., 1964, Presence and effects of microorganisms in tropical peat sediments, *Ann. Inst. Pasteur (Paris)* **107**:303–319.

Volk, B. G., 1972, Everglades Histosol subsidence 1. CO_2 evolution as affected by soil type, temperature, and moisture, *Soil Crop Sci. Soc. Fla. Proc.* **32**:132–135.

Volk, B. G., and Schnitzer, M., 1973, Chemical and spectroscopic methods for assessing subsidence in Florida Histosols, *Soil Sci. Soc. Am. Proc.* **37**:886–888.

Volk, B. G., and Zelazny, L. W., 1973, Infrared analysis of selected Florida Histosols, *Soil Crop Sci. Soc. Fla. Proc.* **33**:132–136.

Wagner, G. H., 1975, Microbial growth and carbon turnover, in: *Soil Biochemistry,* Vol. 3 (E. A. Paul and A. D. McLaren, eds.), pp. 269–305, Marcel Dekker, New York.

Waksman, S. A., and Purvis, E. R., 1932, The microbiological population of peat, *Soil Sci.* **34**:95–113.

Waksman, S. A., and Stevens, K. R., 1929, Contribution to the chemical composition of peat: III. Chemical studies of two Florida peat profiles, *Soil Sci.* **27**:271–281.

Wheatley, R. E., Greaves, M. P., and Inkson, R. H. E., 1976, The aerobic bacterial flora of a raised bog, *Soil Biol. Biochem.* **8**:453–460.

Weir, W. W., 1950, Subsidence of peatlands of the Sacramento–San Joaquin Delta, California, *Hilgardia* **20**:37–56.

Wolf, D. C., and Martin, J. P., 1976, Decomposition of fungal mycelia and humic-type polymers containing carbon-14 from ring and side-chain labeled 2, 4-D and chlorpropham, *Soil Sci. Soc. Am. J.* **40**:700–704.

Terdokimova, N. V., Mostovyy, M. N., and Malyy, Y. I., 1976, Subsidence and biochemical destruction of peat in the Ukrainian Poles'ye, *Sov. Soil Sci.* **1976**:345–347.

Zimenko, T. G., 1972, Activity of the microorganisms and the mineralization of the organic matter in peat soils with varying level of subsoil water, *Izv. Akad. Nauk Kaz. SSR Ser. Biol.* **1972**:846–854 (in Russian).

Zimenko, T. G., Gaurilkina, N. V., Filimonova, T. V., Lishtvan, L. M., Filipshanova, L. I., and Aleichik, L. G., 1974, Changes in the microbiological processes in sandy bog-peat soils, *Fiz. Khim. Geokhim. Microbiol. Protsessy Meliorirovannykh Pochv Poles'ya* **1974**:122–131 (in Russian).

6

The Immunofluorescence Approach in Microbial Ecology

B. BEN BOHLOOL AND EDWIN L. SCHMIDT

1. Introduction

Fluorescent markers appropriately conjugated to antibody proteins provide the basis for a method to visualize those antibodies as they participate in antigen–antibody reactions. The method is referred to as the fluorescent antibody (FA) or immunofluorescence (IF) technique; it has been in widespread and successful use in medical microbiology and in pathology as a highly sensitive and specific cytochemical staining procedure for many years. Most current applications of the technique are for the localization of cellular and viral antigens in tissues and for the rapid detection and identification of infectious agents.

The same features of the FA technique that make possible the detection and identification of microbial pathogens in animal tissues provide its potential for autecological application in microbial ecology. Autecology, as an approach whereby the individual microorganism is studied directly in its natural environment, has been virtually unavailable to microbial ecology because of technical difficulties imposed by the small size and nondescript morphology of microorganisms and by the physical complexities of the natural environment. Direct microscopic examination of the natural environment could become a powerful tool for autecological study if a method were available to recognize one specific microorganism among all others present in the field of view. Such a capability is inherent in the FA technique. Because extremely small amounts of fluorochrome-marked antibody may be detected against a dark background by fluo-

B. BEN BOHLOOL • Department of Microbiology, University of Hawaii, Honolulu, Hawaii 96822.   EDWIN L. SCHMIDT • Departments of Microbiology and Soil Science, University of Minnesota, St. Paul/Minneapolis, Minnesota 55108.

rescence microscopy (Goldman, 1968), microorganisms which bind the antibody may be visualized with great sensitivity. The companion property of specificity results from the great precision of the antigen–antibody reaction. Developments to exploit the unique potential of FA for microbial ecology have come about mostly in the past decade with greater availability of the necessary equipment and facilities and with the resolution of certain technical difficulties.

1.1. Theory and General Considerations

The theory for FA as applied to microbial ecology is simple (see Fig. 1). An isolate of some particular interest is used as an antigen for the preparation of active antiserum. Antiserum to the isolate is labeled with a fluorescent dye, usually fluorescein isothiocyanate (FITC), and the labeled antiserum (FA) is then applied as a stain to a sample of a natural environment. If the microorganism of interest is present in the natural material, the FA will combine specifically with it in an antigen–antibody reaction. When the stained preparation is examined by fluorescence microscopy, the outline of the microorganism of interest is seen by virtue of light emitted from FA bound to its surface. Other organisms present in the same field are devoid of antibody and hence invisible. Figure 2 illustrates how IF can recognize the desired bacteria among all the other microorganisms present in a natural sample. The preparation is a smear from human feces which

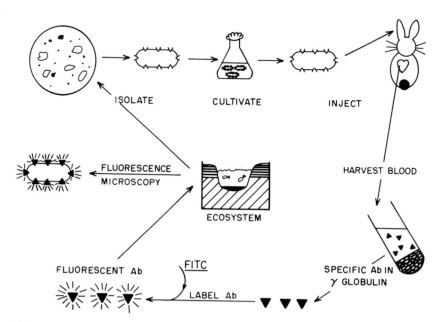

Figure 1. Theory and general considerations in the application of immunofluorescence to autecological studies of microorganisms in natural samples.

was artifically inoculated with *Salmonella typhimurium*. When the smear is stained with a common fluorescent stain such as acridine orange (IA), a wide variety of unidentifiable microorganisms could be seen, but when the specific FA is used, only the organism of interest stains.

Two methods of FA staining are in common usage, the "direct" and the "indirect." For the direct FA, the antibody against the microorganism of interest is itself labeled. The major advantage of the direct method resides in the ability to work with a single antibody reagent. In the indirect or "sandwich" method, the specific antibody is applied unlabeled and is detected on the surface of the cell by a second antibody which is fluorescent and which is directed against the specific antibody. The specific antibody normally is prepared in rabbits injected with the antigen of interest, and the indicator antibody is prepared in goats in response to rabbit globulin. The goat anti-rabbit serum is labeled with a fluorochrome. An obvious advantage of the indirect method is that a single fluorescent antibody can be used to detect the localization of various specific antibodies.

There are a number of background references devoted to the theory, principles, and practical considerations of FA techniques in general. All are oriented to the many applications of FA for use in medical microbiology and pathology. None includes recognition of the much more recent development of the use of FA in microbial ecology. The treatises of Goldman (1968), Kawamura (1977), and Nairn (1975) are the most extensive and should be accessible to the microbial ecologist concerned with the application of FA or the interpretation of FA data. Goldman in particular gives a fine account of the history wherein the central contributions of Albert Coons and co-workers at Harvard University are appropriately highlighted. A small, highly practical, and extremely useful manual published by Cherry *et al.* (1960) unfortunately is out of print, but worth a search on the part of the initiate to FA techniques. Basic immunological procedures starting with the handling of animals are covered in a well-illustrated text by Garvey *et al.* (1977).

The fluorescence microscope is a conventional light microscope equipped with a light source of suitable intensity and wavelength and with necessary filters. Information on the various components of suitable fluorescence microscopy systems is readily available from technical representatives of the major microscope manufacturers.

Appropriate controls are important in all FA procedures, but particularly so when applied to ecological questions. Microbial ecology may call for antibodies to microorganisms whose serology is unknown, and for staining of natural materials of great physical, chemical, and biological diversity. Controls relating to the specificity of the FA reagents and other types of controls as they pertain to applications in microbial ecology were outlined by Schmidt (1973).

1.2. Development of FA for Microbial Ecology

The first paper clearly indicating the potential of FA for microbial ecology was that of Hobson and Mann (1957). They demonstrated that several bacteria isolated from rumen could be detected *in situ* by staining rumen contents with appropriate fluorescent antibodies. Surprisingly, this interesting work was not extended by the authors or other rumen microbiologists for many years. Development of FA for the most complex of natural ecosystems, those of terrestrial environments, began with the paper by Schmidt and Bankole (1962). This and subsequent papers (Schmidt and Bankole, 1963, 1965) were concerned with detection of the fungus *Aspergillus flavus* in soil. Isolates of the fungus were cultured in the vicinity of buried glass microscope slides and subsequently observed on those contact slides by immunofluorescence. The specificity of the FA was found to be adequate for autecological study of *A. flavus,* and background fluorescence could be controlled by the use of certain filters.

Whereas the autecological study of certain fungi by FA appeared feasible for terrestrial habitats, such was not the case for bacteria. The major problem was that of nonspecific adsorption of the FA to soil materials (nonspecific staining). Due to the small size of bacteria, the amount of FA deposited on the cell is very low relative to that localized on a fungal hypha; consequently, the microscopy demands the most efficient lighting and filter systems possible. Required also is a desirable dark, nonfluorescing background. This is necessary not only because the sites of fluorescence are so small that they must be seen in contrast, but also because nonspecifically fluorescing background materials may obscure or even resemble bacteria. Levels of nonspecific staining that could be controlled for fungi with inefficient filter combinations or tolerated because of their larger size and distinctive morphology could not be tolerated for bacteria.

The problem was avoided to some extent in work reported by Hill and Gray (1967). FA techniques were used for differentiation of *Bacillus subtilis* and *B. circulans* in a very sandy forest soil which did not give excessive background fluorescence. Attempts on their part to improve the background with a variety of reagents were largely unsuccessful. Limitations imposed by nonspecific staining were illustrated in the photomicrographs of Schmidt *et al.* (1968). They characterized an FA for a strain of the symbiotic nitrogen fixer *Rhizobium japonicum.* The FA was shown to be highly specific at the strain level and capable of detecting that strain in unknown, wholly natural field soils. Despite the obvious utility of the approach, it was shown that specifically stained cells were obscured and artifacts were troublesome in certain microscopic fields where the conjugate attached nonspecifically to soil colloids and soil films.

Difficulties with respect to nonspecific adsorption of FA to soil background were satisfactorily resolved by Bohlool and Schmidt (1968). A dilute gelatin solution was partially hydrolyzed at a high pH and applied to a soil preparation before staining with the labeled antibody. Such pretreatment with the gelatin apparently saturated the sites of nonspecific adsorption. Subsequent application of the FA resulted in specific staining of the antigens with no interference from the gelatin and with no adsorption of FA to nonspecific sites. By conjugating the gelatin with a fluorochrome of contrasting color to that used on the antibody, the gelatin not only prevented nonspecific staining but also served as a counterstain which made it possible to see the specifically stained bacteria in relation to the gelatin-labeled portions of the microenvironment on a contact slide. Rhodamine isothiocyanate was the fluorochrome conjugated to the hydrolized gelatin.

Early applications made of FA techniques in microbial ecology were summarized in a review by Schmidt (1973). The number was relatively small, and the nature of the reports reflected a concern primarily for the detection and identification of different microorganisms in their natural habitats. It was noted that FA in microbial ecology was still largely in a developmental phase and that few serious autecological problems had been addressed.

One inherent limitation that hampered more effective use of FA for problems dealing with microbes in terrestrial environments especially was the qualitative nature of the technique. Many problems require information gained from enumeration of microorganisms in relation to a process or to a change in the environment. Enumeration data provide a basis for the estimation of biomass, growth rate, and growth response to environmental variables. Such information is obtainable by other techniques only for extreme environments where species diversity is sharply limited (Brock, 1971); hence, it was highly desirable that the unique specificity of the FA be directed to quantitative examination of the normal habitat with characteristic complexity and diversity.

The basic problem in enumerating FA-stained bacteria is again size related. Because of the magnification needed, the area of the field of view is very small, and the amount of soil that can be examined per field without significant interference is correspondingly small. As a consequence, populations must be high in order to encounter a reasonable number of bacteria. Total direct counts by conventional microscopy, as an example, use a maximum of about 1 mg of soil/cm^2. To count one bacterium per field under such circumstances, at approximately 1000 $\times$ magnification, the cell density must be about 10^6 to 10^7/g soil. The problem is less severe for aquatic systems since microorganisms in a given volume may be concentrated onto a membrane filter and observed with incident light fluorescence microscopy. The maximum of dispersed soil that may be examined in a similar fashion is about that deposited during filtration of 0.1 ml of a 1/10 soil dilution. In order to count 1 bacterium per field under such circumstances, the cell density must be about 2×10^6/g. Quite obviously if quantitative autecology were to be carried out at realistically low population levels for the terrestrial system, it would be necessary to separate the bacteria from soil and concentrate them on a surface for FA staining.

A method devised for quantitative FA examination of soil bacteria was first outlined by Bohlool and Schmidt (1973a) and later presented in more detail by Schmidt (1974). Essential steps in the method involve dispersion of a diluted soil sample to release bacteria into suspension, flocculation to remove soil particulates from the supernatant fluid, filtration of a portion of the supernatant fluid through an appropriately pretreated membrane filter, and finally the microscopic enumeration of the specific bacterium on the FA-stained membrane filter surface. Methodological details may be modified somewhat to accommodate to the special properties of the particular soils under examination. The availability of suitable quantitative FA procedures may be expected to provide a new and workable approach to problems in microbial ecology that are currently refractory.

2. Nitrogen-Fixing Bacteria

Papers on nitrogen fixation and the bacteria associated with the process probably comprise a substantial majority of the literature of microbial ecology. Despite their obvious biogeochemical importance and the research attention devoted to them, the nitrogen-fixing bacteria are little known with respect to their biology and activity in natural environments. Only the symbiotic nitrogen fixers, with emphasis on the root-nodule bacteria of legumes, have been studied intensively in their nitrogen-fixing milieu. The root nodule presents a discrete and manipulatable niche where the bacteria occur in pure culture. Niches for the free-living nitrogen fixers and for free-living stages of symbiotic fixers are in the mixed-culture communities of natural habitats, which are inaccessible to direct study. It is not surprising that aspects of the nitrogen-fixing bacteria were among the first of the ecological applications of FA techniques.

2.1. Rhizobia

Various methods are available to study rhizobia once they are concentrated in their main niche in nature, the legume root nodule. FA is a useful adjunct to these methods for the rapid identification and characterization of strains in the nodule isolates, but the main promise of FA in *Rhizobium* ecology lies in the study of events prior to nodulation. Prenodulation events are those that involve the ecology of the *Rhizobium* strain as it adapts to the soil, responds to the rhizosphere of the developing legume host, somehow recognizes the appropriate nodulation site on precisely the right legume root, and begins to interact with the plant to form a functional nodule. Such events are of great practical importance to the effective management of *Rhizobium* and its host legume, but little detailed information is available because the complexity of the plant–soil–bacterium interactions has permitted only indirect experimental approaches. The FA technique is of special pertinence to the ecology of free-living rhizobia because it is the only method to provide the potential for direct investigation of the *Rhizobium* in the soil. Its attractiveness is further enhanced for such studies by the existence of a substantial background literature on the serology of the genus (Graham, 1963; Holland, 1966; Vest *et al.*, 1973; Dudman, 1977).

The first paper to report an autecological study of rhizobia was that of Schmidt *et al.* (1968), a paper that was concerned with *R. japonicum*. A number of important aspects emerged from this study. It was found that the FA-staining reaction was highly strain specific: FA prepared against 1 strain did not cross-react with other strains of R. japonicum belonging to at least 6 other serogroups, nor with other rhizobia or 65 unidentified soil bacteria isolated from 12 soils. Specificity at the strain level is highly desirable for study of the rhizobia because the ecological questions are usually asked in terms of a particular strain of interest because of its superior ability to fix nitrogen. Also demonstrated was the ability to detect the antigen strain on contact slides during its growth in an autoclaved soil. The technique moreover detected FA-reacting bacteria on contact slides in a field soil whose rhizobial content was unknown. This presumptive evidence for the natural occurrence of the antigen strain was strengthened when it was found that the same soil could be used as an inoculant of soybeans to produce nodules whose bacteroids cross-reacted with the specific FA. A final aspect of this study was the evidence that FA was useful to detect and identify a bacteroid strain directly in nodule crushes.

Although FA detection in soil of a specific R. japonicum strain was clearly feasible, Schmidt et al. (1968) noted the nonspecific adsorption of FA to soil particulates and soil films. Further modification of the technique to overcome nonspecific staining was successful (Bohlool and Schmidt, 1968). In subsequent experiments, the FA technique as modified to control nonspecific staining was used (Bohlool and Schmidt, 1970) for the detection of two distinct serotypes of R. japonicum in a range of nonsterile field soils.

R. leguminosarum was studied by Zvyagintzev and Kozhevin (1974) following inoculation at 10^7 to 10^8 cells/g in a soddy-podzolic and chernozem soil. A variation of the direct smear technique was used for quantification by indirect IF. No specificity control data were presented, which is unfortunate, for the authors made the interesting observation that the addition of glucose favored the development of R. leguminosarum in the soils. The ability of rhizobia to compete with other bacteria in soil for available substate is a highly important ecological feature, and it is imperative that FA data reporting a free-living growth response of rhizobia include appropriate controls to insure that only the specific Rhizobium was detected.

All of the work cited, together with the development of suitable procedures to permit the enumeration of bacteria in natural habitats by FA (Bohlool and Schmidt, 1973a; Schmidt, 1974), was either preliminary in the sense of assessing the FA technique for ability to detect rhizobia in the complex soil environment, or developmental as was necessary to overcome the limitations that were encountered. None of the ecological questions that have been difficult or impossible to approach by indirect methods had yet been addressed by FA.

Considerable attention has been focused on the inability of desirable inoculant strains of rhizobia to compete with indigenous strains in field soils (Vest et al., 1973). The soil-adapted strains commonly account for the majority of nodules even though high populations of a potentially better nitrogen-fixing strain are added as inoculant. Factors affecting the competitive success of given rhizobia have not been resolved, and studies of competition between strains have had to rely on the end result of competitive interactions—the nodule. Competition studies have not yet taken advantage of FA to examine the obviously important population dynamics of competing strains in the soil and rhizosphere before the race has been won. FA can be used in the terminal analysis of nodule rhizobia for strain identification purposes (Schmidt et al., 1968; Trinick, 1969) and hence constitutes a convenient means to identify the strain or strains that have been successful under various competition conditions (Jones and Russell, 1972; Bohlool and Schmidt, 1973b; Lindemann et al., 1974).

Jones and Russell (1972) carried out a highly artificial study involving two serologically distinctive strains of Rhizobium trifolii, separately or in a pure-culture mixture, and aseptically grown clover seedlings. FA was used to serotype the nodules. A more extensive but still preliminary study of competition was reported by Bohlool and Schmidt (1973b). They measured the ability of R. japonicum strain USDA 110 to nodulate soybeans in competition with the resident rhizobia of a silt loam soil. Varying numbers of strain 110 were added to soil containing a low but constant population of indigenous R. japonicum, and the soil was used to inoculate soybeans. Relatively high inoculant/resident ratios were needed to overwhelm the resident populations. When the log of the numbers of introduced strain 110 was plotted against the percentages of strain 110 nodules, a sigmoidal curve resulted. The authors suggested that curves similarly derived for any given soil might be descriptive of the resident competition barrier to be overcome by an inoculant strain. Strain 110, added to the test soil and then allowed to incubate, apparently failed to persist since it formed progressively fewer nodules in competition with indigenous strains with increasing time.

Competition experiments of another sort were carried out by Lindemann et al. (1974) with detailed IF examination of nodules to provide evidence that two serologically different strains of R. japonicum may occur in a single soybean nodule. Strains USDA 117 and 138 were used in 50:50 mixtures at different population levels to inoculate soybeans grown in Leonard jars. At an inoculation density of 10^8 rhizobia/plant, 32% of the nodules contained both strains, and this double infection declined as inoculum density decreased. Strain 138 generally was more competitive than strain 117. The widely accepted assumption that only a single strain of Rhizobium is to be found in a given nodule was found not necessarily to be the case.

Double infection was also demonstrated in nodules of lentils grown in the field (May, 1979). Seeds pelleted with mixtures of serologically distinct strains of Rhizobium leguminosarum were planted in a Hawaiian inceptisol. IF typing of the resultant nodules revealed that over 30% of nodules could contain two strains, if two highly competitive strains (e.g., Hawaii 5-0 and NZP5400) were used as inoculum. On the other hand, when the pair in the inoculum included a poor competitor (e.g., NZP5400 and 128A12), then only 6% of the nodules were

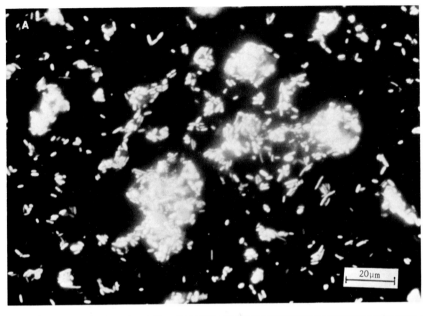

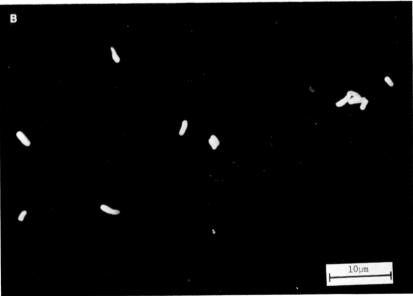

Figure 2. Smear from human feces which was artificially inoculated with *S. typhimurium* Succ-L. A: Stained with acridine orange. B: Stained with FA against *S. typhimurium* Succ-L.

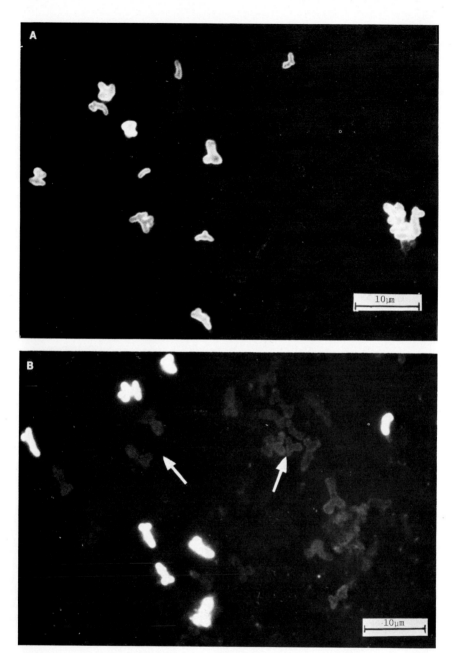

Figure 3. Mixed infection in nodules of lentils grown in the field. Smears from two nodules from the same plant were stained with FA against strain Hawaii 5-0. A: All the bacteroids stain (single strain in the nodule). B: Only a proportion of bacteroids stain (two strains in the same nodule). The unstained bacteroids (arrows) in B are visualized by the use of a double-light system using a dark-field condenser. From May (1979) and May and Bohlool (1979).

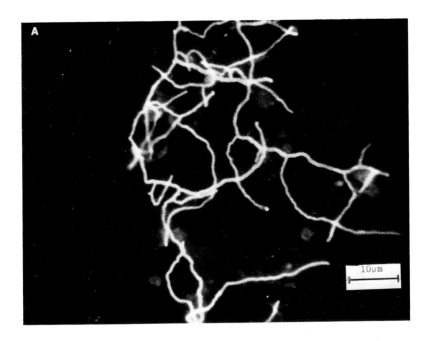

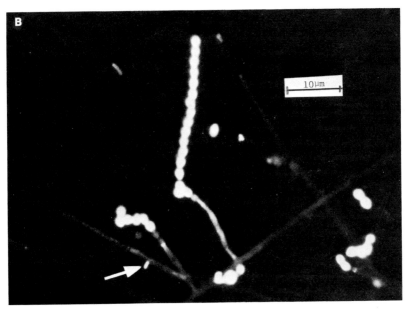

Figure 4. Problems associated with (A) heterologous cross-reaction and (B) nonspecific staining. A: FA against *R. japonicum* strain Nitragin 61A72 cross-reacts with an unidentified soil actinomycete growing in autoclaved soil. From Bohlool and Schmidt (1970). B: FA against *R. japonicum* strain USDA31 nonspecifically stains conidia of a soil *Fusidium* growing, together with USDA31, in autoclaved soil. Arrow depicts a cell of strain USDA31 attached to the fungal filament. From Bohlool and Schmidt (unpublished).

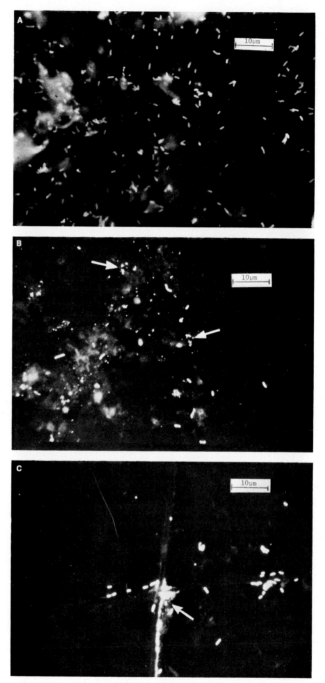

Figure 5. Disintegration of *R. japonicum* USDA110 cells in normal soils. A: Intact cells at 1 day. B: Intact cells and antigen debris (arrows) at 8 days. C: Intact cells and solubilized antigen (arrow) along filament of unidentified soil fungus.

doubly infected. Figure 3 shows IF typing of two nodules from the same lentil plant: A, nodule formed by a single strain and B, nodule containing two strains.

The procedures to use FA to quantify soil bacteria "*in situ*" were developed with *R. japonicum* as a model system and were applied to study the growth rate of that species in soil. Bohlool and Schmidt (1973a) made FA counts of strain 110 during growth of the bacterium in a sterilized silt loam soil. The growth curve was conventional and closely similar to that obtained by plate count, with a mean generation time of 14 hr by either procedure. Growth features of the same strain inoculated into nonsterile soil and followed by FA were strikingly different (Schmidt, 1974). Growth rates were extremely slow, with a mean doubling time of 240 hr in one of the test soils and 365 hr in another. These limited studies merely showed that FA could be used to study growth rates in soil and that growth in natural mixed-culture habitats probably bears little relation to pure-culture growth. The population features of both indigenous and introduced rhizobia in soil make up a highly important segment of rhizobial ecology which should be pursued.

The ecological question of *Rhizobium* response to the rhizosphere of developing legumes is one that should profit from direct FA examination. Indirect evidence by plant-dilution assay has indicated that rhizobia in the rhizosphere of host legumes increase greatly in numbers as compared to nonrhizosphere soil (Rovira 1961; Tuzimura and Watanabe, 1961; Robinson, 1967). Such data have been interpreted as "specific stimulation" by a number of authors (Nutman, 1965; Brown *et al.*, 1968; Vincent, 1974), as indicating that a host legume somehow selectively enhances growth of its symbiont over other bacteria and other rhizobia. Such a rhizosphere effect may merely mean that rhizobia are good rhizosphere bacteria. The nature of the rhizosphere response and its significance to nodulation are unresolved problems.

The first direct study of the population dynamics of a *Rhizobium* in rhizospheres was carried out with FA specific for the serotype *R. japonicum* USDA 123 (Reyes and Schmidt, 1979). Indigenous populations of strain 123 were only a few hundred to a few thousand per gram of field soil. Rhizosphere effects from field-grown soybean plants were modest, reaching a maximum of about 10^4 to 10^5 cells of strain 123/g inner rhizosphere soil. Comparably slight rhizosphere effects were observed with field corn. Pot experiments with the same soil gave data comparable to that derived from field soils. No evidence suggestive of the specific stimulation theory was found in any of these experiments.

Vidor and Miller (1980a) used IF techniques to study four strains of *R. japonicum* heavily inoculated (10^7/g) into two normal Ohio soils. All strains introduced into a *R. japonicum*-free soil dropped to about 10^6/g level, which was maintained for 180 days. Strains inoculated into a soil with an indigenous *R. japonicum* population died off rapidly as compared to the other soil, and evidence suggestive of a lytic bacteriophage was presented. This work is of particular interest in that it illustrates the feasibility of a direct approach in the evaluate of saprophytic competence in soil of specific strains and some of the factors involved in rhizobia–soil population interactions. In a companion study, Vidor and Miller (1980b) followed population changes by FA in the same *R. japonicum* strains in soils supporting soybean plants. Populations of strain 110 increased substantially from 10^3 to 10^6/g soil, whereas cell densities of the other strains increased only slightly. Release of rhizobia from nodules to augment soil populations as observed by Reyes and Schmidt (1979) was also noted in this study.

2.2. Asymbiotic Nitrogen-Fixing Bacteria

The capacity for nitrogen fixation by free-living procaryotes is scattered among numerous genera. Recent listings (Knowles, 1977; LaRue, 1977) attest to the considerable biological and ecological diversity of forms with this potential. But the extent to which the nitrogen-fixing capacity of these organisms is actually expressed in nature is an unresolved problem that is crucial to an understanding of the nitrogen economy of the biosphere. Despite advances that have been made in the detection and assay of the nitrogen-fixation process, process chemistry will provide data of little predictive value unless related to the microbiology responsible for the process. Asymbiotic nitrogen fixation is tied to growth, and the nitrogen fixed may be suspected to vary in magnitude and rate with the growth response of the bacteria responsible. The problem thus involves the identification of the causative organisms in a nitrogen-fixing habitat and the quantification of their growth responses in relation to the atmospheric nitrogen they assimilate. The contribution of FA to asymbiotic nitrogen fixation will be in direct proportion to its capacity to identify the nitrogen fixers in a natural environment and to measure their growth. Very little has been done to date, and all of that has been highly preliminary.

2.2.1. Azotobacter

A great many media with highly selective properties have been devised for the isolation of specific microorganisms; however, those that may be used to quantify a specific microorganism in its natural environment with adequate efficiency are exceptional. Culture techniques fail to mimic the natural environment well enough or lack precision, or both. The genus *Azotobacter* should be well suited to selective plating based on simple, nitrogen-free medium, and such media have served to follow the fate of *Azotobacter* inoculated at high densities relative to the soil population, as with *Azotobacter paspali* (Brown, 1976). Selective plating is less effective at sizes of *Azotobacter* populations normally encountered in soil, since plates at low dilutions develop colonies of oligonitrophiles which obscure the *Azotobacter* (Duncan and Rosswall, 1974). As ideally suited for selective plate enumeration as this genus may appear to be, there remains an urgent need for a reliable method of enumerating natural populations of *Azotobacter* and other nitrogen-fixing bacteria (Postgate, 1972).

The only published reports dealing with FA study of *Azotobacter* are those of Tchan and DeVille (1970) and DeVille and Tchan (1970). They prepared an anti-*Azotobacter vinelandii* conjugate with FITC and an anti-*Azotobacter chroococcum* conjugate with rhodamine and mixed the two in equal proportions to detect the two *Azotobacter* species in inoculated soil smeared on slides. Distinction between the two species was excellent with the bivalent antibody stains. Control of both autofluorescence and nonspecific adsorption was achieved with methylene blue treatments. In the second of these papers (DeVille and Tchan, 1970), the authors report a procedure to quantify the occurrence of *Azotobacter* in soil by FA. They prepared serial dilutions of soils, smeared a known volume of soil dilution on a slide, and stained with FA. After noting the presence or absence of *Azotobacter*-stained cells in each smear, they used MPN (most-probable-number) tables to obtain estimates of the population. The validity of this approach is questionable on both practical and theoretical grounds. In practice, cell densities must be unusually high in order to detect a specific microorganism in a soil dilution (Schmidt, 1974). MPN statistical estimates are based on dilution to extinction so that some of the inocula for replicate tubes have one bacterium per unit volume. In the MPN procedure, the bacterium develops to a detectable level in positive tubes; if the inoculum itself is examined directly, as proposed by DeVille and Tchan, the likelihood of finding one bacterium in the 5000 or more microscope fields constituting a 1-cm^2 smear is extremely small. Failure to detect fluorescent bacteria in such endpoint smears could not, in theory, be considered as negative unless all 5000 fields were examined.

Specificity data were not included in the above reports, but the implication of the data was that the *Azotobacter* FAs were species specific. This was found not to be the case for an *A. chroococcum* FA which cross-reacted with certain strains of *A. vinelandii* (Schmidt, unpublished data). The serology of the genus has received little attention, and data are conflicting. Petersen (1959) found that *A. chroococcum* and *A. vinelandii* had agglutinogens in common and had serological affinities to *Beijerinckia* species. Norris and Kingham (1968) reported that antisera to strains of *A. chroococcum* and *A. vinelandii* did not cross-react with 31 strains representative of 5 heterologous species of *Azotobacter*. Both agglutination and precipitin reactions were restricted to homologous species, but considerable strain specificity was reflected in the data for *A. chroococcum*, whose serum agglutinated only the homologous strain and none of six others of the same species. Review of the Russian work by Rubenchik (1963) indicated both serological cross-reactivity between *A. chroococcum* and *A. vinelandii* and strain specificity for *A. chroococcum* serum. Serological relationships are equally unclear among other azotobacters.

Nonspecific staining is a factor that should receive particular attention in the ecological study of *Azotobacter* by FA. It has been observed (Schmidt, 1973) that certain soil organisms, notably some fungus spores, have been termed "universal acceptors," because they cross-react with any conjugate. Cysts of *A. chroococcum* were shown by Good (1972) to be included among the universal acceptors. Vegetative cells of *A. chroococcum* only stained specifically, but once out of log-phase growth, cyst occurrence followed in culture, and cysts stained indiscriminately with all conjugates used. Nonspecific staining of *A. chroococcum* cysts was readily controlled by application of the rhodamine-gelatine conjugate of Bohlool and Schmidt (1968) without interference with specific staining. Cyst formation is also a feature of *A. vinelandii*, *A. beijerinckia*, and *A. paspali*, so that nonspecific adsorption of antibody must also be considered in any immunodetection system involving these species.

Extremely interesting and significant problems remain concerning the kind, distribution, and growth features of *Azotobacter* as related to the nitrogen-fixing process in soil and rhizosphere. The consensus has been that *Azotobacter* probably adds little to the supply of fixed nitrogen in the biosphere, but the disclaimer, in the absence of adequate technology, has been the possibility of the specialized niche as illustrated by the *A. paspali–Paspalum* association (Dobereiner *et al.*, 1972). FA appears to have excellent potential for the exploration of the specific *Azotobacter* in the special niche, but clearly its application must be preceded by more preliminary work in which careful attention is paid to controls.

2.2.2. Beijerinckia

Ecological questions as to the significance of *Beijerinckia* to nitrogen fixation are very similar to those of *Azotobacter*. The genus has been isolated occasionally from temperate soils, but it occurs commonly in tropical soils, where its special habitat for nitrogen fixation could be the rhizosphere (Knowles, 1977). Leaf surfaces of tropical plants may also provide a selective environment conducive to nitrogen fixation since Ruinen (1961) reported high populations of *Beijerinckia* on rain-forest vegetation in Indonesia.

The two reports of FA study of *Beijerinckia* were both concerned with a temperate-soil isolate obtained from the rhizosphere of rice (Diem *et al.*, 1977, 1978). The first of these was introductory, documenting the specificity of the FA as appropriate for detection of the antigen strain. None of 4 species of *Beijerinckia* cross-reacted, nor did any of 6 species of *Azotobacter* or any of 44 unidentified rhizosphere bacteria. Occurrence of the isolate in sterilized and nonsterile soil after inoculation was demonstrated indirectly by plating and subsequent FA examination of characteristic colonies. Colonization of the rice rhizosphere by the *Beijerinckia* isolate was followed by FA examination of plants grown in soil underlaid with agar in the laboratory. Roots imbedded in agar at the soil–agar interface were observed by direct IF, and the isolate could be seen distributed in scattered sites on root surfaces and in adjacent rhizospheres. Little ecological information was obtained in this study beyond the demonstration of at least brief persistence of *Beijerinckia* in the vicinity of rice roots. The feasibility of the approach was not fully resolved in these experiments in view of the artificial nature of the root environment used and the interfering autofluorescence noted for the roots.

2.2.3. Azospirillum

A report by Dobereiner and Day (1976) noted high acetylene-reduction activities in the rhizospheres of tropical grasses as a function of the occurrence of a spiral-shaped nitrogen-fixing bacterium. The bacterium was found to be concentrated in loosely organized subepidermal patches of root tissue identified as the site of nitrogen fixation. The plant–bacteria relationship was described as associative nitrogen fixation. The bacteria now have been designated as making up the new genus Azospirillum, with two species A. lipoferum and A. brasilense (Tarrand et al., 1978).

The considerable interest that has focused on associative symbioses involving Azospirillum has brought attention to the relevant ecological questions: the nature of the niche in the root, the events involved in establishing the niche, the behavior of Azospirillum in the soil and rhizosphere apart from the niche, and the fate of Azospirillum inoculants. All these are autecological questions immediately suggestive of FA approaches.

Schank et al. (1979) have issued the first report of FA detection of Azospirillum. Antibodies prepared against four strains of A. brasilense were applied by indirect IF. Specificity of the one FA examined most critically appeared to be excellent with respect to absence of cross-reaction with known cultures, including A. lipoferum, Azotobacter paspali, and numerous random soil isolates. Azospirillum isolates obtained from Brazil, Florida, Venezuela, and Ecuador generally cross-reacted with one or more of the four antisera, but four such cultures failed to stain. The data indicate the specificity, as in most systems, is at the strain level, although many strains appear to share antigens of more than one serotype. The authors call attention to one antiserum, UM125A2, as highly strain specific, but their data do not support such a conclusion. Cross-adsorption studies are obviously needed to arrive at serogroups, and the use of adsorbed serum would probably be advisable in future studies if other workers encounter similar cross-reactivity at the strain level. Preliminary attempts to detect Azospirillum in root and rhizosphere preparations were generally encouraging. Cross-reactive bacteria were found unevenly distributed in the root mucigel, but no localizations or rudimentary root structures indicative of a niche were observed. Problems with respect to root autofluorescence and nonspecific binding of the fluorescent goat anti-rabbit reagent were noted.

3. Nitrifying Bacteria

FA techniques are particularly appealing for ecological study of the nitrifying bacteria because such studies have been almost entirely blocked by the complexities of natural habitats and the specialized biology of the nitrifiers. The bacteria are slow-growing chemoautotrophs refractory to the usual microbiological approaches for examination, enumeration, and isolation. As a result, essentially nothing is known with respect to two basic aspects of nitrifying populations: the diversity of nitrifiers within a given population and the *in situ* nitrifying activities of the individual components of a population. It has only been possible to arrive at rough statistical estimates of the overall nitrifying population by means of the MPN method; the method has high statistical uncertainty, requires many weeks' incubation, and is selective only for certain components of the population (Belser and Schmidt, 1978b). The peculiar problems of the microbiology of the nitrifiers was discussed by Schmidt (1978).

More thorough understanding of the microbiology underlying nitrification is needed to control the process in its many applications such as nitrogen management in field crops, nitrogenous waste disposal in sewage systems, and prevention of ground-water and lake contamination with nitrate. With adequate information, it will be possible to construct models with useful predictive capabilities. Prospects for this are still somewhat remote, but there are indications that FA will make unique and significant contributions toward that goal.

3.1. Nitrite Oxidizers

Bacteria responsible for the second stage of nitrification, the oxidation of nitrite to nitrate, were the first of the nitrifiers to be studied by FA. Fliermans *et al.* (1974) prepared antisera against *Nitrobacter agilis* and *N. winogradskyi* and evaluated them as to specificity. Low-level cross-reactions between the sera were removed by cross-adsorption. All 15 *Nitrobacter* isolates that had been obtained from a variety of natural environments were found to react with one or the other of the 2 adsorbed FAs, but no cross-reaction was found in extensive tests that included 5 isolates of ammonia-oxidizing autotrophs and 668 heterotrophic bacteria from soil, water, and sewage. Staining of the *Nitrobacter* isolates with the highly specific FAs indicated that nitrite oxidizers were composed of only two serotypes, with the implication that *Nitrobacter* FAs were possibly species specific and that diversity among nitrite oxidizers was surprisingly limited. Both implications have proved not to hold, since *Nitrobacter* isolates obtained more recently by the nonenrichment procedures of Belser and Schmidt (1978b) were of several different serotypes than those cited above (Stanley *et al.*, 1979). Aside from the application of FA to the problem of diversity, this study included ecological data in the form of the detection of *Nitrobacter* in various ecosystems. A subsequent refinement of the FA technique, using autoradiography combined with FA to distinguish between metabolically active and inactive cells, was developed with *Nitrobacter* as a model (Fliermans and Schmidt, 1975). A soil known to contain *Nitrobacter agilis* by FA detection in the Fliermans *et al.* (1974) study was treated with nitrite and incubated to enrich for *Nitrobacter*. The soil was then exposed to $^{14}CO_2$ briefly and processed for combined FA-autoradiography analysis. *Nitrobacter* cells were detectable by their specific fluorescence, and about 40% of these cells had incorporated the $^{14}CO_2$ as evidenced by silver-grain development in the autoradiographic emulsion directly above the cells.

Development of procedures to enumerate specific bacteria in soil (Bohlool and Schmidt, 1973a; Schmidt, 1974) was particularly pertinent to the autecology of the nitrifiers as a possible alternative to the MPN method. Schmidt (1974) inoculated a partially sterilized soil with a strain of *N. winogradskyi* originally isolated from that same soil, added nitrite as substrate, and followed *Nitrobacter* growth by FA along with nitrate formation. The data showed that the log plot of the population paralleled the log plot of nitrate formation and pointed to the success in this modified natural system of relating the dynamics of a bacterial population to the dynamics of the process which it carries out. This kind of approach offers for the first time the possibility of linking the nitrification process to its microbiological base in terms of growth constants and substrate conversion rates—an ecological problem of first-order significance to control of the nitrification process and the development of useful predictive models. Subsequent extension of this application of FA to wholly natural systems has not been fully successful because of technical problems in recovering nitrifiers from soil and the existence of multiple serotypes in many nitrifying populations (Belser and Schmidt, 1978a).

Growth of *Nitrobacter* in soil following nitrite addition was observed by Fliermans and Schmidt (1975) with a bivalent FA. Numbers increased from 4.8×10^3 to $3.6 \times 10^6/g$. Other direct FA enumerations of *Nitrobacter* in natural habitats were reported by Rennie *et al.* (1977) in a study relating to the controversy of whether roots are or are not nitrifier inhibitors. The rhizospheres of soybeans and wheat examined in this study manifested no inhibitory effects on *Nitrobacter*. *Nitrobacter* counts in various soils with a combined *N. winogradskyi-N. agilis* FA preparation were all in the 10^4-$10^5/g$ range (Rennie and Schmidt, 1977a). Stanley *et al.* (1979) applied standard quantitative FA protocols to enumerate *Nitrobacter* in lake-water columns. The ability to quantify a microorganism in nature greatly expands the variety of autecological questions that may be posed and the extent to which they may be probed. Use of FA to enumerate *Nitrobacter* in soils, water, sediments, and sewage is clearly possible if the *Nitrobacter* serotypes therein are ones for which FA reagents are available; what is needed now is more information on the serotypic diversity of the nitrite oxidizers and the distribution of the serotypes in nitrifying environments. Little such data are now available.

3.2. Ammonia Oxidizers

Detailed study of the ammonia oxidizers by FA was begun somewhat later than that of the nitrite oxidizers. Belser and Schmidt (1978b) looked at three of the four genera of terrestrial ammonia oxidizers. Most of the strains examined were isolates of *Nitrosomonas;* these fell into four serotypes. The 11 isolates of *Nitrosolobus* were included in 3 serotypes, and 6 of 7 isolates of *Nitrosospira* fit into 2 serotypes. No intergeneric cross-reactions were observed. The obvious conclusion was that the ammonia oxidizers are a much more diverse group than has been generally recognized. With additional isolates and additional FAs, the number of serotypes increased (Belser and Schmidt, 1978c).

In subsequent work, Belser and Schmidt (1978b) examined the question of diversity in ammonia oxidizers with emphasis on the nitrifying population of a single soil. MPN media of several kinds were used for microscopic examination, isolation, and enumeration. Depending on the MPN medium used, either *Nitrosomonas* or *Nitrosospira* was the most abundant genus in that soil. *Nitrosolobus* was seen occasionally in MPN tubes at the lower dilutions. This was the first report of the coexistence of multiple genera of ammonia oxidizers in the same environment, and the finding points out clearly that the widely accepted textbook convention of equating the process of ammonia oxidation with the term *Nitrosomonas* is a myth. The point is significant because the various genera differ in growth rate, biomass, yield, and substrate-oxidizing activity (Belser and Schmidt, manuscript in preparation).

Isolates obtained from the single soil cited above were reacted with 16 different FAs for ammonia oxidizers (Belser and Schmidt, 1978c). There were 7 *Nitrosomonas*, 12 *Nitrosospira*, and 1 *Nitrosolobus* isolate. Of these, five of the seven *Nitrosomonas* isolates were stained effectively by the use of four FAs; six of the *Nitrosospira* were identifiable with four FAs. The one *Nitrosolobus* was stained well by only one of the three FAs used. A greater degree of cross-reactivity would certainly be more convenient to allow for a genus-specific, strain-comprehensive FA reagent. From the above results, it would appear that at least five FAs would be needed to cover the *Nitrosomonas* present in this soil and five or more FAs for the *Nitrosospira* component. Thus, composites involving perhaps a dozen FAs would be required for this particular soil assuming all serotypes had been isolated. The existence of a large number of serotypes poses obvious practical problems for achieving the ideal of detection and enumeration of an entire nitrifying population. Further experiments are needed to evaluate the perhaps more reasonable objective of determining the genera and strains that predominate in numbers and activities and to develop the appropriate FA reagents. The specter of infinite serological variability appears not to be real since FAs prepared against both ammonia oxidizers and nitrite oxidizers isolated from Minnesota soils were shown to stain isolates from diverse parts of the world (Fliermans *et al.*, 1974; Schmidt, unpublished).

4. Sulfur- and Iron-Oxidizing Bacteria

The chemolithotrophic bacteria that use reduced forms of sulfur and iron as sources of energy and reducing power are widespread in nature. They play an important role not only in biogeochemical cycling of these elements but also in acid and iron pollution of lakes and rivers.

4.1. *Thiobacillus* in Coal-Refuse Material

One of the most abundant substances associated with coal deposits is pyrite, which occurs in a variety of forms. One of these, iron pyrite, consists of both iron and sulfur in the reduced form. It therefore provides an ideal substrate for sulfur- and iron-oxidizing bacteria of the genus *Thiobacillus*. These bacteria can oxidize either sulfur (*T. thiooxidans*) or both sulfur and iron (*T. ferrooxidans*) and produce large quantities of sulfuric acid and ferric hydroxide (Lundgren et al., 1971). The consequences of these microbial processes in coal-refuse disposal are highly acid mine drainage waters and metal precipitates ("yellow-boy"), which constitute a major water-pollution problem in coal-producing regions.

Iron pyrite oxidizes in air spontaneously, but only at pH values above 5.0. Abiotic acid production therefore becomes self-limiting once the pH of the refuse reaches this value. Below this pH, biological oxidation, mainly carried out by thiobacilli, becomes the major contributing factor to acid production in coal-refuse piles. For a description of a pH-dependent succession of iron bacteria in pyrite material, see Walsh and Mitchell (1972).

Both the iron- and sulfur-oxidizing bacteria can be isolated from coal-refuse piles and acid mine drainage waters (Colmer et al., 1950; Tuttle et al., 1969). However, few quantitative studies have examined these bacteria *in situ*.

Belly and Brock (1974) used $^{14}CO_2$ uptake by coal-refuse material as a measure of total autotrophic activity and showed that it was related to the most-probable-number of iron-oxidizing bacteria. These results, however, only represent the overall $^{14}CO_2$-fixing potential of the sample, and the data do not relate directly to the autecology of the organisms involved.

Apel et al. (1976) are the only authors who have used immunofluorescence (indirect) for detection and enumeration of *T. ferrooxidans* in acid mine environments. The specificity of the FA used in these studies was reported to be adequate in that their reagent did not react with a variety of other microorganisms, including other members of the genus *Thiobacillus*. Apparently, however, it was necessary to adsorb the FA with a number of unrelated bacteria in order to achieve the level of specificity reported.

The authors were not able to observe stained bacteria directly colonizing the coal-refuse particles. This they attributed to low numbers of bacteria and intense fluorescence of the particles. They did, nevertheless, show that when the bacteria were washed from the particles and concentrated by centrifugation, *T. ferrooxidans* could be seen and enumerated. Bohlool and Brock (1974a), on the other hand, had very little difficulty distinguishing *Thermoplasma acidophilum* (a thermophilic acidophilic mycoplasma) colonies in similar samples. They used the direct immunofluorescence technique, with the rhodomine–gelatin conjugate to suppress background fluorescence. Brightly fluorescing cells of *T. acidophilum* could be seen colonizing particles and films on microscope slides buried in the field in contact with coal-refuse material or immersed in the stream draining the refuse pile.

4.2. *Sulfolobus* in Hot Springs

Sulfolobus acidocaldarius is a sulfur-oxidizing facultative autotroph that inhibits sulfur-rich environments that are extremely hot, 55-92°C, and acid, pH 0.9-5.0 (Brock *et al.* 1972; Brock, 1978). These bacteria have been shown to have an optimum temperature for growth of 70-75°C with a lower temperature limit of 55°C. They are also obligately acidophilic, with a growth optimum of pH 2-3. In nature, the major mode of growth for this organism is probably by chemolithotrophic oxidation of elemental sulfur, although its ability to grow heterotrophically has been demonstrated (Brock, 1978).

Questions about the evolution, biogeography, and ecology of *Sulfolobus* in such extremes of temperature and pH are both intriguing and significant from an evolutionary point of view. Immunofluorescence has been useful in gaining some understanding of the origin, distribution, and activity of this organism in its natural environment.

4.2.1. Distribution and Biogeography

Sulfolobus has a wide geographical distribution. It has been found in acidic geothermal areas in North America, El Salvador, Dominica, Italy, Japan, and New Zealand (see Brock, 1978).

Bohlool and Brock (1974b) used immunofluorescence and immunodiffusion to classify into serogroups *Sulfolobus* isolates, "serostrains," cultured from different hot-spring environments. Distribution of serologically distinguishable populations was then studied quantitatively using the membrane-filter immunofluorescence technique. It was shown that several specific "serostrains" could exist in the same spring. This was compatible with the results of Mosser *et al.* (1974b), who had reported isolating different "temperature-strains" from the same source. One of the "serostrains" was found in all the springs sampled, but their relative numbers varied greatly. Numbers as high as 10^7-10^8 cells/ml of a specific strain could be found in several springs. In flowing springs, the majority of the cells were located in the "sediment" of the flowing channel, with only a few in the water while in nonflowing pools, large numbers were suspended in the water.

In a later study, Bohlool (1975) found that several *Sulfolobus* isolates from New Zealand hot springs cross-reacted with FAs against Yellowstone organisms. The antigenic relatedness of New Zealand and Yellowstone populations raises interesting questions of dispersal and evolution of *Sulfolobus* (for further discussion, see Brock, 1978).

4.2.2. Growth Rate in Nature

The stable and steady-state nature of hot springs makes it relatively easy to measure *in situ* growth rates of the populations. Mosser *et al.* (1974a) applied a chemostat approach to measure the steady-state growth rate of *Sulfolobus* in several Yellowstone hot springs. Several springs were enriched with sodium chloride, and the rate of dilution of the chloride ion was determined over a period of time. Since the springs maintained a constant volume, temperature, pH, and *Sulfolobus* numbers, the authors suggested that chloride dilution rates in the spring could be used as an estimate of the rate the bacterial populations have to grow to maintain their numbers. To ascertain that the bacteria were being diluted at the same rate as the chloride ion, formaldehyde-fixed cells of a serologically distinct isolate of *Sulfolobus* were also added to several of the springs at the time of NaCl addition, and the decrease in their numbers was measured using immunofluorescence. The loss rates for formaldehyde-fixed cells were shown to agree reasonably well with those for chloride dilution. Half-times for cell and chloride dilution, presumably equivalent to bacterial doubling time, were estimated at about 10–20 hr for springs ranging in volume from 20 to 2000 liters, but 30 days or more for 2 larger springs of about 1 million liters' capacity.

5. Bacteria Involved in the Methane Cycle in Lakes: Methanogens and Methylotrophs

Decomposition of organic matter in most anaerobic environments leads to the formation of methane gas. The bacteria (methanogens) that are responsible convert the fermentation products (CO_2, H_2, formate, and ethanol) of other anaerobes in these environments to methane. They belong to a highly specialized obligately anaerobic group of bacteria that occupies the terminus of the food chain in the sediment.

In aquatic systems such as lakes and lagoons, the methane gas, being poorly soluble in water, escapes into the overlaying water column, where it serves as a source of carbon and energy for the aerobic and microaerophilic methane-oxidizing bacteria (methylotrophs).

5.1. Methanogens in Lake Sediments

Strayer and Tiedje (1978) used immunofluorescence to study the distribution of *Methanobacterium formicicum* in several lake sediments, anaerobic sewage sludge, and bovine rumen fluid. FA-reactive bacterial cells could be detected directly in some of the samples. FA enumeration of the specific methanogen in one lake sediment revealed numbers as high as 3.14×10^6/g dry sediment, which was at least one order of magnitude greater than MPN counts of the total methanogenic population. The specificity of the FA used in the above studies was satisfactory, since the FA did not react with 9 other methanogens, including *M. formicicum* strains, and 24 heterotrophic bacteria, many of which had been isolated from the study site.

5.2. Methylotrophs in Lake Sediments and Waters

Using direct immunofluorescence, Reed and Dugan (1978) studied the distribution and abundance of two different methylotrophs in the water column and sediments of Cleveland Harbor. *Methylomonas methanica* was detected at all the sites examined. Its numbers were found to increase with depth, with the largest population found in the sediment. *Methylosinus trichosporium*, on the other hand, could only be detected in two of the stations and only in the sediments and 1 m above. These findings are consistent with other published observations that methane oxidation takes place only directly above the sediments where the O_2 concentration is less than 1.0 mg/liter. The antisera prepared for these studies proved quite specific in that they reacted only with their homologous bacteria and not with any of 25 other bacteria, some of which were methylotrophs isolated from Cleveland Harbor.

6. Anaerobic Bacteria

Methods for enumeration of specific anaerobic bacteria from natural samples often involve tedious anaerobic culture techniques and prolonged incubation. In most instances, proper media have not been formulated for selective enrichment and isolation of the desired anaerobe. Immunofluorescence provides a possible rapid shortcut in detection and enumeration of a specific anaerobe directly in the sample of interest. An excellent example has already been noted (Section 5.1) in the recent study of Strayer and Tiedje (1978) on the distribution of the obligate anaerobe *M. formicicum* in lake sediments.

One of the first applications of immunofluorescence in microbial ecology concerned the detection and identification of anaerobic bacteria in the rumen fluid (Hobson and Mann, 1957). In a later study, Hobson *et al.* (1962) reported successful detection of *Selenomonas ruminantium* directly in stained smears of sheep-rumen contents. Their FA prepared against the *S. ruminantium* isolate cross-reacted with another rumen organism, *Veillonella alcalescens;* however, the FA could be made specific by adsorbing with the cross-reactive organisms.

Garcia and McKay (1969) and Garcia *et al.* (1971) have applied immunofluorescence for detection and identification of two anaerobic bacteria, *Clostridium septicum* and *Sphaerophorus necrophorus*, in natural and clinical specimens.

6.1. *Clostridium* in Soil

Clostridia are spore-forming anaerobic bacteria which are widely distributed in soils and are known to be involved in anaerobic decomposition of plant and animal constituents. Many clostridia are also causative agents of a variety of diseases of man and animals. In normal, unamended soil, clostridia are thought to occur mostly in the spore stage. However, Garcia and McKay (1969) combined qualitative immunofluorescence with viable counts to show that vegetative cells added to soils did not sporulate immediately and were capable of growth and multiplication in normal soils. The results of these studies were limited to qualitative FA assessment, for no attempts were made at quantification.

6.2. *Sphaerophorus* in Soil and Animal Tissues

S. necrophorus, a gram-negative nonsporing obligate anaerobe, is known to cause hepatic abscesses in cattle and necrotic gangrenous lesions in man and animals, and it is often associated with enzootic foot-rot infections in livestock. It was shown to survive well in soils incubated anaerobically or moistened to 80% of water-holding capacity. The species-specific fluorescent antibody against an isolate of *S. necrophorus* facilitated the detection of the organisms in liver abscesses, viscera, ruminant contents, and soils (Garcia *et al.*, 1971).

7. Public Health and Water Pollution Microbiology

7.1. Bacteriological Indicators of Fecal Pollution

Most human diseases that are transmitted by water originate from human feces. Common among these are diseases caused by strains of enteropathogenic *Escherichia coli*, *Salmonella*, *Shigella*, *Vibrio*, *Leptospira*, enteric viruses, and a host of parasitic protozoal cysts and worm larvae.

Assessment of the sanitary quality of water has traditionally relied on quantitative detection of indicator organisms rather than on specific pathogens. Part of the reason is that, even in badly polluted waters, pathogens are ordinarily present in very low numbers. Furthermore, monitoring of every pathogenic agent in every water supply becomes an inefficient and costly endeavor.

7.1.1. Enterobacteriaceae

The "coliform" group of bacteria—those aerobic and facultatively anaerobic, gram-negative, non-spore-forming rods capable of lactose fermentation with gas formation within 48 hr at 35°C (American Public Health Association, 1975)—are the most commonly used indicators of fecal pollution. However, the "fecal coliforms—the subgroup of coliforms that can ferment lactose with gas production at 44.5°C (mainly *E. coli*)—are more characteristic of feces of man and other warm-blooded animals and thus, are, more reliable indicators of fecal contamination.

Aside from the reliability considerations in the use of the indicator concept, another factor of prime importance in a water-quality monitoring system is the rapidity and ease with which large numbers of samples can be processed. Application of immunofluorescence for presumptive mass screening and monitoring of water supplies offers a practical, and perhaps economic, alternative to conventional techniques. It has been field-tested for use in rapid detection of enteropathogenic *E. coli* (Cherry *et al.*, 1960; Pugsley and Evison, 1974; Abshire, 1976), *Shigella flexneri* and *S. sonnei* (Thomason *et al.*, 1965), and *Salmonella typhi* (Thomason and Wells, 1971) in fecal smears and for identification of typhoid carriers (Thomason and McWhorter, 1965). Using polyvalent FAs for salmonellae, Cherry *et al.* (1972) screened enrichments of several surface waters and found 60% more positive samples than they could detect by conventional culture methods. These applications were later extended to the detection of *Salmonella* in foodstuff and water (Cherry *et al.*, 1975).

The results of all of these applications have suggested that IF could be a valuable tool in early diagnosis and control of some diarrheal diseases and epidemic conditions. However, as had been emphasized by Cherry and Moody (1965), FA detection of certain members of Enterobacteriaceae must only be considered presumptive and can in no way replace isolation and definitive identification of the etiologic agent. The major limitation to exploitation of FA is the well-known antigenic relatedness and cross-reactivity among the various genera of this family.

The introduction of nonfluorescent membrane filters has made the quantitative detection of low numbers of specific bacteria in water a reality. Cells from a large volume of water can be concentrated on black membrane filters and the desired bacteria identified by staining with appropriate FAs. The approach was first used to quantify low numbers of *E. coli* and *Shigella guarrabara* artificially inoculated into water (Danielsson, 1965; Danielsson and Laurell, 1965; Guthrie and Reeder, 1969), but this potential has never been field-tested directly in water-quality assessment.

7.1.2. Fecal Streptococci

Another normal inhabitant of the intestine of warm-blooded animals which has often been suggested as an indicator of fecal contamination is the group of gram-positive bacteria collectively termed fecal (or Lancefield's Group D) streptococci.

Abshire and Guthrie (1971) and Pavlova *et al.* (1972, 1973) have shown that IF may be used to identify Group D streptococci isolated from contaminated waters. Pugsley and Evison (1975) used Group D-specific FAs to count colonies of fecal streptococci on nonfluorescent membrane filters preincubated on enrichment media for 10 hr. However, reports of direct enumeration of these organisms from natural waters are difficult to find.

Whereas the fecal coliform test does not differentiate between human and animal fecal contamination, it has been suggested that the ratio fecal coliform : fecal streptococci (FC:FS) could indicate this distinction (Geldreich, 1970). The explanation advanced is that in human feces and domestic sewage, the FC:FS ratio is often greater than 4, while in farm-animal feces and effluents from packing houses, feedlots, and dairy farms, the same ratio is generally less than 1. The main objection to the use of the FC:FS concept as an indicator of human vs. animal source of fecal pollution arises from the fact that the two organisms survive differently once they are discharged into water. Therefore, viable counts downstream from the effluent discharge do not reveal the true bacteriology at the source. IF, on the other hand, can overcome this limitation because it can identify viable and nonviable cells alike. More effort should be directed toward developing improved FA reagents and techniques so that the indicator organisms can be quantified directly in the suspected sample.

7.2. Legionnaires' Disease Bacteria (*Legionella*) in Nonepidemic-Related Habitats

The etiologic agent of Legionnaires' disease, *Legionella pneumophila*, has been isolated from a number of man-made and natural environments. However, the organisms have been studied primarily in clinical specimens or pure cultures obtained from infected hosts. Isolates from air-conditioning cooling towers have been shown to be associated with four different outbreaks of Legionnaires' disease (Center for Disease Control, 1978).

Cherry *et al.* (1978) have developed a direct IF procedure for detection of Legionnaires bacteria in lung scrapings, histological sections, and fresh lung tissue obtained at biopsy or autopsy. Their preliminary data indicated that serologically identical or related organisms were common in soils and soil animals from several locations, but none of the cross-reacting organisms was isolated or identified.

Recently, Fliermans *et al.* (1979) have been successful in applying IF for quantitative detection of *Legionella* in samples of lake water concentrated by continuous centrifugation. Confirmatory evidence was obtained by injecting the same water samples into guinea pigs and isolating organisms identified as *L. pneumophila* by morphological, cultural, physiological, and serological characteristics.

Quantitative IF has also provided the tool to study the growth of this organism in association with blue-green algae in the laboratory (Tison *et al.*, 1979). It was shown that a *Legionella* isolate (serogroup 1) from an algal mat community in a man-made thermal effluent (45°C) was capable of rapid multiplication when grown in association with the blue-green alga *Fischerella* sp. in mineral salts' medium.

8. Plant Pathogenic Bacteria

Although the potential of immunofluorescence for applications in plant pathology was recognized quite early (Paton, 1964), progress in this field has been rather slow. Published information on such applications are limited to a few scattered and preliminary reports.

Paton (1964) demonstrated that the technique was effective in locating *Pseudomonas syringae* growing intra- and extracellularly in leaves and tubers of various plants. Autofluorescence of plant tissue was a problem but could be controlled by either acetone extraction followed by treatment with dimethylformamide or by treating with a saturated mercuric chloride solution followed by several rinses in EDTA. Nonspecific staining of plant tissue was reported to be minimized if the serum was preadsorbed with rat-liver homogenate and the slides counterstained with lissamine rhodamine-conjugated preimmune serum prior to FA staining. Some evidence was presented indicating that the bacteria grown in culture were not antigenically identical to those *in vivo*. Acid treatment of *in vivo* bacteria slightly improved the staining reaction of FA reagent produced against culture-grown cells, suggesting that an antigenic layer had been removed from the bacterial surface. Likewise, antibodies produced against *in vivo* cells reacted better with cells *in vivo* than *in vitro*.

Auger and Shalla (1975) were able to identify the causative bacteria of Pierce Disease (PD) of grapevines in host plant tissue and insect vector. The conventional means of diagnosing PD in vines involves transmission, via grafting or insect vector, to another susceptible plant followed by prolonged incubation (24–55 days) for the development of symptoms. Immunofluorescence proved to be a specific and rapid alternative for the diagnosis of PD in infected plants and contaminated insect vectors.

Allan and Kelman (1977) found that FAs prepared against a strain of *Erwinia carotovora* var. *atroseptica* (Eca) was highly specific against Eca in that it reacted only with isolates characterized as *E. carotovora* var. *atroseptica* and not with any of a large number of strains belonging to *E. carotovora* var. *carotovora*, *E. chrysanthemi*, *E. ariodeae*, *E. carnegiana*, and *E. amylovora*. They were able to use the reagent to detect *E. carotovora* var. *atroseptica* in artificially inoculated soils, infected potato tubers and leaves, and fruit flies and maggot flies which had been feeding on pure cultures of bacteria in the laboratory or infected potatoes in the field.

Other applications of immunofluorescence in plant sciences have involved the identification of a nitrogen-fixing strain of *Chromobacterium lividum* in various tissues of leaf-nodulated plants (Bettelheim *et al.*, 1968) and detection of a bacterial contaminant in nodule squashes of some leguminous plants (Van der Merwe *et al.*, 1972).

9. Filamentous Fungi

A large variety of techniques have been developed for the study of fungi from soil and other habitats (for a review, see Parkinson, 1973). Among these, the plate count has been the most frequently used method for quantitative estimation of fungal biomass. This technique has been shown, however, to have serious limitations in that it is, for the most part, selective for those fungi that occur as spores. Direct methods also have intrinsic limitations but nevertheless often afford a realistic view of microorganism *in situ*. The usefulness of direct observation is greatly enhanced when the identity of the microorganism can be established. IF provides such an opportunity.

Schmidt and Bankole explored the potential of IF in recognizing strains of *Aspergillus flavus* grown in sterile soil in the presence of five other soil fungi (Schmidt and Bankole, 1962, 1963, 1965). The FA prepared against one strain of *A. flavus* reacted strongly with all other 14 strains of *A. flavus* but not with 17 isolates of species of *Aspergillus* other than *A. flavus*. Four other aspergilli, particularly *A. sydowi*, cross-reacted intensely. Most other fungi tested reacted weakly or not at all.

Schmidt *et al.* (1974) found IF to be especially promising in the study of certain ectomycorrhizae. FAs prepared against cultures of two ectotrophic mycorrhizal fungi, *Thelephora terrestris* and *Pisolithus tinctorius*, were effective and specific microscopic stains for identification of the homologous fungus on contact slides and in "Hartig Nets" in the mycorrhizal structure. *T. terrestris* FA gave 0 and *P. tinctorius* FA gave only 5 cross-reactions of moderate intensity with 31 diverse nonmycorrhizal fungi. Most of the heterologous cross-reactions could be eliminated by adsorption with the cross-reacting fungus, without significantly affecting the activity of the homologous system.

Eren and Pramer (1966) have used IF to visualize the nematode-trapping fungus, *Arthrobotrys conoides*, in artifically inoculated soils.

Many of the other applications of IF in mycology are either for diagnosis of mycotic infections (for a review, see Goldman, 1968) or for detection of specific fungi in samples of infested grains (Warnock, 1971, 1973).

Questions about the ecology of important members of the fungal flora–i.e., plant pathogens and decomposers–have not been approached using direct techniques. IF provides an attractive opportunity, but its potentials and limitations have not been fully explored.

10. Other Applications

There are only a few other noteworthy examples of the use of IF in ecologically oriented studies.

Gray and his co-workers have used IF for studies on the distribution, growth, and spore formation and germination of *Bacillus subtilis* in soil (Hill and Gray, 1967; Siala *et al.*, 1974; Siala and Gray, 1975). They were able to differentiate between the spores and the vegetative cells of this organism *in situ*, using FAs prepared against the respective antigens. It was shown that *B. subtilis* is present mainly in the vegetative form in an acid horizon of a forest soil but as spores in the alkaline horizon of the same soil profile. Vegetative cells on glass slides in contact with an acid forest soil initially declined in numbers but later grew after fungal hyphae were established on the glass slides. Growth did not occur in sterile soil, nor in alkaline soil, until fungal growth was stimulated. Likewise, spore germination did not occur in the same acid soil unless fungi were also growing. Roots of *Pinus sylvester* growing in the soil was shown to inhibit both vegetative growth and spore germination.

The distribution of *Leptospira* in field soils and waters was studied by Henry *et al.* (1971) using direct IF. Microscope slides in contact with soil or water in the field were colonized within 4 to 14 days with leptospiral-like cells that reacted with fluorescent antibodies against a saprophytic leptospira, serotype *patoc* Patoc I. The presence of leptospires at various sites, as detected by IF, correlated well with data which had been obtained earlier by culture methods.

The thermophilic acidophilic mycoplasma, *Thermoplasma acidophilum*, was studied in spontaneously heated coal-refuse piles in strip-mining regions by Bohlool and Brock (1974a) using IF. FA-reactive cells could be found colonizing refuse particles on glass slides buried in the field. Although some degree of cross-reaction was found for most of the stains tested, IF/cross-adsorption revealed several antigenically distinct serogroups. Survey of their geographical distribution showed that different serogroups could inhabit the same site, whereas the same serogroup could be found in piles from different geographical areas.

IF was used by Farrah and Unz (1975) to establish the role of a strain of *Zoogloea ramigera* in the formation of flocs and finger-like zoogloeae in microbial films that develop on the surface of stored samples of activated sludge. It was shown that some of the fingerlike projections consisted almost entirely of cells of one IF-reactive type, whereas others were devoid of reactive cells.

Fliermans and Schmidt (1977) have applied IF for an autecological study of the unicellular cyanobacterium, *Synechococcus cedrorum*. Their FA reacted brightly and specifically and was thus effective in specifically enumerating *S. cedrorum* cells in water samples from the water column of a stratified lake.

IF has also been applied to study the distribution of *Aeromonas hydrophila* in thermal effluents (Hazen and Fliermans, 1979) and its association with the ciliate *Epistylis* in the epizootic red-sore lesions in large-mouth bass (Hazen *et al.*, 1978).

11. Problems and Limitations

In general, the effectiveness of any technique cannot surpass the competence of the technician. There are, however, intrinsic limitations and possible sources of error for all techniques that the user should be thoroughly familar with in making sound interpretations and conclusions. This is especially true of IF because the results are often based on subjective evaluation of microscopic observations, and, particularly so, when dealing with highly complex and heterogeneous materials, such as soils and sediments. For these reasons, controls of various kinds are essential in all applications of IF. These were outlined in a previous review by Schmidt (1973).

The most important considerations in the application of IF to natural samples relate to (1) the specificity of the antibody to be used and problems with nonspecific staining (NSS), (2) the interference from autofluorescence of, or nonspecific adsorption (NSA) of FA to, the background, (3) the stability of the antigen under different growth conditions and environments, (4) the inability to distinguish between live and dead cells, and (5) the efficency of recovery of the desired cells from the sample for quantification.

11.1. Specificity

The FA-specificity spectrum is one of the first criteria that must be established after the preparation of an FA reagent. The question of specificity must be raised for every new system to be used. In some instances, i.e., Enterobacteriaceae and *Rhizobium*, extensive serological data are already available from published records and can be valuable in selecting the right strain for study. Even for such systems, however, the specificity features of the new reagent must be confirmed for the application planned.

The degree of specificity desired varies with the type of ecological question under study; there could be *too much* or *not enough* specificity in a particular system. For instance, if attention is focused on a specific strain for any particular reason (i.e., a highly effective *Rhizobium* strain or an enteropathogenic strain of *E. coli*), then an FA that will react only with that strain is highly desirable. If, on the other hand, the study involves organisms responsible for a specific process, e.g., methanogens or chemosynthetic ammonium or nitrite oxidizers, then a species-specific reagent is needed. In general, FAs can be made more specific by adsorption with cross-reacting organisms; an example of removal of cross-reaction by this means is found in the study of Fliermans *et al.* (1974). Likewise, strain-specific FAs may be pooled to obtain a more species-specific FA reagent, but the effectiveness of this approach is limited by the number of serotypes that may be involved in the natural population under study, since only about 6–8 active FAs may be pooled because of dilution effects.

Although examples of cross-reaction between totally unrelated organisms are rare, it deserves some attention, nevertheless, since a few examples have been reported. The cross-reaction of OX-19, OX-2, and OX-K strains of *Proteus vulgaris* with typhus and spotted-fever-group rickettsiae is a well-documented case and forms the basis for the Weil–Felix serologic test for diagnosis of these diseases (see Joklik and Willett, 1976). Heidelberger and Elliot (1966) reported cross-reaction of pneumococcal antisera with exopolysaccharides of *Rhizobium* and *Xanthomonas*. Dudman and Heidelberger (1969) later found the *Rhizobium* cross-reaction to be due to a pyruvate subgroup on the polysaccharide moiety.

More pertinent to ecological applications of IF is the observation by Bohlool and Schmidt (1970) that an FA prepared against a strain of *R. japonicum* stained a soil actinomycete isolate specifically (Fig. 4a). Cross-reactions of this type, involving antigens shared by unrelated forms, have not been reported in any other ecological study. Continued attention should be given to this possibility though, for actinomycetes may fragment in pieces that resemble single-cell bacteria.

Another potential problem related to specificity is that some microorganisms (usually fungal spores) are "universal acceptors" (Schmidt, 1974) in that they will stain with any FA, including the preimmune control. Figure 4b illustrates this phenomenon. The soil fungus *Fusidium* and a strain of *R. japonicum* (USDA 31) were grown together in sterilized soil and contact slides stained with USDA 31-FA. Notice that only the *R. japonicum* cells (one attached to a fungal mycelium) and the fungal conidia stained, while the mycelia are negative. It is likely that the surfaces of many fungus spores are highly absorptive and trap any sort of protein material. The fact that *A. chroococcum* cysts also act as universal acceptors of FA and that this appears to be a trapping reaction which may be blocked by gelatin–rhodamine has been mentioned (Section 2).

11.2. Autofluorescence and Nonspecific Adsorption

Autofluorescence of material in the sample and nonspecific attachment of FA to the background can mask the specifically stained cells and seriously limit IF observations. Autofluorescence is specially troublesome with plant material, but it varies greatly with the type of plant used. Hughes *et al.* (1979) had difficulty in using fluorescein-labeled material to stain soybean root hairs because of autofluorescence of root hairs in the fluorescein region of the spectrum. In our experience, most of the interfering autofluorescence of soybean roots is associated with the main root and not root hairs; furthermore, it can be controlled by the use of various extractors and the gelatin–RhITC counterstain (Bohlool and Schmidt, 1968). When autofluorescence presents a serious problem to IF investigation, a variety of modifications can be tried, such as the use of other fluoro-chromes and filter systems, treatment with extractors, and the application of counterstains.

Problems of NSS encountered in the application of IF to soil systems has been discussed in an earlier section (1.2). A counterstain, consisting of hydrolyzed gelatin conjugated to rhodamine, was highly effective in overcoming NSS in a variety of soils and also in some animal and human tissue (Bohlool and Schmidt, 1968). This was also found to be effective in reducing NSS with mycorrhizal sections (Schmidt et al., 1974), other plant material (unpublished), and coal-refuse material (Bohlool and Brock, 1974a).

11.3. Antigen Stability

A major concern in the application of any marker technique in ecological studies is the stability of the marker under different conditions and environments. Does the organism growing in nature maintain its antigenic integrity sufficiently so that it could be recognized by antibodies prepared against cells from artificial media? Unapproachable as this question seems, there exists enough indirect evidence to suggest that the environment has little if any effect in modifying structural antigens. FAs against cultured rhizobia have been shown to react equally well with cells grown in four different media (Bohlool, 1968), bacteria growing in various sterilized soils (Schmidt et al., 1968; Bohlool and Schmidt, 1968, 1973a), and bacteriods in nodules (Schmidt et al., 1968; Trinick, 1969; Bohlool and Schmidt, 1973b; Lindemann et al., 1974; May, 1979). Antigen stability is the premise underlying the widely used practice of serotyping nodules by quick agglutination tests (Means et al., 1964).

Further indirect evidence of antigen stability comes from the finding that certain specific strains of Rhizobium could be detected in natural soils using FA prepared from rhizobia grown in culture medium (Bohlool and Schmidt, 1970). Presumptive evidence for the presence of two strains of R. japonicum in a range of natural soils was further strengthened by the observation that nearly every soil which contained FA-reactive cells, when used to inoculate soybeans, caused nodule formation, and some proportion of the nodules contained bacteroids which also reacted with the same FA. Conversely, when a particular FA failed to detect reactive bacteria in a soil, that soil either did not nodulate soybeans or, if it did, none of the nodules contained bacteriods that reacted with the FA.

More definitive evidence of the stability of rhizobial antigens in soils over a 12-year period was obtained by Diatloff (1977) in Australia. Two serologically distinct strains of Rhizobium for Lotononis bainesii were introduced into an isolated field station where Lotononis-pangola grass pastures were established but no indigenous Lotononis rhizobia were previously present. Stabilities of four rhizobial characters—colony color, effectiveness, antibiotic sensitivity, and antigens—were assessed from nodule isolates. It was shown that antigens and colony characteristics were not changed over a period of 5–12 years of residence in the soil, whereas effectiveness and antibiotic sensitivity had undergone slight modifications.

Yet another line of evidence attesting to the remarkable stability of the antigens of bacteria stems from the world wide distribution of many cross-reacting strains. A few examples will suffice: Fliermans *et al.* (1974) found that FA prepared against a *Nitrobacter agilis* strain from Minnesota stained nitrite-oxidizing chemosynthetic bacteria from soils of Kentucky, Morocco, and Iceland; strains of *Nitrosomonas* and *Nitrosospira* isolated from a Minnesota soil yielded FAs which stained cells in ammonia-oxidizing samples of a tidal bay in New Zealand as well as isolates of ammonia oxidizers isolated from those samples (Belser and Schmidt, unpublished); Bohlool (1975) found that *Sulfolobus* FAs prepared against hot-spring isolates from Yellowstone National Park, U.S.A., reacted with *Sulfolobus* isolates from hot springs in New Zealand; and the same isolate of *Nitrobacter* was found to be entirely constant with respect to FA-staining properties whether grown in strictly autotrophic conditions or in strictly heterotrophic medium in the absence of nitrite (Stanley and Schmidt, unpublished). Despite great differences in geographical distribution, environmental pressures, and growth conditions, it appears that major antigenic components of bacteria remain essentially unchanged.

11.4. Viability

IF cannot differentiate living from dead cells because both will stain brightly and specifically. Dead microorganisms, however, are not expected to persist in nature very long, for they would soon be decomposed and consumed by other organisms. Some evidence for this was obtained by Bohlool and Schmidt (1973b), who observed that heat-killed (65°C for 3 h) *R. japonicum* cells were decomposed in soil and completely cleared within 1–2 weeks. Live cells (from culture) were also degraded in the soil, but it took much longer for them to decline in numbers. Figure 5 illustrates how intact cells are degraded into pieces of fluorescent debris, which are cleared completely after a period of time. The duration of persistence will perhaps vary considerably with the type of cell, their numbers, and the environment. Figure 5c illustrates degradation of cells by an unidentified soil fungus.

Escherichia coli is one of the few bacteria that can be enumerated in normal soil by selective plating. Although *E. coli* does not grow in soil, the die-off rate of a population added to soil may be observed. Data reported by Schmidt (1974) in a comparison of selective plate counts with FA counts of *E. coli* as a function of time in contact with a soil population reflect the relatively rapid clearance of dead bacteria from soil. Between 3 and 10 days there was a large decline in the number of viable *E. coli* cells (plate count), but only a small percentage of the dead cells was sufficiently intact to be recognizable by IF.

Even more rapid disappearance of *Nitrobacter* cells occurred in several soils once added nitrite was exhausted by an actively nitrifying population (Rennie and Schmidt, 1977b). Cell numbers peaked at about 8 days after addition of substrate to the soils, at which time the nitrite had been completely oxidized; between 8 and 14 days, the numbers of FA-detectable cells dropped almost to initial levels. These data together with the common sense realization that dead cells do not pile up in most natural environments suggest that inclusion of non-viable cells in counts is unlikely to be a major source of error in IF enumeration.

11.5. Quantification

Problems associated with the enumeration of FA-stained bacteria in natural materials are most pronounced in the case of soil populations. The microscopic field is extremely small in area, and only a limited quantity of soil may be tolerated in the field lest bacteria be buried or masked. As a consequence, populations must be large in order to encounter a reasonable number of cells from a tolerable deposition of soil. Problems related to microscopic field size and cell densities may be handled by passing a volume of sample through a suitable membrane filter, one that is adequate to trap a convenient number of cells per field of filter. Practical limitations are dictated by inverse relationships between the size of the specific population and the amount of suspended particulates in the water. To quantify realistically low population levels in soils or sediments, it is necessary to separate cells from soil and concentrate them on a surface for FA staining. Development and evaluation of a procedure for the enumeration of a specific population in natural habitats were reported by Bohlool and Schmidt (1973a) and Schmidt (1974).

Limitations associated with IF quantification of soil organisms derive largely from the efficiency with which cells can be released from soil particulates and retained in a supernatant fluid following removal of those particulates. Trapping of cells below the surface in the fibrous matrix of cellulose acetate membranes and masking by residual soil particulates are additional, but likely less important, sources of error. Estimates of the efficiency with which bacteria can be recovered from soil are not easily obtained, for there are no means to obtain fully reliable standards. Enumeration by plate count is the best standard for evaluation of recovery by quantitative FA, but adequate selective plating procedures are available for very few soil organisms. A reasonable compromise is to introduce a known inoculum of the organism of interest into sterilized soil and to estimate recovery by both plating and IF as the pure culture grows. In all such studies, the many limitations of the plate count itself, especially with respect to clumping and statistical variability, must be kept in mind. Recovery estimates obviously are still difficult for organisms, such as the nitrifiers, which are poorly amenable to plating.

The few recovery data thus far available are from procedures based on the protocol of Bohlool and Schmidt (1973a). It is clear that recovery varies with the microorganism and with the soil so that modifications in procedural details may be necessary. Bohlool and Schmidt (1973a) reported recoveries for *R. japonicum* USDA 110 of 25–130% at various growth stages in a sterilized soil. Recovery of viable *E. coli* cells from nonsterile soil after 1 and 3 days was 89 and 64%, respectively, relative to a plate count reported by Schmidt (1974), who also observed that attempts to recover *A. chroococcum* from normal soil (selective plating) were less successful than for *E. coli. R. japonicum* strain USDA 123 was recovered from a Waukegan soil with an estimated 30% efficiency (Reyes and Schmidt, 1979). Vidor and Miller (1980a), using the same procedure but 1% $CaCl_2$ as flocculant, recovered *R. japonicum* strain USDA 110 from a Rossmoyne soil with 80% efficiency, but with only 20% efficiency from a Miami silt loam. Recoveries such as those reported do not pose serious limitations so long as they are consistent for a given system and for populations in the 10^5 or 10^6/g range. Low rates of recovery, even if consistent, become a technical limitation when the cell of interest is present in the soil at densities below 10^5/g.

Some tropical soils which fix added rhizobia rapidly and irreversibly with respect to the usual soil-release procedures of IF enumeration were encountered by Kingsley and Bohlool (unpublished). A strain of chick-pea *Rhizobium* was studied in a sand perlite mixture from which it could be released and counted by FA with more than 90% efficiency. Addition of only small amounts of a

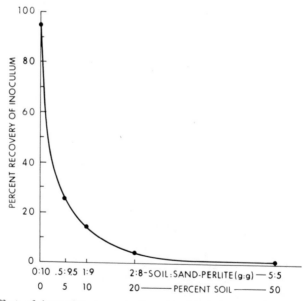

Figure 6. Effect of increasing concentration of a tropical soil (a Hawaiian Oxisol) on recovery of chick-pea *Rhizobium* strain ICRISAT 3889 from a sand–perlite mixture. From Kingsley and Bohlool (unpublished).

Hawaiian Oxisol to the mixture decreased recoveries dramatically to less than 1% (Fig. 6).

New techniques as well as modifications to enhance recovery from refractory soils or of refractory organisms are in order for situations where recovery is found to be a limitation. Work underway with rhizobia-soil systems (Donaldson, Bezdicek, and Sharma, unpublished; Wollum and Miller, unpublished) has focused on density gradient centrifugation procedures as a means to achieve consistently high recoveries. Preliminary results from both groups indicate recovery rates near 100% in tests with heavily inoculated soil.

ACKNOWLEDGMENTS. Much of the published and unpublished data cited by the authors in connection with their own studies was supported by National Science Foundation Grants GB8413, GB29636, DEB76-19518, and BMS75-14020, Minnesota Agricultural Experiment Station Project 25-072, and USDA/SEA Grant 701-15-60. The authors thank Sheila N. May and Mark T. Kingsley for contributing unpublished data from their thesis research.

References

Abshire, R. L., 1976, Detection of enteropathogenic *Escherichia coli* strains in wastewater by fluorescent antibody, *Can. J. Microbiol.* 22:364-378.

Abshire, R. L., and Guthrie, R. K., 1971, The use of fluorescent antibody technique for detection of *Streptococcus faecalis* as an indicator of faecal pollution of water, *Water Res.* 5:1089-1097.

Allen, E., and Kelman, A., 1977, Immunofluorescent stain procedures for detection and identification of *Erwinia carotovora* var. *atroseptica*, *Phytopathology* 67:1305-1312.

American Public Health Association, 1975, *Standard Methods for the Examination of Water and Wastewater*, 14th ed., American Public Health Association, Washington, D.C.

Apel, W. A., Dugan, P. R., Filppi, J. A., and Rheins, M. S., 1976, Detection of *Thiobacillus ferrooxidans* in acid mine by indirect fluorescent antibody staining, *Appl. Environ. Microbiol.* 32:159-165.

Auger, J. G., and Shalla, T. A., 1975, The use of fluorescent antibodies for detection of Pierce's disease bacteria in grapevines and insect vectors, *Phytopathology* 65:493-494.

Belly, R. B., and Brock, T. D., 1974, Ecology of iron-oxidizing bacteria in pyritic materials associated with coal, *J. Bacteriol.* 117:726-732.

Belser, L. W., and Schmidt, E. L., 1978a, Nitrification in soils, in: *Microbiology 1978* (D. Schlessinger, ed.), pp. 348-351, American Society of Microbiology, Washington, D.C.

Belser, L. W., and Schmidt, E. L., 1978b, Diversity in the ammonia-oxidizing nitrifier population of a soil, *Appl. Environ. Microbiol.* 36:584-588.

Belser, L. W., and Schmidt, E. L., 1978c, Serological diversity within a terrestrial ammonia-oxidizing population, *Appl. Environ. Microbiol.* 36:589-593.

Bettelheim, K. A., Gordon, J. F., and Taylor, J., 1968, The detection of a strain of *Chromobacterium lividum* in the tissue of certain leaf-nodulated plants by immunofluorescence technique, *J. Gen. Microbiol.* 54:177-184.

Bohlool, B. B., 1968, Fluorescent Antibody for Ecological Studies of *Rhizobium japonicum*, M.S. Thesis, University of Minnesota, Minneapolis.

Bohlool, B. B., 1975. Occurrence of *Sulfolobus acidocaldarius*, an extremely thermophilic

philic bacterium, in New Zealand hot springs: Isolation and immunofluorescence characterization, *Arc. Microbiol.* 106:171-174.

Bohlool, B. B., and Brock, T. D., 1974a, Immunofluorescence approach to the study of the ecology of *Thermoplasma acidophilum* in coal refuse material, *Appl. Microbiol.* 28: 11-16.

Bohlool, B. B., and Brock, T. D., 1974b, Population ecology of *Sulfolobus acidocaldarius.* II. Immunoecological studies, *Arch. Microbiol.* 97:181-194.

Bohlool, B. B., and Schmidt, E. L., 1968, Nonspecific staining: Its control in immunofluorescence examination of soil, *Science* 162:1012-1014.

Bohlool, B. B., and Schmidt, E. L., 1970, Immunofluorescent detection of *Rhizobium japonicum* in soils, *Soil Sci.* 110:229-236.

Bohlool, B. B., and Schmidt, E. L., 1973a, A fluorescent antibody technique for determination of growth rates of bacteria in soil, *Bull. Ecol. Res. Commun. (Stockholm)* 17: 336-338.

Bohlool, B. B., and Schmidt, E. L., 1973b, Persistence and competition aspects of *Rhizobium japonicum* observed in soil by immunofluorescence microscopy, *Soil Sci. Soc. Am. Proc.* 37:561-654.

Brock, T. D., 1971, Microbial growth rates in nature, *Bacteriol. Rev.* 35:39-58.

Brock, T. D., 1978, *Thermophilic Microorganisms and Life at High Temperatures*, Springer-Verlag, New York.

Brock. T. D., Brock, K. M., Belly, R. T., and Weiss, R. L., 1972, *Sulfolobus*: A new genus of sulfur-oxidizing bacteria living at low pH and high temperature, *Arch. Mikrobiol.* 84: 54-68.

Brown, M. E., 1976, Role of *Azotobacter paspali* in association with *Paspalum notatum*, *J. Appl. Bacteriol.* 40:341-348.

Brown, M. E., Jackson, R. M., and Burlingham, S. K., 1968, Growth and effects of bacteria introduced into soil, in: *The Ecology of Soil Bacteria* (T. R. G. Gray and D. Parkinson, eds.), pp. 531-551, Liverpool University Press, Liverpool.

Center for Disease Control, 1978, Isolation of organisms resembling Legionnaires disease bacterium, *Ga Morbid. Mortal. Weekly Rep.* 27:415.

Cherry, W. B., and Moody, M. D., 1965, Fluorescent antibody technique in diagnostic bacteriology, *Bacteriol. Rev.* 29:222-250.

Cherry, W. B., Goldman, M., and Carski, T. R., 1960, Fluorescent Antibody Techniques in the Diagnosis of Communicable Diseases, Public Health Service Publication No. 729, U.S. Government Printing Office, Washington, D.C.

Cherry, W. B., Thomason, B. M., Pomales-Lebron, A., and Ewing, W. H., 1961, Rapid presumptive identification of enteropathogenic *Escherichia coli* in fecal smears by means of fluorescent antibody, *Bull. W.H.O.* 25:159-171.

Cherry, W. B., Hanks, J. B., Thomason, B. M., Murlin, A. M., Biddle, J. W., and Croom, J. M., 1972, Salmonellae as an index of pollution of surface waters, *Appl. Microbiol.* 24(3):334-340.

Cherry, W. B., Thomason, B. M., Gladden, J. B., Holsing, N., and Murlin, A. M., 1975, Detection of Salmonellae in foodstuffs, feces and water by immunofluorescence, in: *Fifth International Conference on Immunofluorescence and Related Techniques* (W. Hijmans and M. Schaeffer, eds.), *Ann. N.Y. Acad. Sci.* 254:350-369.

Cherry, W. B., Pittman, B., Harris, P. P., Hebert, G. A., Thomason, B. M., Thacker, L., and Weaver, R. E., 1978, Detection of Legionnaires disease bacteria by direct immunofluorescent staining, *J. Clin. Microbiol.* 8:329-338.

Colmer, A. R., Temple, K. L., and Hinkle, M. E., 1950, An iron-oxidizing bacterium from the acid drainage of some bituminous coal mines, *J. Bacteriol.* 59:317-328.

Danielsson, D., 1965, A membrane filter method for the demonstration of bacteria by the fluorescent antibody technique. 1. A methodological study, *Acta Pathol. Microbiol. Scand.* 63:597-603.

Danielsson, D., and Laurell, G., 1965, A membrane filter method for the demonstration of bacteria by the fluorescent antibody technique. 2. Application of the method for detection of small numbers of bacteria in water, *Acta Pathol. Microbiol. Scand.* 63: 604–608.

DeVille, R., and Tchan, Y. T., 1970, Etude quantitative de la population azotobacterienne du sol par la methode d'immunofluorescence, *Ann. Inst. Pasteur (Paris)* 119:492–497.

Diatloff, A., 1977, Ecological studies of root-nodule bacteria introduced into field environments. 6. Antigenic stability in *Lotononis* rhizobia over a 12-year period, *Soil Biol. Biochem.* 9:85–88.

Diem, H. G., Godbillon, G., and Schmidt, E. L., 1977, Application of the fluorescent antibody technique to the study of an isolate of *Beijerinckia* in soil, *Can. J. Microbiol.* 23:161–165.

Diem, H. G., Schmidt, E. L., and Dommergues, Y. R., 1978, The use of the fluorescent antibody technique to study the behaviour of a *Beijerinckia* isolate in the rhizosphere and spermosphere of rice, *Ecol. Bull. (Stockholm)* 26:312–318.

Dobereiner, J., and Day, J. M., 1976, Associative symbiosis in tropical grasses: Characterization of microorganisms and di-nitrogen fixing sites, in: *Proceedings, First International Symposium on Nitrogen Fixation* (W. E. Newton and C. J. Nyman, eds.), Vol. 2, pp. 518–538, University of Washington Press, Pullman.

Dobereiner, J., Day, J. M., and Dart, P. J., 1972, Nitrogenase activity and oxygen sensitivity of the *Paspalum notatum–Azotobacter paspali* association, *J. Gen. Microbiol.* 71: 103–116.

Dudman, W. F., 1977, Serological methods and their application to dinitrogen-fixing organisms, in: *A Treatise on Dinitrogen Fixation* (R. W. F. Hardy and A. H. Gibson, eds.), Sect. IV, pp. 487–508, John Wiley & Sons, New York.

Dudman, W. F., and Heidelberger, M., 1969, Immunochemistry of newly found substituents of polysaccharides of *Rhizobium* species, *Science* 164:954–955.

Duncan, L. K., and Rosswall, T., 1974, Taxonomy and physiology of tundra bacteria in relation to site characteristics, in: *Soil Organisms and Decomposition in the Tundra* (A. J. Holding, O. W. Heal, S. F. MacLean, Jr., and P. W. Flanagan, eds.), pp. 79–92, Tundra Biome Steering Committee, Stockholm.

Eren, J., and Pramer, D., 1966, Application of immunofluorescent staining to studies of the ecology of soil microorganisms, *Soil Sci.* 101:39–45.

Farrah, S. R., and Unz, R. F., 1975, Fluorescent-antibody study of natural finger-like Zoogleae, *Appl. Microbiol.* 30:132–139.

Fliermans, C. B., and Schmidt, E. L., 1975, Autoradiography and immunofluorescence combined for autecological study of single cell activity with *Nitrobacter* as a model system, *Appl. Microbiol.* 30:676–684.

Fliermans, C. B., and Schmidt, E. L., 1977, Immunofluorescence for autecological study of a unicellular blue-green alga, *J. Phycol.* 13:364–368.

Fliermans, C. B., Bohlool, B. B., and Schmidt, E. L., 1974, Autecological study of the chemoautotroph *Nitrobacter* by immunofluorescence, *Appl. Microbiol.* 27:124–129.

Fliermans, C. B., Cherry, W. B., Orrison, L. H., and Thacker, L., 1979, Isolation of *Legionella pneumophila* from nonepidemic-related aquatic habitats, *Appl. Environ. Microbiol.* 37:1239–1242.

Garcia, M. M., and McKay, K. A., 1969, On the growth and survival of *Clostridium septicum* in soil, *J. Appl. Bacteriol.* 32:362–370.

Garcia, M. M., Neil, D. H., and McKay, K. A., 1971, Application of immunofluorescence to studies on the ecology of *Sphaerophorus necrophorus*, *Appl. Microbiol.* 21:809–814.

Garvey, J. S., Cremer, N. E., and Susdorf, D. H., 1977, *Methods in Immunology*, 3rd ed., W. A. Benjamin, Reading, Massachusetts.

Geldreich, E. E., 1970, Applying bacteriological parameters to recreational water quality, *J. Am. Water Works Assoc.* 62:113–120.

Goldman, M., 1968, *Fluorescent Antibody Methods*, Academic Press, New York.

Good, H. L. B., 1972, Nonspecific Staining in the Use of Fluorescent Antibody for Microbial Ecology, M.S. Thesis, University of Minnesota, Minneapolis.

Graham, P. H., 1963, Antigenic affinities of the root-nodule bacteria of legumes, *Antonie van Leeuwenhoek J. Microbiol. Serol.* 29:281-291.

Guthrie, R. K., and Reeder, D. J., 1969, Membrane filter-fluorescent-antibody method for detection and enumeration of bacteria in water, *Appl. Microbiol.* 17:399-407.

Hazen, T. C., and Fliermans, C. B., 1979, Distribution of *Aeromonas hydrophila* in natural and man-made thermal effluents, *Appl. Environ. Microbiol.* 38:166-168.

Hazen, T. C., Raker, M. L., Esch, G. W., and Fliermans, C. B., 1978, Ultrastructure of red-sore lesions on largemouth bass (*Mieropterus salmoides*): Association of the ciliate *Epistylis* sp. and the bacterium *Aeromonas hydrophila*, *J. Protozool.* 25:351-355.

Heidelberger, M., and Elliot, S., 1966, Cross-reactions of Streptococcal Group N techoic acid in antipneumococcal horse sera of type VI, XIV, XVI and XXVII, *J. Bacteriol.* 92:281-283.

Henry, R. A., Johnson, R. C., Bohlool, B. B., and Schmidt, E. L., 1971, Detection of *Leptospira* in soil and water by immunofluorescent staining, *Appl. Microbiol.* 21:953-956.

Hill, I. R., and Gray, T. R. G., 1967, Application of the fluorescent antibody technique to an ecological study of bacteria in soil, *J. Bacteriol.* 93:1888-1896.

Hobson, P. N., and Mann, S. O., 1957, Some studies on the identification of rumen bacteria with fluorescent antibody, *J. Gen. Microbiol.* 16:463-471.

Hobson, P. N., Mann, S. O., and Smith, W., 1962, Serological tests of a relationship between rumen selenomads *in vitro* and *in vivo*, *J. Gen. Microbiol.* 29:265-270.

Holland, A. A., 1966, Serologic characteristics of certain root-nodule bacteria of legumes, *Antonie van Leeuwenhoek J. Microbiol. Serol.* 32:410-418.

Hughes, T. A., Lecce, J. G., and Elkan, G. H., 1979, Modified fluorescent technique, using rhodamine, for studies of *Rhizobium japonicum*-soybean symbiosis, *Appl. Environ. Microbiol.* 37:1243-1244.

Joklik, W. K., and Willett, H. P., 1976, *Zinsser's Microbiology*, 16th ed., p. 725, Appleton-Century-Crofts, New York.

Jones, D. G., and Russell, P. E., 1972, The application of immunofluorescence techniques to host plant/nodule bacteria selectivity experiments using *Trifolium repens*, *Soil Biol. Biochem.* 4:277-282.

Kawamura, A., Jr., 1977, *Fluorescent Antibody Techniques and Their Applications*, 2nd ed., University Park Press, Baltimore.

Knowles, R., 1977, The significance of asymbiotic dinitrogen fixation by bacteria, in: *A Treatise on Dinitrogen Fixation* (R. W. F. Hardy and A. H. Gibson, eds.), Sect. IV, pp. 33-83, John Wiley & Sons, New York.

LaRue, T. A., 1977, The bacteria, in: *A Treatise on Dinitrogen Fixation* (R. W. F. Hardy and W. S. Silver, eds.), Sect. III, p. 1962, John Wiley & Sons, New York.

Lindemann, W. C., Schmidt, E. L., and Ham, G. E., 1974, Evidence for double infection within soybean nodules, *Soil Sci.* 118:274-279.

Lundgren, D. G., Vertal, J. R., and Tabita, F. R., 1971, The microbiology of mine drainage pollution, in: *Water Pollution Microbiology* (R. Mitchel, ed.), pp. 69-88, Wiley-Interscience, New York.

May, N., 1979, Ecological Studies on Lentil Rhizobia: Competition and Persistence in Some Tropical Soils, M.S. Thesis, University of Hawaii, Honolulu.

May, S. N., and Bohlool, B. B., 1979, Mixed-infection in lentil nodules and its relationship to competitiveness of *Rhizobium* strains, Proceedings of the 7th North American *Rhizobium* Conference, p. 49, June 17-21, 1979, Texas A & M.

Means, U. M., Johnson, H. W., and Date, R. A., 1964, Quick serological method of classifying strains of *Rhizobium japonicum* in nodules, *J. Bacteriol.* 87:547-553.

Mosser, J. L., Bohlool, B. B., and Brock, T. D., 1974a, Growth rates of *Sulfolobus acidocaldarius* in nature, *J. Bacteriol.* 118:1075-1081.

Mosser, J. L., Mosser, A. G., and Brock, T. D., 1974b, Population ecology of *Sulfolobus acidocaldarius*. I. Temperature strains, *Arch. Microbiol.* 97:169-179.

Nairn, R. C., 1975, *Fluorescent Protein Tracing*, 4th ed., Churchill Livingstone, London.

Norris, J. R., and Kingham, W. H., 1968, The classification of the azotobacters, in: *Festkrift til Hans Laurits Jensen* (L. A. Henriksen, ed.), Statens Planteavls-Laboratorium, Vejle, Denmark.

Nutman, P. S., 1965, The relation between nodule bacteria and the legume host in the rhizosphere and in the process of infection, in: *Ecology of Soil-Borne Pathogens* (K. F. Baker and W. C. Snyder, eds.), pp. 231-247, University of California Press, Berkeley.

Parkinson, D., 1973, Techniques for the study of soil fungi, *Bull. Ecol. Res. Commun. (Stockholm)* 17:29-36.

Paton, A. M., 1964, The adaptation of the immunofluorescence technique for use in bacteriological investigations of plant tissue, *J. Appl. Bacteriol.* 27:237-242.

Pavlova, M. T., Beauvais, E., Brezenski, F. T., and Litsky, W., 1972, Fluorescent-antibody technique for the identification of Group D streptococci: Direct staining method, *Appl. Microbiol.* 23:571-577.

Pavlova, M. T., Beauvais, E., Brezenski, F. T., and Litsky, W., 1973, Rapid assessment of water quality by fluorescent antibody identification of fecal streptococci, in: *Sixth International Conference on Water Pollution Research, Jerusalem, 1972* (S. H. Jenkins, ed.), Pergamon Press, Oxford.

Petersen, E. J., 1959, Serological investigations on *Azotobacter* and *Beijerinckia*, Royal Veterinarian and Agricultural College Yearbook (Copenhagen), pp. 70-90.

Postgate, J., 1972, *Biological Nitrogen Fixation*, Merrow Publishing, Watford, Hertfordshire, England.

Pugsley, A. P., and Evison, L. M., 1974, Immunofluorescence as a method for the detection of *Escherichia coli* in water, *Can. J. Microbiol.* 20:1457-1463.

Pugsley, A. P., and Evison, L. M., 1975, A fluorescent antibody technique for the enumeration of faecal streptococci in water, *J. Appl. Bacteriol.* 38:63-65.

Reed, W. M., and Dugan, P. R., 1978, Distribution of *Methylomonas methanica* and *Methylosinus trichosporium* in Cleveland Harbor as determined by an indirect fluorescent antibody-membrane filter technique, *Appl. Environ. Microbiol.* 35:422-430.

Rennie, R. J., and Schmidt, E. L., 1977a, Immunofluorescence studies of *Nitrobacter* populations in soils, *Can. J. Microbiol.* 23:1011-1017.

Rennie, R. J., and Schmidt, E. L., 1977b, Autecological and kinetic analysis of competition between *Nitrobacter* strains in soil, *Ecol. Bull.* 25:431.

Rennie, R. J., Reyes, V. G., and Schmidt, E. L., 1977, Immunofluorescent detection of the effects of wheat and soybean roots on *Nitrobacter* in soil, *Soil Sci.* 124:10.

Reyes, V. G., and Schmidt, E. L., 1979, Population densities of *Rhizobium japonicum* strain 123 estimated directly in soil and rhizospheres, *Appl. Environ. Microbiol.* 37:854.

Robinson, A. C., 1967, The influence of host on soil and rhizosphere populations of clover and lucerne root nodule bacteria in the field, *J. Aust. Inst. Agric. Sci.* 33:207-209.

Rovira, A. D., 1961, *Rhizobium* numbers in the rhizosphere of red clover and paspalum in relation to soil treatment and numbers of bacteria and fungi, *Aust. J. Agri. Res.* 12:77-87.

Rubenchik, L. I., 1963, Azotobacter and Its Use in Agriculture, Israel Program of Scientific Translations, National Science Foundation, Washington, D.C.

Ruinen, J., 1961, The phyllosphere. I. An ecological neglected milieu, *Plant Soil* 15:81-109.

Schank, S. C., Smith, R. L., Weiser, G. C., Zuberer, D. A., Bouton, J. H., Quesenberry, K. H., Tyler, M. E., Milam, J. R., and Littell, R. C., 1979, Fluorescent antibody technique to identify *Azospirillum brasilense* associated with roots of grasses, *Soil Biol. Biochem.* 11:287-295.

Schmidt, E. L., 1973, Fluorescent antibody technique for the study of microbial ecology, *Bull. Ecol. Res. Commun. (Stockholm)* 17:67-76.

Schmidt, E. L., 1974, Quantitative autecological study of microorganisms in soil by immunofluorescence, *Soil Sci.* 118:141-149.

Schmidt, E. L., 1978, Nitrifying microorganisms and their methodology, in: *Microbiology 1978* (D. Schlessinger, ed.), pp. 288-291, American Society of Microbiology, Washington, D.C.

Schmidt, E. L., and Bankole, R. O., 1962, Detection of *Aspergillus flavus* in soil by immunofluorescent staining, *Science* 136:776-777.

Schmidt, E. L., and Bankole, R. O., 1963, The use of fluorescent antibody with the buried slide technique, in: *Soil Organisms* (J. Doeksen and J. Van Der Drift, eds.), pp. 197-204, North-Holland, Amsterdam.

Schmidt, E. L., Bankole, R. O., 1965, Specificity of immunofluorescent staining for study of *Aspergillus flavus* in soil, *Appl. Microbiol.* 13:673-679.

Schmidt, E. L., Bankole, R. O., and Bohlool, B. B., 1968, Fluorescent antibody approach to the study of rhizobia in soil, *J. Bacteriol.* 95:1987-1992.

Schmidt, E. L., Biesbrock, J. A., Bohlool, B. B., and Marx, D. H., 1974, Study of mycorrhizae by means of fluorescent antibody, *Can. J. Microbiol.* 20(2):137-139.

Siala, A., and Gray, T. R. G., 1974, Growth of *Bacillus subtilis* and spore germination in soil observed by a fluorescent-antibody technique, *J. Gen. Microbiol.* 81:191-199.

Siala, A., Hill, I. R., and Gray, T. R. G., 1974, Populations of spore-forming bacteria in an acid forest soil, with special reference to *Bacillus subtilis*, *J. Gen. Microbiol.* 81: 183-190.

Stanley, P. M., Gage, M. A., and Schmidt, E. L., 1979, Enumeration of specific populations by immunofluorescence, in: *Symposium on Native Aquatic Bacteria, Enumeration, Activity and Ecology*, ASTM STP 695 (J. W. Costerton and R. R. Colwell, eds.), pp. 46-55, American Society of Testing and Materials, Philadelphia.

Strayer, R. F., and Tiedje, J. M., 1978, Application of the fluorescent-antibody technique to the study of a methanogenic bacterium in lake sediments, *Appl. Environ. Microbiol.* 35:192-198.

Tarrand, J. J., Krieg, N. R., and Dobereiner, J., 1978, A taxonomic study of the *Spirillum lipoferum* group with descriptions of a new genus, *Azospirillum* gen. nov. and two species, *Azospirillum lipoferum* (Beijerinck) comb. nov. and *Azospirillum brasilense* sp. nov., *Can J. Microbiol.* 24:967-980.

Tchan, Y. T., and DeVille, R., 1970, Application de l'immunofluorescence a l'etude des azotobacter du sol, *Ann. Inst. Pasteur (Paris)* 118:665-673.

Thomason, B. M., and McWhorter, A. C., 1965, Rapid detection of typhoid carriers by means of fluorescent antibody technique, *Bull. W.H.O.* 33:681-685.

Thomason, B. M., and Wells, J. G., 1971, Preparation and testing of polyvalent conjugates for fluorescent antibody detection of salmonellae, *Appl. Microbiol.* 22:876-884.

Thomason, B. M., Cowart, G. S., and Cherry, W. B., 1965, Current status of immunofluorescence technique for rapid detection of shigellae in fecal specimens, *Appl. Microbiol.* 13:605-613.

Tison, D. L., Pope, D. H., Cherry, W. B., and Fliermans, C. B., 1979, Growth of *Legionella pneumophila* (Legionnaires' Disease Bacterium) in association with blue-green algae, *Science*, submitted.

Trinick, M. J., 1969, Identification of legume nodule bacteroids by the fluorescent antibody reaction, *J. Appl. Bacteriol.* 32:181-186.

Tuttle, J. H., Dugan, P. R., Macmillan, C. B., and Randles, C. I., 1969, Microbial dissimilatory sulfur cycle in acid mine water, *J. Bacteriol.* 97:594-602.

Tuzimura, K., and Watanabe, I., 1961, The growth of *Rhizobium* in the rhizosphere of the host plant. Ecological studies (part 2), *Soil Sci. Plant Nutr. Tokyo* 8:19-28.

Van der Merwe, S. P., Strijdom, B. W., and Van Rensburg, H. J., 1972, Use of the fluorescent antibody technique to detect a bacterial contaminant in nodule squashes of leguminous plants, *Phytophylactica* **4**:97–100.

Vest, G., Weber, D. F., and Sloger, C., 1973, Nodulation and nitrogen fixation, in: *Soybeans; Improvement, Production, and Uses* (B. E. Caldwell, ed.), pp. 353–390, American Society of Agronomy, Madison, Wisconsin.

Vidor, C., and Miller, R. H., 1980a, Relative saprophytic competence of *Rhizobium japonicum* strains in soils as determined by quantitative fluorescent antibody technique, *Soil Biol. Biochem.* (in press).

Vidor, C., and Miller, R. H., 1980b, Influence of the soybean host on the population of *Rhizobium japonicum* in soil, *Soil Biol. Biochem.* (in press).

Vincent, J. M., 1974, Root nodule symbiosis with *Rhizobium*, in: *The Biology of Nitrogen Fixation* (A. Quispel, ed.), pp. 265–341, North-Holland, Amsterdam.

Walsh, F., and Mitchell, R., 1972, A pH-dependent succession of iron bacteria, *Environ. Sci. Technol.* **6**:809–812.

Warnock, D. W., 1971, Assay of fungal mycelium in grains of barley, including the use of the fluorescent antibody technique for individual fungal species, *J. Gen. Microbiol.* **67**:197–205.

Warnock, D. W., 1973, Use of immunofluorescence to detect mycelium of *Alternaria, Aspergillus*, and *Penicillium* in barley grains, *Trans. Br. Mycol. Soc.* **61**(3):547–552.

Zvyagintzev, D. G., and Kozhevin, P. A., 1974, Dynamics of the population of *Rhizobium leguminosarum* in soils studied by immunofluorescence, *Mikrobiologiia* **43**:888–891. (in Russian).

Index